ETHOPHARMACOLOGY OF AGONISTIC BEHAVIOUR
IN ANIMALS AND HUMANS

TOPICS IN THE NEUROSCIENCES

Neuronal Control of Bodily Function: Basic and Clinical Aspects

Ethopharmacology of Agonistic Behaviour in Animals and Humans

edited by

B. OLIVIER
Department of Pharmacology, Duphar B.V., Weesp, The Netherlands

J. MOS
Department of Pharmacology, Duphar B.V., Weesp, The Netherlands

P.F. BRAIN
Department of Zoology, University College of Swansea, Swansea, Wales, United Kingdom

1987 **MARTINUS NIJHOFF PUBLISHERS**
a member of the KLUWER ACADEMIC PUBLISHERS GROUP
DORDRECHT / BOSTON / LANCASTER

Distributors

for the United States and Canada: Kluwer Academic Publishers, P.O. Box 358, Accord Station, Hingham, MA 02018-0358, USA
for the UK and Ireland: Kluwer Academic Publishers, MTP Press Limited, Falcon House, Queen Square, Lancaster LA1 1RN, UK
for all other countries: Kluwer Academic Publishers Group, Distribution Center, P.O. Box 322, 3300 AH Dordrecht, The Netherlands

Library of Congress Cataloging in Publication Data

```
Ethopharmacology of agonistic behaviour in animals
   and humans.

   (Topics in the neurosciences)
   1. Agonistic behavior in animals.  2. Aggressiveness
(Psychology)  3. Psychopharmacology.  I. Olivier,
Berend.  II. Mos, J. (Jan)  III. Brain, Paul F.
IV. Title: Ethopharmacology of agonistic behavior
in animals and humans.  V. Series.
QL758.5.E85  1987        155.2'32        87-22089
```

ISBN-13: 978-94-010-8009-5   e-ISBN-13: 978-94-009-3359-0
DOI: 10.1007/978-94-009-3359-0

Copyright

Preface

Aggression research is in a rapid state of development. The accelerating knowledge of neurotransmitter systems in the brain, their behavioural functions and the development of drugs which may specifically affect systems related to attack and defence is fruitfully combined with studies in which basic ethological observation and quantification techniques are used more routinely.

Moreover, much of the experimental effort has finally applied some order to the initial chaos which afflicted the various experimental aggression models used in pharmacological, physiological and ethological research. This highly desirable trend not only leads to a better understanding of the phenomena studied and the terminologies employed, but it increases our awareness of the multiplicity of factors that are important, making it difficult to allow over hasty and simple generalizations.

This book is a compilation of studies presented at the International Society for Research on Aggression meeting in Chicago 1986, in which leading investigators were invited to cover aspects of ethopharmacological aggression research in a wide variety of species, including studies on humans. The level to which ethological techniques have been incorporated into the various areas of research differs, as well as the knowledge and understanding of the neurotransmitter and experimental drug action on brain functioning in mammalian species. This naturally results in data which are not always easy to compare or to extrapolate between species but useful indicators are starting to emerge.

In this book the emphasis in the different chapters either lies on developments in behavioural studies or on pharmacological actions, as these basis sciences are cornerstones to breakthroughs in the multidisciplinary assessment of aggression and its physiological bases. Developments in behavioural studies focus around experimental aggression models in humans and the importance of factors such as social status and cooperation. Other aspects emphasize the influence of frustration, flight and defense strategies and adaptive reactions to novelty or defeat to our understanding of conflict behaviour.

Intriguing pharmacological aspects dealt with include the renewed interest in the effects of alcohol on aggression – in relation to hormones, degree of domestication and perinatal effects. The conflicting data on the benzodiazepines is also presented which clearly outlines the strength of the ethological approach. Interactions by recently–discovered drugs with serotonin receptors, and animal studies with clinically–used drugs point to an urgent need for integration between animal and human data.

The above distinction between emphases on behavioural and/or pharmacological topics does not imply a sparcity of attempts to integrate animal and human findings and delineate future areas of research. One can perhaps maintain that this volume strongly suggests that consideration should be given to the way in which we apply ethological considerations to the analysis of laboratory situations. This book suggests that interesting areas will emerge at interfaces between neuroscience and behaviour which need careful ethopharmacological study to gain an understanding of adaptive aggression as well as to control pathological aggression.

The organization of the symposium and the preparation of this book were greatly aided by contributions in funds and time from Duphar B.V. (Weesp, The Netherlands).

The efficient help of Mrs. Marijke Mulder was invaluable in giving this book its present typographical layout. We are also greatly indebted to Dr. Paul Bevan for providing us with the means to produce this volume.

Berend Olivier, Jan Mos, Paul F. Brain.

TABLE OF CONTENTS

ETHOPHARMACOLOGICAL STUDIES OF AGGRESSION AND DOMINANCE IN MONKEYS

PSYCHOPHARMACOLOGICAL STUDIES OF HUMAN AGGRESSION

CONTRIBUTORS

Berger, Barry D.
    Department of Psychology, University of Haifa, Haifa 31999, Israel
    Co-author: Richard Schuster

Blanchard, Robert J.
    Department of Psychology, University of Hawaii, Honolulu, HI 96844,
    U.S.A.
    Co-author: D. Caroline Blanchard

Brain, Paul F.
    Biomedical and Physiological Research Group, Biological Sciences,
    University College of Swansea, Swansea, SA2 8PP, United Kingdom
    Co-authors: Jamaan S. Ajarem and Vesselin V. Petkov

Cherek, Don R.
    Department of Psychiatry, Louisiana State University, School of
    Medicine, Shreveport, LA 71115, U.S.A.
    Co-author: Joel L. Steinberg

Dantzer, Robert
    INRA-INSERM U259, Rue Camille St.Saens, 33077 Bordeaux Cedex, France

Dixon, A. Keith
    Sandoz Research Institute Berne Ltd., Postbox 2173, 3001 Berne,
    Switzerland
    Co-author: Hans-Peter Kaesermann

Kruk, M.R.
    Ethopharmacology Group, Dept. of Pharmacology, Sylvius Laboratories,
    Leiden University, Wassenaarseweg 72, 2333 AL Leiden, The Netherlands
    Co-authors: A.M. Van der Poel, J.H.C.M. Lammers, Th. Hagg,
    A.M.D.M. De Hey and S. Oostwegel

McGuire, Michael T.
    Department of Psychiatry-Biobehavioral Sciences School of Medicine,
    University of California at Los Angeles, Los Angeles, CA 90024, U.S.A.
    Co-author: Michael J. Raleigh

Mos, Jan
    Dept. of Pharmacology, Duphar B.V., P.O. Box 2, 1380 AA Weesp, The
    Netherlands
    Co-author: Berend Olivier

Olivier, Berend
    Dept. of Pharmacology, Duphar B.V., P.O. Box 2, 1380 AA Weesp, The
    Netherlands
    Co-authors: Jan Mos, Jan van der Heyden, Jacques Schipper, Martin Tulp,
    Bas Berkelmans and Paul Bevan

X

Panksepp, Jaak
    Department of Psychology, Bowling Green State University, Bowling
    Green, OH 43403, U.S.A.
    Co-authors: Larry Normansell, James F. Cox, Loring J. Crepeau and
    David S. Sacks

Poshivalov, Vladimir P.
    Division of Pharmacology, Pavlov Medical Institute, Leningrad 197089,
    U.S.S.R.

Rodgers, R. John
    Pharmacoethology Laboratory, School of Psychology, University of
    Bradford, Bradford BD7 1DP, United Kingdom
    Co-author: Jill I. Randall

Sheard, Michael H.
    Department of Psychiatry, Yale University Medical School, New Haven, CT
    06508, U.S.A.

Winslow, James T.
    Department of Psychology, Tufts University, Medford, MA 02155, U.S.A.
    Co-authors: Joseph F. DeBold and Klaus A. Miczek

Yoshimura, H.
    Department of Pharmacology, Ehime University School of Medicine, Ehime
    791-802, Japan

FRUSTRATION, AGGRESSION AND DRUGS

Robert Dantzer. INRA–INSERM U259, Rue Camille St–Saens, 33077 Bordeaux Cedex, France.

INTRODUCTION

Frustration is a hypothetical central state elicited by the omission of an expected reward or an inability to reach a desired goal because of some physical or psychological obstacle in the environment. It is generally believed that frustration leads to violence. For example, if my dog tries to bite me when I attempt to remove its food, one may assume that it is frustrated. In the same manner, a common theme in the media is that the present rebellion in South Africa townships finds its roots in frustration engendered by many years of apartheid.

The concept of a causal relationship between frustration and aggression was formalized in a very influential book written by Dollard, Doob, Miller, Mowrer and Sears in 1939. This concept inspired an extensive series of studies conducted mainly during the sixties and the early seventies and aimed at assessing aggressive responses toward conspecifics or inanimate objects in animals placed in frustrating situations. The lack at the time of accurate and objective measurement of aggressive behaviour and the apparent ease with which aggression could be elicited and quantified in frustrated animals prompted the use of frustration–induced aggression as a baseline for studying the effects of drugs on aggression.

Most of the findings obtained in studies of frustration–induced aggression appear a priori to support the concept of a causal relationship between frustration and aggression. However, the question of the specificity of the aggressive response observed in frustrated animals and the mechanisms by which frustration leads to aggression have rarely been addressed. Nor has there been an attempt to systematically assess the mechanisms by which drugs affect frustration–induced aggression. In particular, it is unknown whether drugs that modify frustration–induced aggression do so because of their effects on frustration, on aggression or on other processes involved in the appearance and development of frustration–induced aggression.

The aim of the present paper rather than reviewing the literature on the many different facets of frustration–induced aggression in animals and human beings, is to determine the extent to which aggression is elicited by frustration and to critically examine how the effects of drugs on frustration–induced aggression can be interpreted.

FRUSTRATION AND AGGRESSION

Methodological aspects

There are several ways of inducing frustration. A common technique is to subject animals, previously trained to obtain food by giving an appropriate response (e.g., pressing a lever in a Skinner box or running down an alley), to an extinction session during which the food reward is not presented. Another technique is to use schedules of food reinforcement requiring many responses including periods of non–reinforcement for responding (time–out procedures), intermittent schedules of food reinforcement or simply a high ratio requirements in fixed ratio reinforcement schedules.

Birds (pigeons, chickens) and laboratory mammals (rodents, pigs and primates) have been the most common subjects for studies of frustration–induced aggression. In birds, attack behaviour is usually measured by counting the number of pecks directed against a stuffed or live bird during periods of non–reinforcement in a multiple schedule of food reinforcement (Azrin et al. 1966). Live target birds are restrained in an open top box exposing their head, neck and upper breast. The restraining box is mounted on a stabilimeter that enables recording of frontal attacks. In general, live targets are found to generate a stronger attack and a higher frequency of pecking among experimentally naive animals than alternative targets. However, since the live target can be seriously injured and its defensive movements may activate the recording device, stuffed targets or mirrors have been preferred. Responding on a mirror has been claimed to be functionally equivalent to responding on a live or stuffed bird (Cohen and Looney 1973). However, an obvious limitation of the mirror target is that aggression consists only of beak–to–beak attacks whereas attacks against live or stuffed targets may be directed to the comb (in chickens) and other areas of the body also. Macurik et al. (1978) recently proposed the shielded live target as an alternative to mirror and stuffed targets. In this technique, a transparent screen separates the target bird from the attacker and pecks directed against the screen are recorded. In a comparison with other targets, the live protected target generates the highest rate of attack over all subjects. Another variation of the same technique is to project on a screen the colored image of a conspecific (Looney et al. 1976). However, only 50–70% of the birds exposed to this target developed sustained attack (Looney et al. 1976; Yoburn and Cohen 1979; Yoburn et al. 1981).

In rats, extinction–induced aggression has been measured by the duration of fighting taking place between a satiated target animal and a food deprived animal with a prior history of reinforced operant behaviour (Thompson and Bloom 1966) or between two animals with a previous history of contingent reinforcement (Davis and Donenfeld 1967). Following the demonstration that rats attack other rats in a straight alley on extinction trials (Gallup 1965), Miczek (1974) developed an ingenious procedure for studying frustration–induced aggression in rats. Food deprived rats were first individually trained to run from one compartment to the other through a tube connecting the two compartments, in order to be reinforced by a food pellet. On the day of test, one rat was placed in each compartment with no food available and entrances to the tube were raised simultaneously. Since the width of the tube did not allow more than one rat to move from one compartment to the other, one of the rats invariably forced the other back into its starting compartment, an activity accompanied by intense agonistic interactions. Within a few sessions, a stable hierarchy was evident, with the subordinate displaying typical defensive–submissive postures in response to threats and attacks from the dominant animal. In a replication of this work, however, stable dominance subordinance relationships were observed in only half the pairs (Lagarde 1981). Observational studies involving detailed analyses of agonistic patterns in extinction situations have also been carried out in pigs (Arnone and Dantzer 1980a; Dantzer et al. 1980) and in chickens (Duncan and Wood–Gush 1971).

In monkeys, frustration–induced aggression has mainly been studied by recording biting responses directed toward a pneumatic hose when animals were submitted to fixed ratio schedules of reinforcement (Hutchinson et al. 1968).

Characteristics of frustration–induced aggression

If aggression is causally related to frustration, then aggressive behaviour should be a positive function of the reinforcement value of the frustrated goal response, the degree of frustration of this goal response and the number of frustrated response sequences. Parametric studies of frustration–induced aggression have mainly been carried out on aggression displayed by pigeons during extinction or fixed ratio components of multiple schedules of food reinforcement (cf. Looney and Cohen 1982, for a recent review of aggression elicited by intermittent schedules of reinforcement). Working with pigeons conditioned to peck a key under a procedure that alternated periods of food reinforcement with periods of extinction, Azrin et al. (1966) reported that the duration of extinction–induced aggression displayed against a nearby conspecific, was an inverse function of time since the last reinforcement and a direct function of the number of reinforcements. In addition, prior satiation reduced the probability of attack. In another study, attack developed by pigeons submitted to an intermittent food delivery schedule was found to be directly related to deprivation level (Dove 1976). The probability of attack against a restrained target pigeon was found to be related to the fixed ratio (FR) requirement in pigeons trained to key peck for food on a multiple schedule of reinforcement including components of continuous and fixed ratio reinforcement. No attack occurred during continuous reinforcement and fixed ratio 15. Attack occurred occasionally during FR 25 and 40, and frequently during FR 60 and 120 (Knutson 1970). In the same study, the probability of aggression that developed during FR or extinction, components of the multiple schedule of food reinforcement decreased as a function of time since the last reinforcement, i.e. more aggression occurred immediately after the reinforcement than later on.

The temporal characteristics of frustration–induced aggression have been studied more accurately in response–independent food schedules. Staddon (1977) has suggested that activities occurring in the inter–food intervals of an intermittent food delivery schedule belong to three functional categories: (1) interim activities which occur immediately after eating and which are schedule–induced in the sense that they are excessive and directly related to food rate; (2) terminal activities which occur before the delivery of food and are related to anticipation of food, and (3) facultative activities which occur in between and are not induced by the intermittent food schedule. To qualify as schedule–induced, an attack must decrease in probability as a function of time to the next food delivery, be excessive, vary as a function of the animal's body weight and be differentially affected by reinforcement frequency (Falk 1977; Roper 1981; Staddon 1977). These rules are built on the model of schedule–induced polydipsia. Excessiveness of schedule–induced drinking has been assessed by comparing drinking displayed during exposure to the intermittent food schedule with that which occurs during a baseline session when all the scheduled food is delivered at the beginning of an equal duration session. A schedule–induced activity is typically less intense during baseline conditions than during the intermittent food schedule. In contrast with schedule–induced drinking, a no–food baseline has been commonly used to evaluate the role of the intermittency of food in inducing attack in pigeons. The typical finding is that the amount of attack associated with intermittent food delivery exceeds that during the no–food sessions (Azrin et al. 1966; Gentry 1968). Using a massed food delivery, Yoburn and Cohen (1979) confirmed that attack rates during intermittent food delivery schedules were higher relative to both the massed food and no–food baselines.

The relationship between attack rates and inter–food intervals has been found to follow an inverted U shape similar to the one described for schedule–induced drinking (Cherek et al. 1973; Cohen and Looney 1973; Flory 1969). However, some conflicting results have also been reported (Yoburn and Cohen 1979).

It would appear that the available findings are generally consistent with the proposition that attack observed in animals submitted to intermittent food delivery schedules is just another example of schedule–induced or adjunctive activities, i.e. activities that typically occur in situations of conflict and thwarted motivation (Falk 1971; 1977). The validity of this proposition can be tested by the extent to which schedule–induced attack can substitute for other forms of schedule–induced behaviour. This is not easy to test since the probability of appearance and development of different types of schedule-induced activity are very variable. In pigeons, for example, the prototypical schedule-induced activity, drinking, is very difficult to obtain. It is therefore not surprising that Yoburn and Cohen (1979) were unable to observe a similar increase in drinking and attack under intermittent food delivery conditions. In the same manner, schedule-induced drinking has a higher probability of appearance than attack in rats and the activities do not substitute for each other when either of them is prevented (Knutson and Schrader 1975). These results can be interpreted as suggesting that drinking and attack belong to different functional categories or, more likely, that once elicited, each activity becomes dependent on its own controlling factors rather than being just driven by a common energizing factor.

Relationship of frustration-induced aggression to other forms of aggression

The frustration elicited by the intermittent distribution of small food deliveries to highly motivated individuals is considered to have diffuse drive properties. Killeen et al. (1978) proposed that heightened arousal is the main determinant of the excessiveness of schedule-induced behaviour. Hungry animals exposed to periodic delivery of food remain highly aroused when kept in presence of cues predicting food. Because of the small size of the food rewards, bouts of eating behaviour are too short to allow this heightened arousal to dissipate.

Arousal therefore cumulates over the successive food deliveries, if opportunities for alternative consummatory behaviour are not available. According to this interpretation, schedule-induced attack would be one of the means by which animals attempt to cope with the frustrating situation and the development of the former would be critically dependent on its de-arousal properties. There has been no direct test of the possible de-arousal properties of frustration-induced aggression, although there is evidence that other forms of schedule-induced behaviour are accompanied by decreased physiological activation (Brett and Levine 1979; Dantzer and Mormede 1982; Tazi et al. 1986). The point of the present discussion is, however, whether aggression occurring in frustrating situations is different from other forms of aggressive behaviour, such as intermale fighting, predatory aggression or irritable aggression.

A useful approach in that respect is to describe frustration-induced aggression in terms of target areas for pecks or bites and basic strategies of the attacker (Blanchard et al. 1977; Brain 1981). In pigeons, casual observations of schedule-induced aggression suggest that attacks consist mainly of short-duration pecks directed towards the throat and the head, especially the eyes, of the target bird (Looney and Cohen 1982). During these aggressive pecks, pigeons keep their eyes closed and their beaks open. Vocalizations and other behavioural patterns such as bow-cooing and ground pecks have also been noted. However, the similarity of these responses to attack displayed in natural agonistic encounters has not been systematically assessed.

Although there is some evidence that schedule-induced attack is more easily obtained in aggressive strains of pigeons (Looney and Cohen 1982), a direct test of this possibility using cocks selectively bred for aggressiveness has failed to confirm this possibility. More specifically, when submitted to a time-out procedure similar to the one described by Moore et al. (1976), cocks of a commercially available strain did not differ from fighting cocks in their rate of schedule-induced mirror responding, in spite of clear evidence for higher aggression toward conspecifics and humans in the latter animals (Dantzer, unpublished results).

In a systematic study of aggressive behaviour in pairs of food-deprived adult hens exposed to food they could see but not obtain, aggression consisted mainly of threats and pecks to the head and neck regions (Duncan and Wood-Gush 1971). Grips and chases were rarely observed. In every case, only the dominant member of the pair showed any aggression. Birds intermediate in the hierarchy were aggressive when tested with a subordinate bird but not when tested with a dominant one. The same result had been observed in goats several decades ago by Scott (1948). He was therefore able to conclude that frustration per se does not induce aggression, but merely increase the probability of such a behaviour when eliciting stimuli, or releasers, are present (Scott 1975). A similar conclusion was reached by Arnone and Dantzer (1980a) in a systematic comparison of frustration–induced aggression to other forms of aggression. Pigs with a previous history of operant responding for continuous reinforcement were submitted in pairs to food competition tests or extinction sessions and their behavioural and physiological responses to these conditions were studied. During the food competition tests, pairs of animals were put together into the operant conditioning cage with only one response panel and one food bowl available. This resulted in the development of a stable dominance–subordinance relationship, with agonistic behaviours mainly consisting of pushing and biting episodes, without true fighting.

Plasma cortisol levels were not influenced by food competition nor by social rank. In animals exposed in pairs to extinction sessions, the outcome depended on the degree of acquaintance of the members of the pair. In pairs made up with previously acquainted animals, aggression did not increase over the level reached in food competition and plasma cortisol levels were not altered. This lack of response of the pituitary–adrenocortical axis to frustration was in contrast to the increase normally observed in pigs individually subjected to extinction (Dantzer et al. 1980).

In pairs made up with unacquainted animals, aggression was significantly more severe and it was accompanied by elevations of plasma cortisol levels. This increase was more marked in subordinate than in dominant pigs. The behavioural and physiological responses to extinction in pairs of unacquainted animals were essentially similar to those observed in pigs submitted to paired social encounters. In a different experiment, pigs were submitted in pairs to extinction with or without access to the response panel and the food bowl (Dantzer et al. 1980). Pushing, biting and fighting occurred more frequently when a partition prevented the animals from reaching the response panel and food bowl.

The view that frustration does not induce aggression but merely facilitates its development in the presence of appropriate cues is important because it implies that frustration–induced aggression is non–specific. This lack of specificity is best illustrated by the results of a recent study performed by De Waal (1984) on reconciliative behaviour in small groups of rhesus monkeys. The original hypothesis tested by De Waal was derived from the frustration–aggression viewpoint. It postulated that frustration elicited by competition for food should increase aggression and eventually result in a high incidence of positive contact behaviour, an important means by which social cohesion can be maintained in monkeys. However, when competition for food was systematically varied by contrasting monopoly tests (only one piece of apple was made available to all members of the group) with equality tests (several small pieces of apple were thrown into the pen), the amount of aggression remained the same, but the frequency of reconciliative behaviour increased only in the monopoly condition. These results were interpreted by postulating that inequality within the group due to food competition does not cause aggression but rather increases social tension, and that the amount of reconciliative behaviour is a direct function of this tension.

Human studies

The generated conclusion that frustration can facilitate aggression only in the presence of appropriate cues is in contrast to the original formulation of the frustration–aggression hypothesis. In their influential monograph on frustration and aggression, Dollard et al. (1939) stated that "the occurrence of aggression always presupposed the existence of frustration, and contrariwise, the existence of frustration always leads to some form of aggression". It was soon recognized that the second part of the statement was misleading since it could be taken to imply that frustration has no consequence other than aggression and that other responses can dominate and actually inhibit aggression. Miller therefore proposed that "frustration produces instigation to aggression, but this is not the only type of instigation that it may produce. Responses incompatible with aggression may, if sufficiently instigated, prevent the actual occurrence of acts of aggression" (Miller 1973).

There is little evidence to support the frustration–aggression hypothesis in human studies. At best, it applies only to the vigour of the response, but not to actual aggressive behaviour. For example, Haner and Brown (1955) did a study in which children playing a game had to push a plunger to start the game again when a loud buzzer was turned on. The results showed that the closer the children were to completing the game, the harder they pushed the plunger. In other words, frustration is more likely to occur when interruption happens close to the goal than far away from it. More recently, Kelly and Hake (1970) studied frustrative reactions of teenagers who pulled a knob for monetary reward and concurrently pressed a button or punched a padded cushion with a higher force requirement to escape or avoid periodic presentations of an aversive tone. Button pressing was normally the preferred avoidance response.

However, after discontinuation of the monetary reinforcement, most of the subjects chose the non–preferred punching response. When the pushing response was replaced by another non–preferred, but non-aggressive response (twisting a door knob), neither this response nor button pressing increased during non–reward. Although these findings are presented as evidence that frustration does not enhance responding in a diffuse way but specifically increases the response having aggressive components, they are likely to be biased by the strong habit we have developed to punch or kick coin machines that do not work properly!

Attempts have been made to assess the relationships between frustration and aggression in a variety of studies differing in the way that frustration is elicited, the nature of the aggressive response (from tearing apart dolls to apparently delivering electric shocks to the experimenter's confederate) and the availability of "aggressive" cues (e.g. weapons or films of violence). Generally, even in children, non reward leads to increased performance vigour in only certain instrumental tasks and the reaction to frustration appears dependent on several individual variables (Ryan and Natson 1968). Frustration is believed to instigate aggression only when there are cues associated with direct threat and when aggression has an instrumental value in helping to reach the desired goal. Both Bandura (1973) and Berkowitz (1974) have suggested that aversive stimuli facilitate aggression through their general arousing properties. When emotional arousal is labeled as anger, it leads to aggressive behaviour primarily directed to the provoker. If aggression cannot be directed against the source of frustration, then it may be redirected against targets associated with that source. For example, prison inmates frustrated by psychological and material deprivations, are unable to express their aggression against prison guards and are therefore likely to divert their aggression toward fellow inmates (Kosewski 1979).

Whether anger or arousal leads to aggression directed primarily against the source of annoyance or against associated targets, the underlying theoretical position is the same and is some variant of what is known as the cognition–arousal theory (Schachter 1964). This theory sees emotion as the product of an interaction between physiological arousal (characterized as heightened sympathetic activation) and view of the cause of that arousal. An extension of this theory, the excitation transfer theory (Zillman 1971), emphasizes the possibility of an addition of residual arousal from prior situations with arousal induced in the specific situation. This combined arousal intensifies both emotional experience and emotional behaviour in the later context and is misattributed to the emotional stimuli present there. The excitation transfer theory has been helpful in explaining, for instance, how residual sexual arousal can promote aggression and vice–versa (Zillman 1979).

However, its foundations are as shaky as the Schachter's theory (cf. Leventhal and Tomarken 1986, for a recent account of this theory).

In conclusion, both animal and human studies provide no support for the view that frustration has a direct influence on aggression, but suggest it may facilitate such expression in certain individuals in the presence of appropriate cues. The mechanisms by which these facilitatory influences take place are still controversial.

DRUGS AND FRUSTRATION–INDUCED AGGRESSION

Although it would be desirable to organize this section according to the different classes of psychotropic drugs, the methodologies used for studying effects of drugs on frustration–induced behaviour are so variable that this is impossible.

The Skinnerian approach

Attempts to assess the effects of drugs on aggression using a baseline attack elicited by intermittent reinforcement schedules is very traditional, in the sense that it is mainly descriptive, relies heavily on the use of measures of attack rate as a convenient dependent variable and disregards other behavioural elements.

Aggression induced by intermittent reinforcement schedules was initially believed to be a very useful technique for studying the effects of drugs on aggression. The reasons for this view were twofold: (1) the methods of schedule–induced aggression facilitated and automated objective and quantitative assessment of aggressive behaviour, and (2) it was possible to compare the concurrent effects of drugs on aggressive behaviour and food–reinforced responding, in the same animal during the same session.

This latter point is important in attempts to determine the specificity of drug action on aggression. Viewed in this way, the methods of schedule–induced aggression held the same promises for testing potentially anti–aggressive drugs as the approach–avoidance conflict procedure for assessing anxiolytic drugs. For example, Cherek and Thompson (1973) reported that schedule–induced attack of a live target in pigeons was suppressed to a greater degree by delta 1 – tetrahydrocannabinol (THC) than the rate of key–pecking. This was the case even when the rates of the two behaviours (key–pecking and attack) were balanced by appropriate parametric manipulation of the reinforcement schedule.

Surprisingly other drugs tested in the same paradigm acted in exactly the same way as THC. Chlordiazepoxide, for example, produced a marked decrease in schedule–induced mirror responding at doses that had little or no effect on key–pecking for food (Moore et al. 1976). Cocaine had the same effect on extinction attack responses against a stuffed target pigeon (Hutchinson et al. 1977) and on schedule–induced mirror responding (Moore and Thompson 1978). High doses of cocaine disrupted food–reinforced responding and chronic administration resulted in tolerance to this effect, but there was no tolerance to the effect on aggressive behaviour.

There has been no attempt to determine why schedule–induced aggression in pigeons is so easily suppressed by such a variety of psychotropic drugs. It is tempting to assume that the decreases in attack rates reflect some non–specific factor such as state–dependency rather than a truly anti–aggressive action. Although it would also be useful to compare the effects of drugs on schedule–induced attack with their effects on other schedule–induced behaviours, this task appears hopeless because most of the research on the pharmacology of schedule–induced polydipsia has been carried out in rats while most pharmacological studies on schedule–induced attack used pigeons.

The ethological approach

More useful information about the mechanisms of action of drugs on behaviour has been gained from studies relying on detailed behavioural analyses of the agonistic activities displayed by frustrated animals. Using the extinction–induced aggression paradigm previously described, Miczek investigated the effects of several drugs on the pattern of offence displayed by the dominant rat and the patterns of submission–defence displayed by the subordinate rat. The drugs under study were administered to the dominant or to the subordinate animal. This approach can be examplified by the results of delta 9 – THC treatment (Miczek and Barry 1974). Subordinates treated with THC spent less time in the submissive supine posture but more time in the immobile crouch posture. As a result, they were more severely attacked by undrugged dominants and suffered more injuries. In contrast, treated dominants attacked less frequently and displayed fewer threats, so that opponents engaged significantly less frequently in the mutual upright posture. In spite of the reduced attack by the dominant, the undrugged subordinate animal did not show increased aggressiveness. In a different series of experiments, low doses of chlordiazepoxide or amphetamine significantly increased offensive postures in the dominant animal (Miczek 1974; 1977). Chlordiazepoxide prolonged submissive and defensive postures in the subordinate animal. Another benzodiazepine, clorazepate, was found to enhance offensive postures in both dominant and subordinate animals, when either were treated (Lagarde 1981). This variation from Miczek's results could be due to a weaker differentiation between the dominant and the subordinate members of the pairs under study. Similar effects of drugs on various components of attack and defence have been described in other paradigms, such as the resident–intruder model (Miczek and Krsiak 1979). These findings suggest that agonistic patterns are a critical determinant of drug action. A similar conclusion has been reached in a study of the effects of diazepam on extinction–induced aggression in pigs (Arnone and Dantzer 1980b). As mentioned earlier, pigs with a previous history of continuous reinforcement for food were tested by pairs in extinction under two different conditions, with or without access to the response panel and the food bowl. In the first case, when access to the response panel and food bowl was permitted, diazepam–treated animals spent longer trying to get access to the food bowl and pushed the response panel for a longer duration than placebo–treated animals. Aggression which consisted mainly of pushing and biting episodes, was not modified by diazepam. In contrast, when access to the response panel and the food bowl was not permitted, diazepam increased the severity of aggression between animals. Diazepam has been shown to increase resistance to extinction in pigs submitted individually to the extinction procedure (Dantzer 1977a) and to enhance attack in paired encounter tests (Arnone 1979). These results may be interpreted as meaning that diazepam does not act on frustration or aggression per se, but strengthens the animal's prevailing behaviour at the time of the test. The way such an effect is obtained and its relationship with other behavioural effects of benzodiazepines have been discussed elsewhere (Dantzer 1977b; 1978; 1986).

The relationships between competition for food, frustration and aggression were examined earlier in this chapter, when reference was made to the finding that food competition enhances reconciliative behaviour in small grousp of monkeys (De Waal 1984). There has been no systematic comparison of the effects of drugs on aggression during competition for food with other forms of aggressive behaviour. Chronic treatment of pigs with lithium carbonate gradually decreases intra–group aggression displayed during monopoly tests. The same treatment did not prevent aggression and fighting that occurred after mixing pigs from different origins (Dantzer and Mormede 1979). The finding that lithium did not attenuate inter–group aggression unless administered at toxic doses was confirmed by McGlone et al. (1981). Rather than suggesting a selective effect of lithium on frustration, these results are consistent with the hypothesis that lithium is more effective on behaviours under weak rather than strong stimulus control (Johnson 1979).

CONCLUSION

Frustration-induced aggression is a research subject with a rich past but an impoverished present. The same is true for pharmacological studies of frustration-induced aggression. The Skinnerian perspective to the effects of drugs on aggression occurring in frustrating conditions has contributed little to our understanding of the mechanisms of action of drugs and has failed to generate useful concepts or methodologies for the field of behavioural pharmacology. Methodological improvements in the analysis of agonistic patterns and parametric studies of the effects of drugs on extinction-induced aggression have, in contrast, generated data that have been influential for the current thinking on psychopharmacology of aggression. However, the present emphasis on ethological approaches in pharmacology of social behaviour has made obsolete any concern for the issue of drug effects on frustration-induced aggression.

Papers on schedule-induced attack referenced in Animal Behaviour Abstracts have appeared at the rate of 1-2 per year during the last three years, and there seems no reason why this trend should change in the near future. This is unfortunate, because some of the issues we have discussed need a thorough reevaluation. This is particularly the case for the instrumental or coping value of aggression in frustration situations (Ursin 1981). However, the contribution of psychopharmacology to this issue and, conversely, the importance of these concepts for the psychopharmacological analysis of aggression are doubtful.

Acknowledgement: This paper was prepared while the author was on sabbatical leave at the Department of Animal Sciences, University of Illinois, Urbana, Il 61801.

REFERENCES

Arnone M (1979) Etude comportementale et pharmacologique des conduites agressives chez le porc. Unpublished PhD Thesis, University Paul Sabatier, Toulouse, pp 1-108

Arnone M, Dantzer R (1980a) Does frustration induce aggression in pigs? Appl Anim Ethol 6: 351-362

Arnone M, Dantzer R (1980b) Effects of diazepam on extinction-induced aggression in pigs. Pharmacol Biochem Behav 13: 27-30

Azrin NH, Hutchinson RR, Hake DF (1966) Extinction-induced aggression. J Exp Anal Behav 9: 191-204

Bandura A (1973) Aggression: A social learning analysis. Prentice Hall, Englewood Cliffs, New Jersey

Berkowitz L (1974) Some determinants of impulsive aggression: role of mediated associations with reinforcement. Psychol Rev 81: 165-176

Blanchard RJ, Blanchard DC, Takahashi LK (1977) Reflexive fighting in the albino rat: aggressive or defensive behavior? Aggr Behav 3: 145-155

Brain PF (1981) Differentiating types of attack and defense in rodents. In: Brain PF, Benton D (eds) Multidisciplinary approaches to aggression. Elsevier, Amsterdam, pp 53-77

Brett LP, Levine S (1979) Schedule–induced polydipsia suppresses pituitary–adrenal activity in rats. J Comp Physiol Psychol 93: 946–956

Cherek DR, Thompson T (1973) Effects of 1–tetrahydrocannabinol on schedule–induced aggression in pigeons. Pharmacol Biochem Behav 1: 493–500

Cherek DR, Thompson T, Heistad GT (1973) Responding maintained by the opportunity to attack during an interval food reinforcement schedule. J Exp Anal Behav 19: 113–123

Cohen PS, Looney TA (1973) Schedule–induced mirror responding in the pigeon. J Exp Anal Behav 19: 395–408

Dantzer R (1977a) Effects of diazepam on behaviour suppressed by extinction in pigs. Pharmacol Biochem Behav 6: 157–161

Dantzer R (1977b) Behavioural effects of benzodiazepines: a review. Biobehav Rev 1: 71–86

Dantzer R (1978) Dissociation between suppressive and facilitating effects of aversive stimuli on behavior by benzodiazepines. A review and a reinterpretation. Progr Neuropsychopharmacol 2: 33–40

Dantzer R (1986) Behavioral analysis of anxiolytic drug action. In: Greenshaw AJ, Dourish CT (eds) Experimental approaches in psychopharmacology. Humana Press, Clifton, New Yersey, in press

Dantzer R, Arnone M, Mormede P(1980) Effects of frustration on behaviour and plasma corticosteroid levels in pigs. Physiol Behav 24: 1–4

Dantzer R, Mormede P (1979) Effects of lithium on aggressive behaviour in domestic pigs. J Vet Pharmacol Ther 2: 299–303

Dantzer R, Mormede P (1982) Pituitary–adrenal consequences of adjunctive activities in pigs. Horm Behav 15: 386–395

Davis H, Donenfeld I (1967) Extinction induced social interaction in rats. Psychon Sci 7: 85–86

De Waal FBM (1984) Coping with social tension: sex differences in the effect of food provision to small rhesus monkey groups. Anim Behav 32: 765–773

Dollard J, Doob LW, Miller NE, Mowrer OH, Sears RR (1939) Frustration and aggression. Yale University Press, New Haven, Connecticut

Dove LD (1976) Relation between level of food deprivation and rate of schedule–induced attack. J Exp Anal Behav 25: 63–68

Duncan IJH, Wood–Gush DGM (1971) Frustration and aggression in the domestic fowl. Anim Behav 19: 500–504

Falk JL (1971) The nature and determinants of adjunctive behavior. Physiol Behav 6: 577–588

Falk JL (1977) The origin and functions of adjunctive behavior. Anim Learn Behav 5: 325–335

Flory RK (1969) Attack behavior as a function of minimum interfood interval. J Exp Anal Behav 12: 825–828

Gallup GG Jr (1965) Aggression in rats as a function of frustrative nonreward in a straight alley. Psychon Sci 11: 99–100

Gentry WD (1968) Fixed–ratio schedule–induced aggression. J Exp Anal Behav 11: 813–817

Haner CF, Brown PA (1955) Clarification of the instigation to action concept in the frustration–aggression hypothesis. J Abn Soc Psychol 51: 204–206

Hutchinson RR, Azrin NH, Hunt GM (1968) Attack produced by intermittent reinforcement of a concurrent operant response. J Exp Anal Behav 11: 489–495

Hutchinson RR, Emley GS, Krasnegor NA (1977) The effect of cocaine on the aggressive behavior of mice, pigeons and squirrel monkeys. In: Ellingwood EH Jr, Kilbey MM (eds) Cocaine and other stimulants, Plenum Press, New York, pp 457–480

Johnson FN (1979) The psychopharmacology of lithium. Neurosci Biobehav Rev 3: 15–30

Kelly JF, Hake DF (1970) An extinction–induced increase in an aggressive response with humans. J Exp Anal Behav 14: 153–164

Killeen PR, Hanson SJ, Osborne SR (1978) Arousal: its genesis and manifestation as response rate. Psychol Rev 85: 571–581

Knutson JF (1970) Aggression during the fixed ratio and extinction components of a multiple schedule of reinforcement. J Exp Anal Behav 13: 221–231

Knutson JF, Schrader SP (1975) A concurrent assessment of schedule–induced aggression and schedula–induced polydipsia in the rat. Anim Learn Behav 3: 16–20

Kosewski M (1979) The prison as an aggressive social organization. In Feschbach S, Fraczek A (eds) Aggression and behavior change. Praeger, New York, pp 208–228

Lagarde B (1981) Psychopharmacologie de l'agression: effets d'un sedatif anxiolytique sur les comportements agonistiques du rat. Unpublished Master Thesis, Universite de Bordeaux II

Leventhal H, Tomarken AJ (1986) Emotion: today's problems. Ann Rev Psychol 37: 565–610

Looney TA, Cohen PS (1982) Aggression induced by intermittent positive reinforcement. Neurosci Biobehav Rev 6: 15–37

Looney TA, Cohen PS, Yoburn BC (1976) Variables affecting the establishment of schedule–induced attack on pictorial targets in White King pigeons. J Exp Anal Behav 26: 349–360

McGlone JJ, Kelley KW, Gaskins CT (1981) Lithium and porcine aggression. J Anim Sci 51: 447–455

Macurik KM, Kohn JP, Kavanaugh E (1978) An alternative target in the study of schedule–induced aggression in pigeons. J Exp Anal Behav 29: 337–339

Miczek KA (1974) Intraspecies aggression in rats: effects of D–amphetamine and chlordiazepoxide. Psychopharmacologia 39: 275–301

Miczek KA (1977) Effects of alcohol on attack and defensive–submissive reactions in rats. Psychopharmacology 52: 231–237

Miczek KA, Barry H III (1974) Δ9–tetrahydrocannabinol and aggressive behavior in rats. Behav Biol 11: 261–267

Miczek KA, Krsiak M (1979) Drug effects on agonistic behavior. Adv Behav Pharmacol 2: 87–162

Miller NE (1973) The frustration–aggression hypothesis. In: Marple T, Matheson DW (eds) Aggression, hostility and violence, nature or nurture. Holt, Rinehart and Winston, New York, pp 103–115

Moore MS, Thompson DM (1978) Acute and chronic effects of cocaine on extinction–induced aggression. J Exp Anal Behav 29: 309–318

Moore MS, Tychson RL, Thompson DM (1976) Extinction–induced mirror responding as a baseline for studying drug effects on aggression. Pharmacol Biochem Behav 4: 99–102

Roper TJ (1981) What is meant by the term "schedule–induced" and how general is schedule induction? Anim Learn Behav 9: 433–440

Ryan TJ, Natson P (1968) Frustration and non reward theory applied to children's behavior. Psychol Bull 69: 111–125

Schachter S (1964) The interaction of cognitive and physiological determinants of emotional state. Adv Exp Soc Psychol 1: 49–80

Scott JP (1948) Dominance and the frustration–aggression hypothesis. Physiol Zool 21: 31–39

Scott JP (1975) Aggression, 2nd edition, University of Chicago Press, Chicago

Staddon JER (1977) Schedule–induced behavior. In: Honig WK, Staddon JER (eds) Handbook of operant behavior, Prentice–Hall, Englewood Cliffs, New Jersey, pp 125–152

Tazi A, Dantzer R, Mormede P, Le Moal M (1986) Pituitary–adrenal correlates of schedule–induced polydipsia and wheel running in rats. Behav Brain Res 19: 249–256

Thompson TT, Bloom W (1966) Aggressive behavior and extinction–induced response rate increase. Psychon Sci 5: 335–336

Ursin H (1981) Neuroanatomical basis of aggression. In: Brain PF, Benton D (eds) Multidisciplinary approaches to aggression research. Elsevier, Amsterdam, pp 269–291

Yoburn BC, Cohen PS (1979) Assessment of attack and drinking in White King pigeons on response–independent food schedules. J Exp Anal Behav 31: 91–101

Yoburn BC, Cohen PS, Campagnoni FR (1981) The role of intermittent food in the induction of attack in pigeons. J Exp Anal Behav 36: 101–117

Zillman D (1971) Excitation transfer in communication–mediated aggressive behavior. J Exp Soc Psychol 7: 419–434

Zillman D (1979) Hostility and aggression. Lawrence Erlbaum Publishers, Hillsdale, New Jersey

PHARMACOLOGICAL ASPECTS OF SOCIAL COOPERATION

Barry D. Berger and Richard Schuster. Department of Psychology, University of Haifa, Haifa, 31999, Israel.

At first impression, it may seem paradoxical that a paper on cooperation would be included in a symposium devoted to aggression. Aggression and cooperation are usually depicted as contrasting and unrelated phenomena. Aggression is typically applied to an inherently selfish interaction, whereby one individual gains at the expense of another who loses by virtue of injury, expulsion, or reduced access to a desired resource. In cooperation, the social interaction is usually characterized as positive, involving a joint or collaborative action directed towards obtaining a desired goal for one or both (Hake and Vukelich 1972; Nisbet 1968). This dichotomous approach is supported by a tendency in laboratory research to study aggression or cooperation in isolation from one another. Aggression has been by far the more popular subject. The few studies on cooperation have been aimed at demonstrating whether animals are capable of behaviour meeting the criteria of cooperation. Thus, rats learned to coordinate an exchange of places for feeding without obtaining shock (Daniel 1942), chimpanzees learned a series of food discrimination tasks (Crawford 1941), rhesus monkeys learned to avoid shock by attending to social cues of fear (Miller et al. 1962), and pigeons learned to peck on corresponding keys in near simultaneity (Skinner 1953).

An alternative approach seeks to establish a linkage between aggression and cooperation. If aggression and cooperation are actually opposing phenomena, a reciprocal relationship is implied whereby an increase in one would mean, ipso facto, a decrease in the other. Although such reciprocity between aggression and cooperation may be demonstrated (Schuster et al. 1982), the relationship between aggression and cooperation may turn out to be more complex than simple reciprocity. Field observations of many social species suggest that prolonged coexistence, whether within the framework of a stable social group or reproductive pair bond, involved a mix of cooperative and competitive social interactions and even occasional violence (Schuster et al. 1982). The observed tendency towards stability suggests that aggression and cooperation are not nearly as incompatible as a reciprocal relationship might imply. Based on research with primates, Crook (1971) and Kummer (1971, 1974) proposed that aggression can even be an indespensible factor in the capacity for a cooperative relationship. The suggestion is that aggression, by intensifying social awareness while clarifying status and roles, creates conditions necessary for later cooperation.

Indeed, it may not be far-fetched to suggest that aggression alone, when occurring repeatedly between the same individuals over a span of time, contains many if not all of the ingredients for a cooperative relationship. When defined operationally cooperation requires the collaboration of two (or more) organisms in procuring positive, or removing negative reinforcement for at least one (Hake and Vukelich 1972; Keller and Schoenfeld 1950; Marwell and Schmitt 1975). Sustaining a cooperative relationship requires, minimally, that (1) each organism's action be a discriminative social cue for the other's behaviour, and (2) each organism be reinforced for his particular role.

These criteria could equally describe the aggressive relationship between two humans in a boxing ring, two antelopes on neighboring territories or two baboons in the same hierarchy. All of these situations involve a pair of individuals engaged in repeated and highly coordinated interactions which limit injury and stress through the use of ritualized or conventional means of conflict with clearly defined "rules" governing victory and defeat. While the precise relationship between aggression and cooperation awaits discovery, it is not too soon to propose that they can be interrelated processes in the formation and maintenance of a social relationship.

Our interest in cooperation reflects a more general neglect of the whole area of social interaction which is not exclusive to experimental psychology. In psychopathology, the traditional focus in symptomology, diagnostic criteria and definition of categories has been on the individual. More recently, there has been an upsurge of interest in social factors, particularly in recognizing states occurring uniquely in a social context becoming manifest in specifically sociopathic or anti-social behaviours (McGuire 1982; Lindner et al. 1970; Reid 1978; Spraque and Sleator 1977). Nevertheless, the emphasis upon the individual has been one of the factors steering research away from a social emphasis.

By relying almost exclusively upon the behaviour of individuals, it has been tacitly assumed that social phenomena, regardless of their emergent properties (Lindsley 1966) will not reveal anything new about the underlying behavioural and physiological processes governing action. Skinner (1953), was explicit: "..........a 'social law' must be generated by the behaviour of individuals. It is always an individual who behaves and he behaves with the same body and according to the same processes as in a non-social situation." (p.298)

In this theoretical climate, given the choice between studying behavioural processes by means of individual or social models, the individual has been clearly preferable. Using a single subject, individual action can be controlled by restricting stimuli and behavioural options. In a social paradigm requiring collaborative action, the independent variables – task, cues, reinforcement contingencies – provide only a context within which each pair works out its particular solution (Thibaut and Kelly 1961).

Another more practical advantage of the single subject is reliance upon dimensions of behaviour which can be easily automated. When studying social interaction, important aspects such as communication, aggression, and status require trained observers.

Notwithstanding these theoretical, historical, and methodological obstacles, it is our contention that the study of behavioural and physiological mechanisms of social interaction in general and cooperative behaviour in particular, is important on several levels. First, considering the incidence of social behaviour in its various forms in human and animal behaviour, it is unacceptable that this has been a largely untapped research area. Second, the communality of mechanisms between social and non social behaviour has only been assumed, never really tested and it is possible that the two broad classes involve qualitatively different mechanisms and follow different laws of behaviour. Finally, social behaviours may reveal subtle effects of the new generation of psychoactive drugs or of other specific brain manipulations more easily than the relatively crude and simple techniques using individual behaviours. Indeed, it is possible that mapping studies and drug development can open new subject areas for study and application.

In the studies reported here, an animal (rat) model of cooperative behaviour is described, and recent findings on the effects of psychoactive drugs on various aspects of the social interaction task presented.

THE APPROACH

Animals were adult naive albino (male, unless specified otherwise), Sprague – Dawley derived rats, obtained from a local supplier. They were maintained on

Purina Lab Chow ad libitum with water available for 30 minutes/day. The rats were housed two to a cage (grouped housing) except in specified cases when they were housed alone (isolated housing).

The apparatus (Research Authority, University of Haifa) consisted of a rectangular Plexiglass chamber divided into two compartments (A and B) of unequal size (fig. 1). A feeder with two dippers for the delivery of 0.1 cm³ sweetened water (10 parts water to 1 part sugar by weight) was located at one end of the chamber together with a magazine lamp that was lit during the 6-sec dipper cycle. The stainless steel grid floor was divided into three separate hinged sections, each mounted on microswitches that served as sensors for determining on which of the sections of the chamber the animals were located.

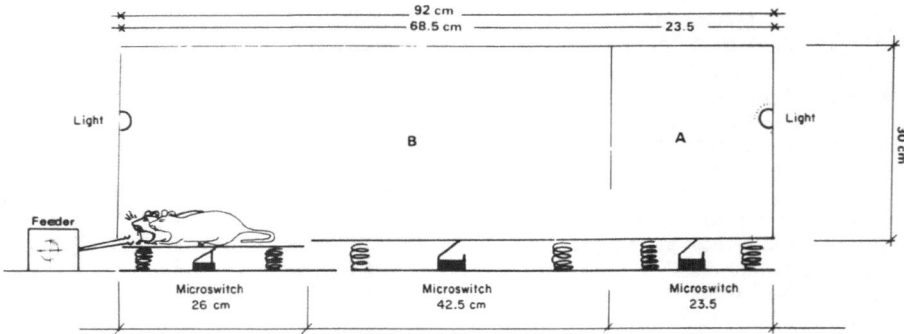

Fig. 1 Schematic diagram of the coordination learning apparatus. Pairs of rats must coordinate shuttling from compartment B to A and back again to B in order to obtain reinforcement.

For the coordination task, the animals were placed in the apparatus in pairs. Sugar water reward was delivered if both rats moved from the dipper end of the apparatus and were present together in the smaller (A) compartment and then moved back again and were present together in the dipper area of the larger (B) compartment in that sequence. At the end of the feeder cycle, the magazine light was turned off to start a new trial. There were fifty such trials per session, which was limited to a maximum duration of one hour (3600 seconds). The location of the two animals was monitored by the microswitches mounted under the three hinged floor sections. Thus, when both animals entered the A-compartment, only the A-compartment microswitches were activated, and when both animals then moved into the feeder area at the other end of the box, only that floor section was activated. These two unique states of the floor positions occurring in sequence represented correct "coordinated behaviour" and fulfilled the criteria for delivery of reinforcement. In a control task not based on coordination or social interaction, single animals, rather than pairs were placed in the apparatus and were run in exactly the same fashion as pairs, the sugar water being presented when the individual rat shuttled into the A-compartment and back into the B-compartment. Thus the sensory, motor, and motivational demands of the single animal task were similar or identical to those of the social task, differing primarily in the coordination demand of the two-animal social taks.

Several behavioural measures were recorded. Total time was defined as the time in seconds for the completion of 50 correct coordinations (in pairs of animals), or time to complete 50 shuttles in single animals. Proximity failure refers to the number of occurrences that the members of a pair were located at opposite ends of the apparatus. This measure provided a rate-independent measure of proximity and coordination. For single animals, the proximity failure was meaningless by definition. Correct response was defined as a trial in which the animal that first entered the A-compartment did not leave the compartment until the partner rat entered. This measure was also relevant only to pairs. In addition to these automated measures, the experimenter also recorded which animal of the pair was the first to enter and leave the A-compartment and which animal was the last, this in an attempt to identify possible leader-follower relations. Finally, any occurring aggression, was noted using the method of Valzelli (1974).

I. Behavioural and methodological studies:

Early in training, pairs of rats are inefficient at obtaining reinforcement using coordinated activity. This is evidenced by long time scores to obtain reinforcement, few errorless trials, and many proximity failures. Moreover, observation of the animals indicates little evidence that the location of a given animal is influenced by the position of its partner. However, as training progresses, performance of the cooperative task improves. Time scores typically reduce from 3000 sec or more in early sessions to 900 sec or less at asymptote, correct trials increase from approximately 20% to approximately 80% and proximity failures reduce from 4 - 5 per reinforcement to one or less. It is tempting to speculate that this improvement in performance is due to the learning of a cooperation strategy. However, it is also possible that other factors are involved e.g. the animals may have learned to pace themselves individually so as to optimize reinforcement in some adventitious fashion. Our early studies were designed to answer these and related methodological questions, and to address ourselves to the question of whether the paradigm in fact satisfies the criteria for cooperation behaviour mentioned earlier (Hake and Vukelich 1972; Keller and Schoenfield 1950; Marwell and Schmitt 1975). The results of these experiments reinforced the idea that the paradigm is indeed based on social cooperation.

Thus, we observed that restricting social contact between the partners by placing an opaque barrier between them in the apparatus or by placing each animal in a physically separate, but electronically yoked apparatus, severely impairs acquisition and performance of the two-animal task (Berger et al. 1980). This is taken as evidence that, under normal conditions, when social cues are available, the animals use each other's behaviour as discriminative stimuli in performing the task. This is the most basic criterion for cooperation behaviour.

Other experiments in our laboratories further revealed the social nature of the task. An indication that pairs develop and utilize specific social strategies, was demonstrated in a study in which the performance of pairs running with their same partners from session to session (fixed partners) was compared to pairs in which the partners were changed from session to session (switched partners). Switching partners either during acquisition or following training impaired performance in the task. In a second study, pairs were first trained in the task following which each member was retrained with a naive partner. The learning curve of the trained-naive pair was facilitated relative to the original learning curve of two naive animals, suggesting that imitation or even active "instruction" may take place (Berger et al. 1980; Berger and Schuster 1982).

In other studies, we have documented that: 1) manipulations such as mild footshock or nonreward that may induce stress and frustration improve the performance of the cooperation task (Schuster et al. 1982); 2) isolated housing impairs the learning and performance of the social task, but not a comparable

18

non-social problem, in male rats, but not in females (Schuster et al. 1982; Swanson et al. 1985); 3) the social task influences other behaviours such as expression of taste aversion (Berger et al. 1979); and 4) various social emergents such as coordination, proximity, communication, and leader-and-follower patterns of behaviour are observed in the course of coordination performance.

We now have evidence that performance of the cooperation task results in a type of pair bonding between the partners (Z. Ackerman, unpublished results). Following extensive training in the cooperation task, one member of the pair (test animal) was placed in a circular open field apparatus. At four locations in this apparatus (north, east, south and west) were cages, respectively containing: 1) its partner rat from the cooperation task; 2) the rat sharing its home cage; 3) a stranger rat and 4) no rat.

A sensing device measured the time spent by the test animal in the area near each of these four cages. In a control session to measure for possible position bias, the test animal was placed in the open field with all four cage locations being empty. Figure 2 summarizes the results. A clear preference was obtained for the animal that had served as its partner in the cooperation task. Interestingly, the test rats tended to prefer a stranger rat to their home cage mates (in agreement with S.File, personal communication), though this tendency was not statistically significant.

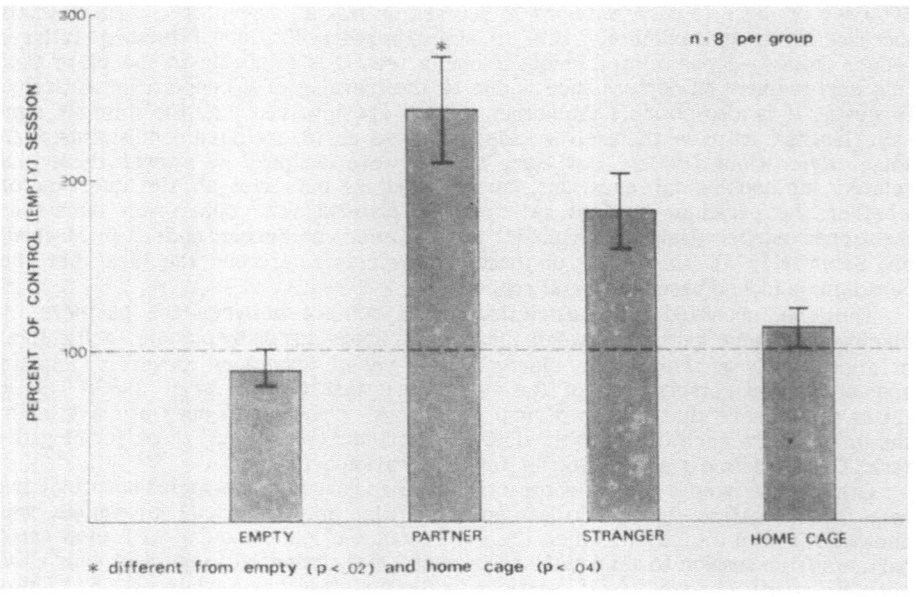

Fig. 2 Preference for pair – partner following training in the cooperation task in a test situation where the time spent near four possible compartments was measured. The compartments either were empty or contained the partner rat from the cooperation learning task, a stranger rat, or the home cage mate.

These results indicate that working together in the social task apparently produced a "bond" between the pair that generalizes to situations other than the training apparatus, and provides further support for the validity of the task as a measure of cooperative behaviour.

II. Drug studies:

The cooperation task is particularly advantageous for the study of the effects of drugs on social behaviour for several reasons. First, it is possible to determine if a given drug exerts a selective action on cooperation behaviour, by comparing its effects in the two-animal paradigm with its effects in the single-animal task. Second, the different objective and observational behavioural measures permit a quantitative and qualitative analysis of drug effects on various aspects of cooperation behaviour. Finally, in instances where normal cooperative behaviour is disrupted by competing responses of aggression (see below), a drug-induced facilitation of cooperation is reflected by an increase in responding, thus eliminating confounding factors of explanation in terms of sedation, a confounding factor common in traditional paradigms of social interaction.

FACILITATION OF COOPERATIVE BEHAVIOUR BY PROPRANOLOL

We have previously reported that the noradrenergic receptor blocker, dl-propranolol, facilitated cooperative behaviour that had been disrupted by aggression (Berger and Schuster 1982, 1985). In one experiment, pairs of rats, housed two animals per cage (grouped housing) were trained to stable performance in the cooperation task. At this point, the training regime was interrupted, and the pairs were isolated from all social contact for 21-days (isolated housing). When retested, the performance of the coordination task was greatly disrupted, and remained so even after 23 subsequent sessions of retraining (fig. 3). Repeated fights were observed between the partners during this period with the result that one animal of the pair typically froze in a crouched, submissive posture in one corner of the apparatus for the entire 60-minute session. The second (dominant) animal would usually remain in the opposite end of the apparatus unless the submissive partner begin moving which stimulated further attack.

A number of drugs were administered to both partners during this isolation-induced disruption of cooperation behaviour. In this experiment and in the following experiments, drugs were administered intraperitoneally, 15 minutes prior to the experimental session. Figure 3 gives an example, in which d-amphetamine, and diazepam failed to facilitate the disrupted performance, but dl-propranolol (5 mg/kg) caused a large improvement in performance in three separate replications (days 27, 28, 34).

Observation of the animals during propranolol treatment suggested several possible explanations for the facilitation of cooperation performance. One impression was that propranolol reduced the number and intensity of fights, in line with several recent reports (Miczek et al. 1984; Weinstock and Weiss 1980), or possibly by reducing the freezing behaviours and crouching postures that follow a fight and interfere with cooperation. Another possibility is that propranolol exerted a more direct effect on social interaction, rather than by an indirect action on aggression. This could include an action on sociability per se, an effect on proximity to the partner rat, on dominant-submissive or on leader-follower relationships. Finally, it is possible that propranolol caused motor, sensory, motivational, or emotional changes, and these effects adventitiously resulted in a facilitation of shuttlebox behaviour.

In order to test these, or other alternative explanations for the finding of facilitation of cooperation performance by propranolol, and in order to replicate and further document the finding, several additional experiments were conducted.

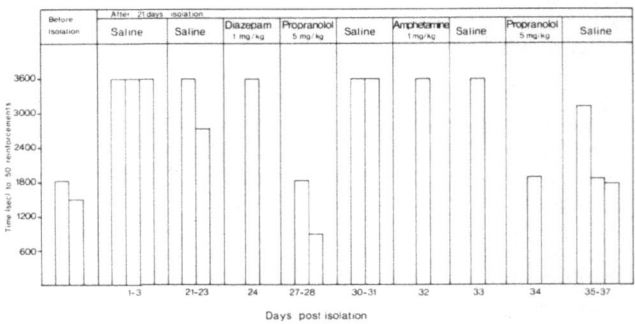

Fig. 3   dl–Propranolol (5 mg/kg, IP) selectively improves cooperation performance disrupted by 21 days of social isolation after acquisition (pair 88–89). A score of 3600 seconds to achieve 50 reinforcements indicates that the pair did not complete the session in the maximum time allotted. Diazepam (1 mg/kg IP) and d–Amphetamine (1 mg/kg IP) did not improve performance.

PROPRANOLOL AND ACQUISITION

This study manipulated housing condition (isolated vs grouped), social or individual learning task (pairs vs singles), and drug condition (saline vs propranolol) on the acquisition phase of cooperation learning. In agreement with previous findings in our laboratory (Schuster et al. 1982), isolated housing severely retarded acquisition of the cooperation task in pairs, but had little or no effect on the learning of the shuttle response in single animals. Treatment with propranolol, completely eliminated the disruptive effects of isolated housing on the acquisition of cooperation behaviour. A large facilitation in acquisition of cooperation was observed in isolated pairs treated with propranolol relative to isolated pairs treated with saline (fig. 4). However, propranolol did not appear to facilitate acquisition in animals housed in grouped conditions (fig. 5). Similarly, propranolol exerted no significant effect in single animals in the shuttlebox task.

These data replicate and extend the initial pilot study indicating that propranolol facilitates cooperation behaviour in animals housed in isolation. The propranolol effect is observed only in the pair and not in the single animal paradigm, suggesting that the drug acts on social features of the task, and not on more general motor, motivational, or sensory aspects. Moreover, the facilitation by propranolol was obtained in this experiment only under conditions of isolated housing. This suggests that the effect of propranolol is on aggression (commonly observed in pairs housed in isolation) rather than on social cooperation per se. Inspite of this, it should be pointed out that acquisition of cooperation in grouped housed animals is relatively rapid, and thus it may be difficult on methodological grounds for facilitative effects of propranolol to be observed. Also, other experiments in our laboratory have occasionally obtained facilitation by propranolol under conditions of grouped housing when aggressive behaviour is rarely observed (Berger and Schuster 1982; and see below). Thus, isolated housing is a sensitive, but not exclusive, condition for demonstrating facilitation of cooperation by propranolol.

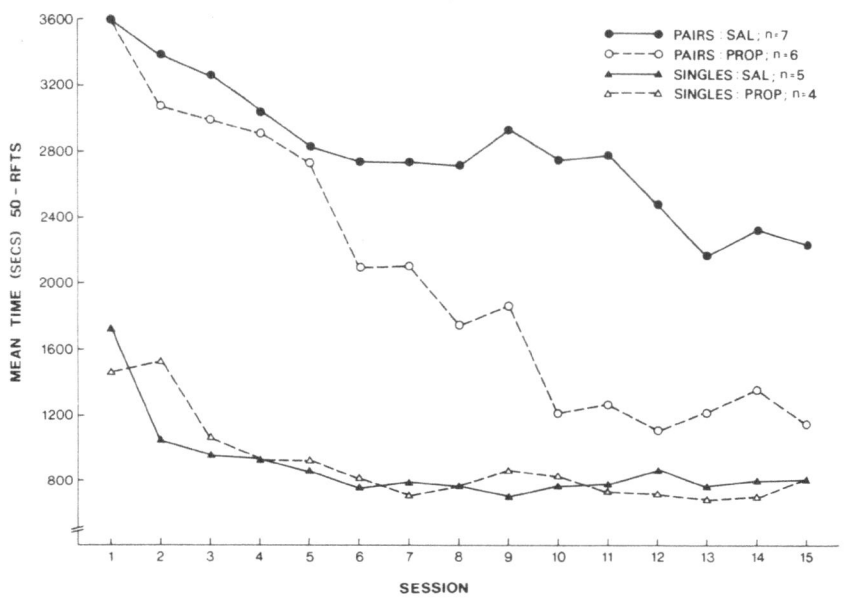

Fig. 4 Isolated Housing: dl–Propranolol facilitates acquisition of cooperation performance of pairs of male rats. Performance of the shuttle task in single animals was not affected by dl–propranolol.

PROPRANOLOL AND PARTNER CONDITION

The purpose of this study was to evaluate the effects of propranolol on cooperation behaviour in pairs of rats running with their same partners from day to day (fixed partners) as opposed to pairs running with a different partner each day (switched partner). Previous research had indicated that imposing the switched partner condition is disruptive to cooperation performance, and it was therefore of interest as to whether dl–propranolol would facilitate cooperation under these conditions as it had under the conditions of isolated housing.

Pairs of rats, housed two to a cage (grouped housing), were trained for 10 sessions in the cooperation apparatus. Half the pairs were trained under the influence of saline and half under the influence of propranolol (5 mg/kg, IP). During this phase, all pairs were run with their same partners from session to session (fixed partners). Following acquisition, training continued for an additional five sessions, but for each session the partners were changed (switched partners). The drug schedule remained the same.

The results are summarized in fig. 6. During the initial acquisition phase with fixed partners, propranolol exerted a facilitatory effect, particularly from session 4. Switching partners disrupted cooperation efficiency in both drug groups somewhat, but the disruption under propranolol was transient and the disturbance under saline was maintained throughout the five days of testing. These results further document the facilitatory effects of propranolol on cooperation learning. In this case the facilitation did not appear to be due primarily to a decrease in aggression, although switching partners is sometimes accompanied by fighting.

22

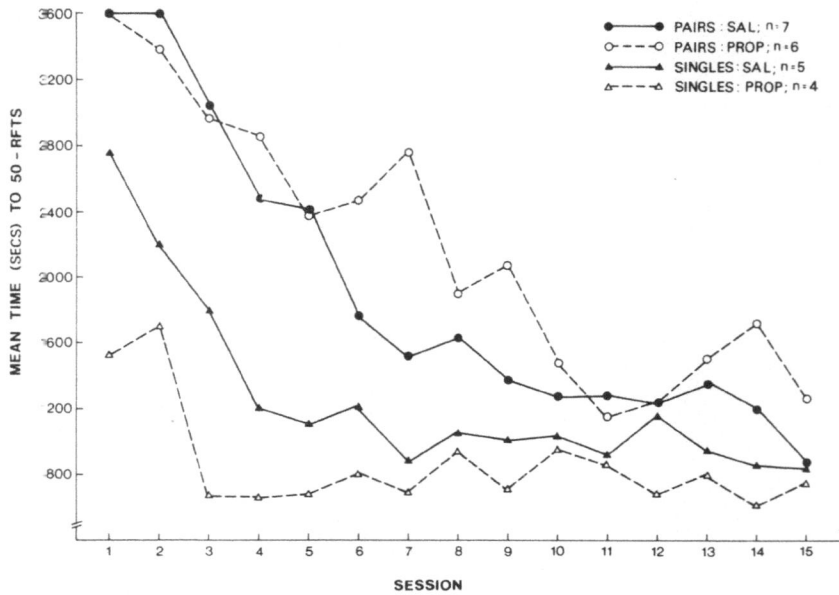

Fig. 5 Grouped Housing: Lack of facilitation by dl–propranolol in the acquisition of the cooperation taks.

Rather, it seemed that propranolol was facilitating the ability of the new partners to focus their actions on performing the coordination task under conditions of new discriminative stimuli. Saline–treated animals appeared much more affected by the stressful effects of relearning the task with a new partner.

PROPRANOLOL AND HIERARCHY

The purpose of this study was to further explore the effects of propranolol in the interactive behaviour of the partner rats in the cooperation task. Specifically, this experiment tested for possible differential effects of propranolol on "dominant" and "submissive" members of a pair. Pairs of rats, housed two per cage, were trained in the cooperation task until the pair successfully completed 50–reinforcements within a time limit of 3600 seconds maximum. In the course of this pretraining, which typically took two or three training sessions, an assessment was made according to defined criteria of the controller (dominant) or the controlled (subordinate) animal of the pair (Berger et al. 1979). Training was then suspended for 21–days during which the animals were placed on an isolated housing regimen. Following this isolation period, cooperation training was continued for 12 additional sessions, during which time the pairs remained in isolated housing conditions. Propranolol (5 mg/kg IP) was administered either to both members of the pair, to the dominant member, or to the subordinate member, 15 minutes prior to each session. In a control group, saline was administered to both members of the pair.

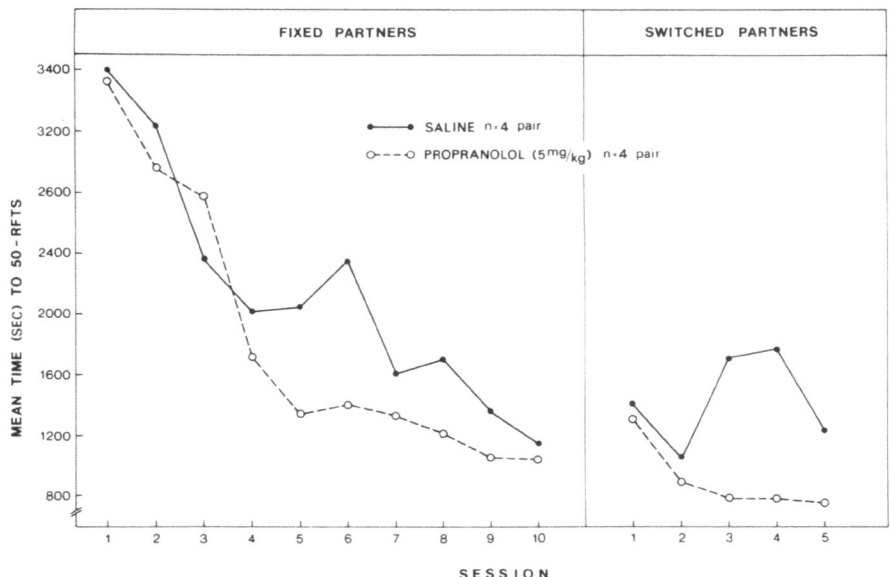

Fig. 6 dl–Propranolol given to fixed partners (housed two to a cage) facilitates performance of the cooperation task. Subsequent testing with switched partners is also facilitated by dl–propranolol.

The results are summarized in fig. 7. Consistent with the studies described above, isolated housing caused an impairment of acquisition of cooperation performance, evidenced by a slow rate of improvement and by relatively poor asymptotic levels of responding. Pairs in which both animals were treated with propranolol, on the other hand, learned the task readily, replicating the finding that propranolol facilitates learning and performance of the cooperation task. Propranolol administered either to the dominant or to the subordinate member exerted an intermediate facilitative effect. No definitive statement can be made as yet as to differential effects of propranolol on dominant or subordinate members.

FACILITATION OF COOPERATIVE BEHAVIOUR BY FLUPRAZINE

The cooperation paradigm is particularly suitable for the study of drug effects on aggression. As mentioned earlier, severe fighting results in responses that are incompatible to coordinated shuttlebox performance in pairs of rats, though the performance of single animals is unaffected.

Traditional methods of measuring aggression may confound what may in fact be a general sedative effect of a drug with what might be interpreted as an anti–aggressive action. Drug–induced reduction in aggression in the cooperation paradigm, on the other hand, results in an _increase_ in coordinated shuttlebox performance, thus controlling for possible sedative effects. Moreover, single animals running in the same shuttlebox under similar sensory, motor, and motivational conditions as pairs but without the social component, serve as a control condition for other non–specific effects of the drug. Finally, we have shown

24

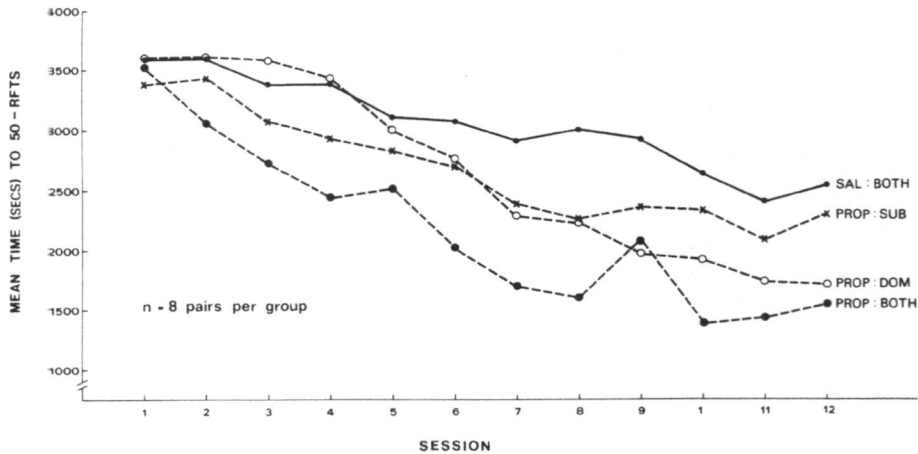

Fig. 7 Facilitation of cooperation performance acquisition by dl–propranolol administered to both partners, to the dominant member, or to the subordinate member.

that mild aggression or stress may result in facilitated performance in the cooperation task (Schuster et al. 1982), and thus it might now be possible, using the cooperation procedure, to measure the effects of drugs on mild stress or in situations that cause only moderate levels of aggression.

It will be recalled that one property of propranolol may be an anti–aggressive action. With the recent interest in anti–aggressive compounds, and in particular with the specific "serenics" (Bradford et al. 1984; Olivier et al. 1986), we evaluated propranolol using the cooperation technique. It has been reported, using ethological techniques, that fluprazine (the prototypic serenic) may have additional behavioural effects, including those on social and non social behaviours (Benton et al. 1983). The cooperation paradigm measures various aspects of social behaviour in addition to aggression. Thus the technique seems suitable for a more inclusive evaluation of fluprazine.

COMPARISON OF FLUPRAZINE AND PROPRANOLOL ON ACQUISITION OF COOPERATION PERFORMANCE

Pairs of rats were trained in the cooperation procedure to a criterion of 50 reinforcements within a maximum 60 min session. The animals were then placed in isolated housing conditions for 7 days, following which training in the cooperation procedure was resumed. The first session post–isolation was conducted without drug and served as a means to constitute matched groups. For the next 6 days, drugs or saline were administered intraperitoneally 15 min prior to each session according to the following groups: saline; dl–propranolol (5 mg/kg); fluprazine (5 mg/kg); combination of dl–propranolol (2.5 mg/kg) + fluprazine (2.5 mg/kg). A parallel study with single animals was conducted exactly as with pairs in order to evaluate the role of the social task on the drug effect.

The results of the experiment on pairs are summarized in fig. 8. As in previous studies, isolated housing resulted in retarded acquisition of the cooperation task.

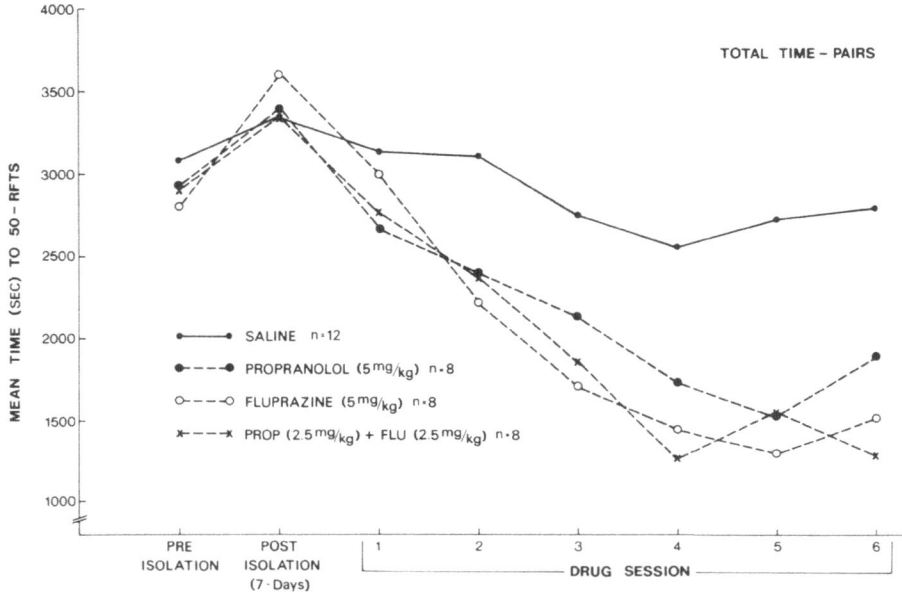

Fig. 8   Fluprazine and dl–propranolol facilitate acquisition of the cooperation task in animals under Isolated Housing conditions.

The performance of the control group receiving saline barely improved over the six–day training period, remaining at scores of 2500–3000 sec to achieve 50 reinforcements. As before, propranolol caused a relative facilitation in pairs with performance in these animals steadily improving to total time levels of 1500–2000 in the 6 day period of drug treatment. Fluprazine also caused a statistically significant improvement in total time scores over the acquisition period, tending to facilitate cooperation performance to an even greater degree than propranolol. Separate analyses of the rate–independent "proximity failures" measure, similarly showed facilitation by fluprazine and propranolol, the fluprazine scores again tending to be better than those of propranolol at equivalent doses. The combination of propranolol and fluprazine at half doses caused a facilitation roughly equivalent to that of either drug at the 5 mg/kg dose alone. This finding suggests that the effects of propranolol and fluprazine may be additive and that they may be acting on a common biochemical or behavioural mechanism.

The facilitative effects of fluprazine and propranolol are absent in the single animal variation of the shuttlebox task (fig. 9). This result indicates that propranolol and fluprazine in pairs act on some interactive feature of the shuttlebox task. One possibility is that the drugs reduce aggression, and thus facilitate cooperation performance. Indeed, the number of fights commonly seen in the first few days of training following isolation was reduced in the drug groups. However, fighting was largely absent in the control group by the end of training; yet pairs given propranolol, fluprazine, or the combination continued to perform more efficiently. Perhaps the continued poor performance of the saline group

26

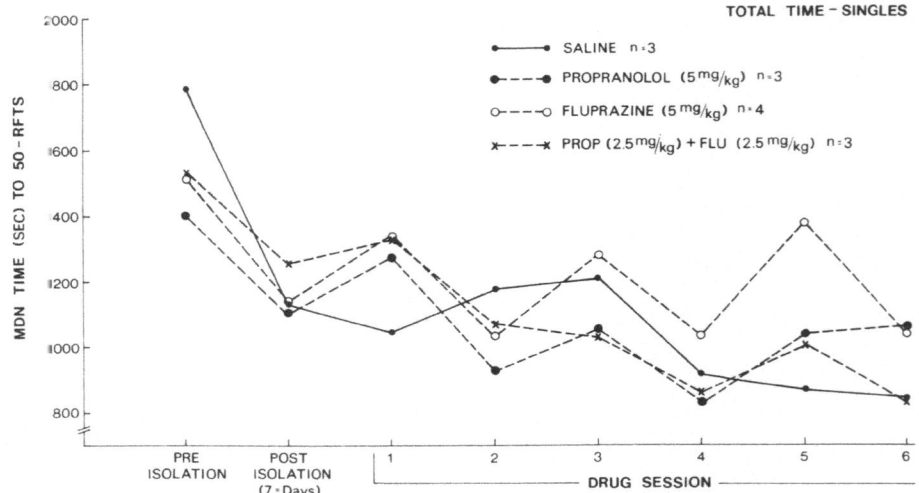

Fig. 9 Fluprazine and dl–propranolol do not affect shuttle performance in single animals.

(pairs) was due to the maintenance of responses such as freezing and crouching originally associated with fights. Propranolol and fluprazine may facilitate cooperation behaviour by attenuating these responses that are incompatible to the cooperation task.

EFFECTS OF FLUPRAZINE, PROPRANOLOL, AND LITHIUM CHLORIDE ON COOPERATION PERFORMANCE FOLLOWING LESIONS OF THE DORSAL AND MEDIAN RAPHE NUCLEI

The purpose of this study was to replicate and extend the facilitation in cooperation performance obtained with fluprazine and propranolol in intact animals to brain damaged preparations. Raphe lesions were of particular interest in part because of the aggression that is seen following lesions in these nuclei (Jacobs and Cohen 1976) and in part because of the possible role of serotonin in aggression, dominance, and other aspects of social interaction (Valzelli 1974). Our decision to evaluate lithium in this study was based on the well documented finding that lithium reduces aggressive behaviours in a range of situations (e.g. Eichelman and Thoa 1973; Sheard 1971, 1983). It therefore was of interest to determine if lithium facilitated cooperation behaviour as do propranolol and fluprazine.

Pairs of rats were trained to stable levels in the cooperation task. They then were lesioned in the dorsal and in the median raphe nuclei according to the stereotaxic coordinates of Jacobs and Cohen (1976). Other pairs were given sham lesions. The animals were housed in pairs. Following recovery of about one week from the surgery, the pairs were reevaluated in the coordination task for 5 additional sessions. The animals were then placed in conditions of isolated housing for 7 days following which they were retested in the coordination task for another 5 sessions. After this test phase, the pairs were given saline, propranolol (5 mg/kg),

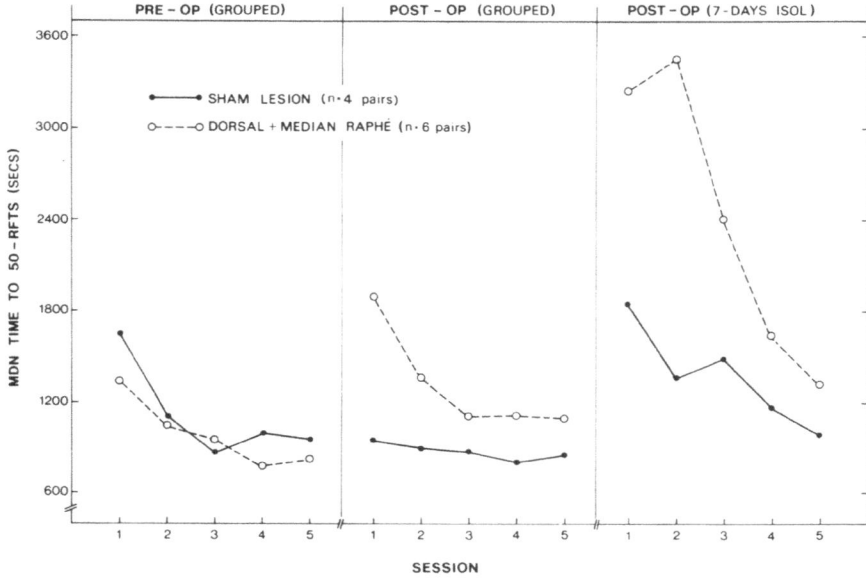

Fig. 10 Lesions of the median and the dorsal Raphe nuclei impair cooperation performance in grouped and in isolated housed pairs, that had previously been trained to criterion.

fluprazine (5 mg/kg), or lithium chloride (14 mg/kg), intraperitoneally 15 min prior to the test session. Respective drug sessions were separated by approximately 4 days, during which time the pairs continued testing with saline.

A moderate impairment in cooperation performance was seen following lesions of the median and the dorsal raphe nuclei (fig. 10). No impairment was observed as the result of the sham surgery. The impairment appeared to be due to a lack of coordinated responding between the members of the pair rather than to an increase in aggression. Indeed, aggression scores in this phase were even lower in the lesion group than in the sham group. The impairment in cooperation in the raphe–lesioned group was exacerbated following the period of isolated housing. Sham lesioned pairs were moderately impaired by isolated housing, whereas a very large and sustained impairment was observed in the lesioned pairs (fig. 10). Recording of aggression indicated a large increase in the incidence of fights that persisted, though gradually diminished, throughout the 5 day test period in the lesioned group. A transient increase in aggression was also observed in the sham lesioned pairs.

The effects of propranolol, fluprazine, and lithium on cooperation performance are summarized in fig. 11, both for individual pairs and in terms of the median score for each drug group. Clear facilitation by all three drugs was observed. Each pair performed the cooperation task more efficiently under drug than under saline conditions (except for lithium in the case of pair # B11). There were no statistically significant differences among the three drugs at the doses tested. It should be pointed out that during this phase of drug testing, fighting was virtually nonexistent even under no drug or saline conditions. Despite this lack of aggression, propranolol, fluprazine and lithium facilitated cooperation performance.

28

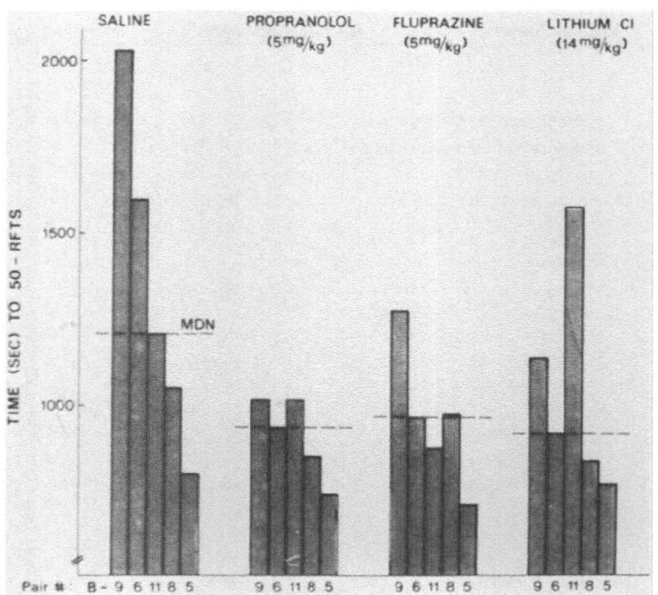

Fig. 11   dl–Propranolol, fluprazine, and lithium chloride, facilitate performance of cooperation performance in pairs of rats lesioned in the median and dorsal Raphe nuclei. Data are presented for 5 individual pairs.

Thus, it again appears that while propranolol, fluprazine, and lithium, may each have anti–aggressive actions, the facilitation of cooperation performance is likely to be due to some other property of the drugs.

EFFECTS OF OTHER PSYCHOACTIVE DRUGS ON PERFORMANCE OF THE COOPERATION TASK

We have evaluated other drugs in the cooperation task in order to determine the generality and specificity of the findings with propranolol and fluprazine, and in an attempt to reveal the pharmacological mechanisms of these effects. It is premature to identify the brain systems involved in cooperation. Nevertheless, one trend arising from our studies is that manipulations that affect serotonin systems also affect cooperation behaviour in our model. Depletion of serotonin synthesis by PCPA impairs acquisition and performance of cooperative behaviour. Similarly, diazepam, which reduces turnover of serotonin also impairs performance in the cooperation task (Berger and Schuster 1982). Raphe lesions, as discussed above, also impair performance in the social task. Fluprazine, a serotonin agonist, facilitates cooperation performance.

Preliminary data suggest that the serotonin precursor, 5–HTP, may also facilitate cooperation performance in isolated housed pairs. Finally, lithium chloride, which may act through serotonin systems in the brain (Broderick and Lynch 1982; Sheard and Aghajanian 1970), also facilitated cooperation performance.

Serotonin may also be involved in effects of propranolol on cooperation. Propranolol may interact with serotonin systems (Weinstock and Weiss 1980), and serotonin influences beta–adrenergic receptors in the brain (Stockmeier et al. 1985). The finding reported here that the combination of fluprazine and propranolol was additive in the cooperation task, suggests that the two drugs may share common mechanisms of action. Recent findings in our laboratory indicate that the effects of propranolol on cooperation are apparently due to central rather than peripheral sites of action, as atenolol, (a beta–receptor blocker with weak central nervous system activity), exerts little effect in the cooperation task. Support for the idea that beta–receptors are involved in the propranolol effects on cooperation, come from preliminary studies that the beta–receptor agonist clenbuterol impairs cooperation and that the beta–receptor antagonist ICI-118551, like propranolol, facilitates cooperation performance.

In addition to these compounds, other drugs that have been tested so far in the cooperation paradigm include haloperidol, d–amphetamine, and scopolamine (Berger and Schuster 1982). These drugs impair cooperation performance in single animals as well as in pairs in the shuttlebox situation. Thus, unlike the compounds discussed above, these drugs apparently exert a nonselective action in the appetitive shuttlebox task.

SUMMARY AND CONCLUSIONS

The overall aim of this research program is to study brain and behavioural mechanisms of operationally defined cooperation performance. Our initial studies have identified drugs that may influence some component intrinsic to the cooperation task. These include propranolol, fluprazine, lithium, and other drug and brain manipulations that affect serotonin systems and central noradrenergic beta–receptors. There are several possible mechanisms by which these drugs can affect cooperation behaviour in our task. For example, if a drug were to influence spatial affinity or social distance to other animals, it could affect performance in our model since efficient performance in our task is determined in part by proximity of one animal to another. Furthermore, we have suggested that aggression and dominance are important in the acquisition and performance of the cooperation task, and brain manipulations that affect these processes, would also affect cooperation behaviour. Also, drugs that affect perception of familiarity of partners and drugs that influence motor patterns of behaviour, would be expected to affect cooperation performance in our model. Finally, one must include the possibility of drug effects on brain systems that are organized for the integration of social interaction. Pathologies of such systems could result in antisocial or sociopathic behaviours or the withdrawal and lack of communication observed in some schizophrenic behaviours (Polsky and McGuire 1980; Rosen et al. 1980, 1981; Widom and Newman 1985).

We propose that our animal model provides an important and sensitive tool for the analysis of complex forms of normal and pathological goal–directed behaviour and its physiological correlates, and furthermore, may have important implications for comparative psychology, drug evaluation programs, and behavioural toxicology.

Acknowledgements: We acknowledge the expert technical assistance of Dalya Teucher and Avram Papo. We also note the following students and staff who conducted the individual studies described in this report: S. del Canho; S.Ilan; D.Kotvitch; F.Landa; S.Landa; T.Shinitsky; D.Teucher; V.Teucher.

REFERENCES

Benton D, Brain P, Jones S, Colebrook E, Grimm V (1983) Behavioural examinations of the anti-aggressive drug fluprazine. Behav Brain Res 10: 325–338

Berger BD, Cwengel P, Peshkin N, Schuster R (1979) Social factors in taste aversion. Neuropharmacology 18: 1003–1006

Berger BD, Mesch D, Schuster R (1980) An animal model of cooperation learning. In: Thompson RF, Hicks LH, Shvyrkov VB (eds) Neural mechanisms of goal–directed behavior and learning, Academic Press, New York, pp 481–492

Berger BD, Schuster R (1982) An animal model of social interaction: Implications for the analysis of drug action. In: Spiegelstein MY, Levy A (eds) Behavioral models and the analysis of drug action. Elsevier, Amsterdam, pp 415–428

Berger BD, Schuster RH (1984) Role of aggression in learning and performance of a task requiring mutual coordination and its modification by drugs. Proc Fifth Eur Winter Conf Brain Res (Abstract)

Bradford LD, Olivier B, Van Dalen D, Schipper J (1984) Serenics: The pharmacology of fluprazine and DU 28412. In: Miczek KA, Kruk MR, Olivier B (eds) Ethopharmacological aggression research, Alan R Liss, New York, pp 191–207

Broderick P, Lynch V (1982) Behavioral and biochemical changes induced by lithium and l–tryptophan in muricidal rats. Neuropharmacology 21: 671–679

Crawford MP (1941) The cooperative solving by chimpanzees of problems requiring serial responses to color cues. J Soc Psychol 13: 259–280

Daniel WJ (1942) Cooperative problem solving in rats. J Comp Physiol Psychol 34: 361–368

Eichelman BS, Thoa NB (1973) The aggressive monoamines. Biol Psychiatry 6: 143–164

Hake DF, Vukelich R (1972) A classification and review of cooperation procedures. J Exp Anal Behav 18: 333–343

Jacobs BL, Cohen A (1976) Differential behavioral effects of lesions of the median or dorsal raphe nuclei in rats: Open field and pain – elicited aggression. J Comp Physiol Psychol 90: 102–108

Keller FS, Schoenfeld WN (1950) Principles of psychology, Appleton, New York, pp 357–358

Kummer H (1971) Spacing mechanisms in social behavior. In: Eisenberg JF, Dillon WS (eds) Man and Beast: Comparative Social Behavior, Smithsonian Institution Press, Washington DC, pp 219–234

Kummer H (1974) Rules of dyad and group formation among captive gelada baboons. In: Kondo S, Kawai M, Ehara A, Kawamura S (eds) Symposium of the 5th Congress of the International Primatological Society, Japan Science Press, Tokyo, pp 129–159

Lindner LA, Goldman H, Dinitz S, Allen HE (1970) Antisocial personality type with cardiac lability. Arch Gen Psychiatry 23: 260–267

Lindsley OR (1966) Experimental analysis of cooperation and competition. In: Verhave T (ed) The Experimental Analysis of Behavior, Appleton-Century-Crofts Inc, New York, pp 1–17

Marwell G, Schmitt DR (1975) Cooperation: An experimental analysis, Academic Press, New York, pp 1–17

McGuire MT, Raleigh MJ, Brammer GL (1982) Sociopharmacology. Ann Rev Pharmacol Toxicol 22: 643–661

Miczek KA, DeBold JF, Thompson ML (1984) Pharmacological, hormonal, and behavioral manipulations in the analysis of aggressive behavior. In: Miczek KA, Kruk MR, Olivier B (eds) Ethopharmacological aggression research, Alan R Liss, New York, pp 1–26

Miller RE, Banks JH, Ogawa N (1962) Communication of affect in "cooperative conditioning" of rhesus monkeys. J Abn Soc Psychol 64: 343–348

Nisbit R (1968) Cooperation. In: International encyclopedia of the social sciences, vol 3, Macmillan and The Free Press, New York, p 384

Olivier B, Van Dalen D, Hartog J (1986) A new class of psychotropic drugs: Serenics. Drugs Future 11: 473–494

Polsky RH, McGuire MT (1980) Observational assessment of behavioral changes accompanying clinical improvement in hospitalized psychiatric patients. J Behav Assessment 2: 207–223

Reid WH (1978) The psychopath, Brunner/Mazel, New York

Rosen AJ, Tureff SE, Daruna JH, Johnson PB, Lyons JS, Davis JM (1980) Pharmacotherapy of schizophrenia and affective disorders: Behavioral correlates of diagnostic and demographic variables. J Abn Psychol 89: 378–389

Rosen AJ, Sussman S, Mueser KT, Lyons JS, Davis JM (1981) Behavioral assessment of psychiatric inpatients and normal controls across different environmental contexts. J Behav Assessment 3: 25–36

Schuster RH, Rachlin H, Rom M, Berger BD (1982) An animal model of dyadic social interaction: Influence of isolation, competition, and shock-induced aggression. Aggr Behav 8: 116–121

Sheard MH (1971) Effect of lithium in human aggression. Nature 230: 113–114

Sheard MH (1983) Aggressive behavior: Effects of neural modulation by serotonin. In: Simmel WC, Hahn ME, Walters JK (eds) Aggressive Behavior: Genetic and neural approaches, pp 167–183

Sheard MH, Aghajanian GK (1970) Neuronally activated metabolism of brain serotonin: Effect of lithium. Life Sci 9: 285–290

Skinner BF (1953) Science and human behavior. The Free Press, New York

Sprague RL, Sleator EK (1977) Methylphenidate in hyperkinetic children: Differences in dose effects on learning and social behavior. Science 198: 1274–1276

Stockmeir CA, Martino AM, Kellar KL (1985) A strong influence of serotonin axons on β–adrenergic receptors in rats brain. Science 230: 323–325

Swanson HH, Scholtens J, Van de Poll NE, Schuster RH (1985) Sex differences in aggression as a modulating influence on social interactions. Proc Fifth Eur Winter Conf Brain Res (Abstract)

Thibaut JW, Kelley HH (1961) The social psychology of groups. John Wiley, New York

Valzelli L (1974) 5–hydroxytryptamine in aggressiveness. In: Costa E, Gessa GL, Sandler M (eds) Advances in Biochemical Psychopharmacology (vol 11) Raven Press, New York, pp 255–263

Weinstock M, Weiss C (1980) Antagonism by propranolol of isolated–induced aggression in mice: Correlation with 5–hydroxytryptamine receptor blockade. Neuropharmacology 19: 653–656

Widom CS, Newman JP (1985) Characteristics of non–institutionalized psychopaths. In: Farrington DP, Gunn J (eds) Aggression and Dangerousness, John Wiley and Son, New York, pp 57–80

ETHOPHARMACOLOGY OF HYPOTHALAMIC AGGRESSION IN THE RAT

M.R. Kruk, A.M. Van der Poel, J.H.C.M. Lammers, Th. Hagg, A.M.D.M. De Hey and S. Oostwegel.

Ethopharmacology Group, Dept. of Pharmacology, Sylvius Laboratories, Leiden University, Wassenaarseweg 72, 2333 AL Leiden, The Netherlands.

INTRODUCTION

An ethopharmacological study of hypothalamic aggression may seem a little paradoxical, as essentially an artificially-induced response is studied using ethological methods, which stress the importance of studying natural behaviour in a natural environment. An attempt will be made to show, however, that an ethopharmacological study of hypothalamic responses is worthwhile for both behavioural pharmacology and ethology.

Drugs, Behaviour and Ethopharmacology

Pharmacology is a branch of science which accepts a variety of new techniques and methodologies to study drug effects as well as the physiological and biochemical mechanisms underlying them. Ethology is one addition to the methodological inventory of pharmacology. But what is ethopharmacology? Ethology was applied to behavioural pharmacology (Mackintosh et al. 1977; Silverman 1966a,b) long before the word 'ethopharmacology' became fashionable (Miczek et al. 1984). The recent popularity of the word suggests that a field has already emerged as a distinct discipline, with its own generally-accepted methods and goals. However, such a feeling may be premature as one must initially ask what ethology can contribute to the fundamental goal in pharmacology namely: the detection, description and explanation of drug effects? In ethology, animal behaviour is usually studied in so-called 'semi-natural' environments which seem relevant to the ecology of the animal studied. Detailed records of postures and actions of animals within such an environment are the basis of the behavioural analysis. One of the claimed advantages of ethology in behavioural pharmacology is that it could increase sensitivity and specificity in testing for psychoactive drugs. Sure enough, ethological methods of observation may draw attention to drug effects which are overlooked in conventional drug screening (see e.g. Dixon et al. 1984; Miczek et al. 1984; Olivier et al. 1984). Microprocessor technology has made detailed behavioural recording more easy but methods to analyse the complex sequential data-structure resulting from ethological records are still in their infancy. Detecting and testing drug effects by comparing sets of sequential data is an even more complicated task. There are currently no generally accepted, effective and powerful methods to analyse such records. As a consequence, much information on drug effects is probably lost. Moreover, almost every research group active in this field seems to have its own methodology (Kruk and Brain 1985). The consequences of such diversity in methodology are hard to assess but it is probable that it affects both results and conclusions. The full potential of ethology in behavioural pharmacology will only become apparent when these problems have been solved.

Drug Mechanisms and Ethopharmacology

Detecting and describing drug effects is only the first step in pharmacological analysis. A fundamental question is whether ethology is also useful in unravelling the brain mechanisms accounting for drug effects on behaviour. Molecular neuropharmacology suggests that the behavioural effects of a specific drug may well result from varied actions on a range of molecular mechanisms at different parts of the brain. It is also apodictic that most centrally–acting drugs affect several aspects of behaviour in ways that often cut across the distinctions made by ethological theory. It is no surprise, therefore, that efforts to link certain classes of drugs with specific behavioural concepts such as aggression are unsuccessful (Allikmets 1974; Eichelman and Thoa 1973). Application of ethological techniques of observation and analysis is not sufficient if the aim is to unravel brain mechanisms underlying drug effects on behaviour. Analysis of the mechanisms by which drugs change behaviour also require anatomical and physiological study of the particular brain mechanisms controlling specific aspects of behaviour. Unfortunately, efforts to localize and map such neural mechanisms have long been derided as a new kind of phrenology. However, brain stimulation studies suggest that different parts of the hypothalamus and the brainstem are involved in varied aspects of behaviour (Kruk et al. 1983; Mos et al. 1982; Roberts 1969; Woodworth 1971).

Classifying Hypothalamic Aggression

It has long been known that electrical stimulation of the hypothalamus in many species induces "aggressive" responses (Hess 1928; Hess and Brugger 1943; Hunsperger and Bucher 1967). Usually several distinct responses can be induced which have been allocated to behavioural categories such as defence, offence or predation, usually on the basis of differences in behavioural topology and accompanying stimulation–induced phenomena. In the cat, for example, two different forms of attack can be induced: 'Quiet Biting Attack' from lateral hypothalamic sites and 'Affective Attack' from medial hypothalamic sites (Flynn et al. 1970). The latter response is accompanied by signs of autonomic arousal, piloerection and hissing and it has accordingly been classified as defensive. The former is preceded and accompanied by exploratory and orienting movements and it has accordingly been classified as part of a predatory response. A similar scheme has been adopted in the rat (Panksepp 1971b). In the past, such ideas have been criticized mainly on theoretical grounds (Kruk and Van der Poel 1980) but recent data casts further doubt on the feasibility of such nice distinctions in the rat. Here evidence is presented suggesting that the idea that hypothalamic stimulation activates a neural and behavioural equivalent of a motivational system could be a misconception. It will be argued that hypothalamic attack differs in many respects from spontaneous aggression, offence, defence or predation leading to the suggestion that hypothalamic attack can be considered as a unique category. Hypothalamic stimulation in the rat seems to activate a system which can best be described as a "general–purpose attack system", which can be equally subservient to offence, defence or predation. Also, the pharmacology of hypothalamic attack responses differs considerably from the pharmacologies of many forms of 'spontaneous' attack behaviour.

A detailed description of hypothalamic attack behaviour can be found in Kruk et al. (1979, 1984a,b). Ethological procedures for the study of territorial aggression towards intruders have been described in Olivier et al. (1982; 1984) and Van der Poel et al. (1984). Procedures and designs to test drug effects on hypothalamic behavioural responses are given in Van der Poel et al. (1982) and Kruk et al. (1984a,b). A preliminary overview of published and unpublished pharmacological and ethological data on hypothalamic attack behaviour is presented here. The potential of ethopharmacology as a discipline facilitating the unravelling of brain mechanisms in behaviour is explored by comparing the ethopharmacology of hypothalamic aggression with territorial aggression.

Similarities and differences

Hypothalamic aggression has many features in common with territorial, maternal, defensive and offensive behaviour (Kruk et al. 1979; Kruk and Van der Poel 1980; Mos et al. 1987). The importance of features present in territorial aggression but absent in hypothalamic aggression has only recently attracted attention. These include differences in topology, sensitivity to environmental constraints and pharmacology.

Topology of Fighting

Electrical stimulation in the 'aggressive' area of the hypothalamus induces several forms of attack (fig. 1). In ascending order of intensity these are relatively mild bite–attacks to the head and neck region of the opponent (fig. 1.2), strong bites on the head neck and on the upper back of the opponent (fig. 1.3), bite and kick attacks on the opponents or attack jump (fig. 1.1). The last two can be followed by a clinch–fight (fig. 1.4). The attack topology depends on the strain of the rat studied, the intensity of the stimulation, and the position of the electrode tip in the hypothalamus.

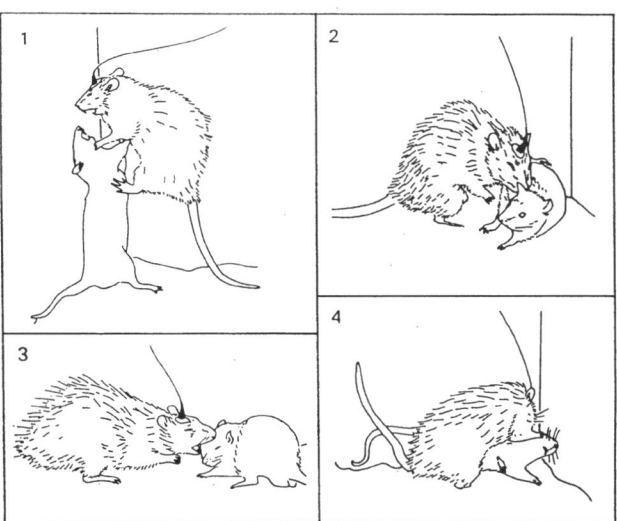

Fig. 1 Hypothalamic attack topology 1: attack jump, 2: mild bite, 3: strong bite, 4: one of the many postures in a clinch–fight. Typical examples: all sorts of topological mixtures and transitions may occur.

There are several reasons to believe that these topological differences derive from driving the same neural system at different intensities. These include:

a) fig. 1 gives only a sample of typical attacks in a continuous series of attack topologies and many intermediate forms occur. Such a topological continuity is ethologically often considered as evidence that these forms belong to the same behavioural system.

b) Stimulation by the same electrode – even at the same intensity may yield more than one of these attack topologies (Kruk et al. 1984b).

c) Increasing the intensity of stimulation with any electrode generally biasses the attack topology towards the more violent forms (Kruk et al. 1984a).

d) Attacks all derive from the same undivided continuous area in the hypothalamus (Kruk et al. 1983; Lammers et al. unpublished results).

e) It is easier to induce violent attack forms at low intensities of stimulation in the centre of the above area. The milder forms are easier to induce at the margins of this area, although at somewhat higher intensities (Kruk et al. 1983).

If the above arguments hold true, it makes little sense to look into different response topologies as markers of neural mechanisms involved in different motivational systems. A further reason why it is not easy to classify hypothalamic attack into categories such as defence, offence or predation on the basis of attack topology is that these attacks are part of the offensive as well as the defensive repertoire of the rat. The attack jump e.g. is an opening to a clinch fight especially if the opponent is in defensive upright in territorial offensive aggression. It is often called "bite-and-kick-attack" there. This form of attack is frequently observed in maternal aggression (Olivier et al. 1986; Van der Poel et al. 1984). In the defence of a cornered rat, however, the same pattern is an opening move in an effort to escape. All these attacks – even the attack jump – are also seen in stimulation–induced mouse killing (Woodworth 1971) whereas the topology of spontaneous mouse killing is rather different (Adamec and Himes 1978).

The family of postures called "sideways threat" or "lateral display" are conspicuously absent in stimulation–induced aggression (Van der Poel et al. 1984). Ethologists have suggested that these patterns are the expression of an ambivalence within the animal between the tendency to attack and to withdraw (Van der Poel et al. 1984). They also differ from overt attack in that, although they may convey an impression of physical strength to the opponent, they do not inflict direct physical damage. These postures make up a substantial part of the repertoire of the resident in territorial aggression in a resident–intruder paradigm but are absent in hypothalamic aggression if the rat is stimulated outside its own territory. This last finding is not due to the intensity of stimulation. When we stimulated 18 WeZob–rats at 4 current intensities around attack threshold, 661 attacks were induced by 720 stimulations each of 10 seconds duration. Video analysis shows that less violent attacks were induced at lower intensities but there was not one lateral threat (Kruk et al. 1984a). If rats are hypothalamically stimulated in their own territory during a resident–intruder conflict, attack is increased but lateral threat reduced (Koolhaas 1978). So, if lateral threat reflects caution or ambivalence, the effect of stimulation is to reduce that ambivalence, biasing the rat's behaviour towards the more violent forms of attack.

Fighting and Other Responses

Attack is often classified into behavioural categories according to the responses accompanying the actual attack. In spontaneous behaviour, aggression is often accompanied by piloerection and/or teeth chattering. Defensive aggression is often accompanied by escape or flight, while offensive aggression is often accompanied by explorative locomotion. It is consequently tempting to conclude that similar responses accompanying hypothalamic attack arise by the activation of the same neural system via which attack is induced. One should keep in mind, however, that given the anatomical intricacy of the hypothalamus, one is likely to concomitantly activate many neural systems, each neural system having its own behavioural consequences. The varied behaviours seem to derive from the activation of independent neural substrates which in some places overlap the hypothalamic aggression area (Kruk et al. 1983, 1984a,b; Lammers et al. unpublished results).

Therefore, it seems doubtful whether accompanying responses can be used to classify the aggressive responses.

Appraisal and Relevance of Fighting

Some of the factors controlling fighting in natural settings and spontaneous aggression do not seem to apply to hypothalamic aggression. In spontaneous fighting, males do not attack females, especially not when the latter are in oestrus, subordinate males do not readily fight dominants and females tend not to attack males usually, except when it is in the defense of pups. These rules seem advantageous to the natural ecology of the species.

In contrast, hypothalamic attack is induced as readily in females as in males (Kruk et al. 1984b) and both sexes will attack animals of both sexes with equal facility. Hypothalamically–stimulated rats even attack dominants (Koolhaas 1978), frozen rats (Kruk et al. 1979) or mice (Woodworth 1971). However, they do not attack rubber rats or other rat–shaped toys (unpublished observations). In hypothalamic aggression, stimulation apparently overrides or bypasses the factors which ultimately determine the value of fighting in natural settings.

Drugs and Aggression

Many drugs affect aggression in a variety of ways (for a recent review see Miczek 1987). A fighting animal probably taxes its sensory, motor, and other CNS facilities to their utmost, as losing a fight may have awesome consequences. Aggression tests may have become increasingly popular in drug testing as any effects on important brain mechanisms, are likely to show up. In spite of this, the pharmacology of hypothalamic aggression in rats has hardly been seriously studied (Panksepp 1971a) although drug effects on responses to such stimulation can easily be quantified. Drug treatment effects may show up in the intensity of stimulation required to induce a particular response (threshold changes), in the time lag of the response since stimulation onset (latency changes) or in the form of the attack (topology changes). Controls for behaviourally nonspecific effects can be obtained by determining the drug effects on concomitant or other behavioural responses to stimulation. These can even be studied in the same electrode placements in the same rat, by aiming at areas in which several responses can be induced by the same electrode (Van der Poel et al. 1982).

Using the above techniques, the selective effects of the serenic fluprazine on hypothalamic aggression in male (Van der Poel et al. 1982) and in female rats (Kruk et al. 1984a) were studied. A preliminary overview of an unpublished study of drug effects on hypothalamic attack and other responses is given in Table 1. Facilitating and inhibiting effects of drugs are given as deviations from vehicle threshold levels at the highest doses tested. Only significant dose-dependent results are shown. The drugs tested here have (with the exception of mianserine and phenytoin) marked effects on spontaneous aggression, either in the resident–intruder paradigm, or in maternal aggression in the same dose range (see Table 2; Olivier and Van Dalen 1982; Van der Poel and Remmelts 1971; see chapters by Mos and by Olivier in this book). However, the majority of these drugs do not affect hypothalamic attack at all, although some do affect other responses.

The effects of drugs fall into three broad categories. Drugs like quipazine and haloperidol apparently affect every hypothalamic response tested. All the stimulation–induced responses are changed as well as the behaviour in between–stimulation trials with the animals seeming to be generally incapacitated. However, drugs strongly interfering with ongoing behaviour do not necessarily suppress hypothalamic responses. Drugs like mianserine, scopolamine, alcohol and amphetamine considerably affect spontaneous behaviour between stimulation periods, but do not affect hypothalamic attack.

Table 1 Drug effects on hypothalamic responses as % of the threshold under vehicle conditions.

Drug	Attack	Loco-motion	Teeth-chatter	Stimu-lation escape	Doses used (mg/kg)	Route of admin-istration
dl-amphetamine	x	−46	x	x	0.5, 1.0, 2.0	IP
scopolamine	x	−43	x	−29	0.25, 0.5, 1.0	IP
chlordiazepoxide	x	x	x	x	5, 10, 20	PO
oxazepam	8	x	x	x	5, 10, 20	PO
alcohol	x	x	x	x	250, 500, 1000, 2000	PO
dl-propranolol	35	x	38	x	5, 10, 20	IP
haloperidol	400	n.d.	37	–	0.5, 1.0, 2.0	PO
naloxone	x	x	x	–	0.1, 1.0, 10	IP
quipazine	48	34	36	80	1.25, 2.5, 5, 10	IP
fluprazine	71	16	45	x	4, 8	IP
TFMPP	83	x	21	45	0.5, 1.0, 2.0	IP
8-OH-DPAT	x	x	x	x	0.05, 0.1, 0.2	SC
mianserine	x	x	−6	x	3, 10, 30	IP
fluvoxamine	20	x	x	39	5, 10, 20	IP
pCPA	−9	−21	–	x	*375	IP
phenytoin	x	x	–	–	100	PO
carbamazepine	x	x	–	–	50	IP

Maximum effects at highest dose tested of drugs on thresholds of responses induced in hypothalamic aggression area. Only significant results are shown (Page's test for multiple doses; Wilcoxon matched pairs for single dose). Data are given as percentage of the vehicle threshold. A negative value indicates facilitation of that response (lower threshold); a positive value indicates inhibition of that response (higher threshold); n.d. = not determinable; – = not tested; x = no significant effect; * pCPA time-course study maximum effect 72 hrs after injection.

Although the animal at first hand seems incapacitated either by compulsive locomotion (scopolamine, amphetamine) or by sedation (mianserine, alcohol) when it is not stimulated, these drugs produce no changes in attack latency, attack threshold, or attack topology. Therefore, the argument that an animal cannot attack as it is in a drugged state, should be used with care. The animal may be perfectly able to fight, but may simply fail to perceive the normal incentives required for fighting. Animals receiving high doses of chlordiazepoxide or naloxone do not seem to be affected between stimulation periods by these drugs and they had no effects on hypothalamic attack thresholds or latencies. However, these animals shift to less violent forms of attack at higher doses. In the case of chlordiazepoxide, this may be due to its muscle relaxant properties. However, oxazepam shows a similar shift and a slight increase in attack threshold. By far the most potent and selective effect on hypothalamic attack behaviour is caused by trifluorometaphenylpiperazine (TFMPP), a serotonin agonist selective for 5-HT$_{1B}$ receptors (Sills et al. 1984) which is structurally related to the serenic fluprazine. In contrast, the selective 5-HT$_{1A}$ agonist 8-OH-DPAT (Peroutka 1985; Sills et al. 1984) does not affect attack or other responses induced by hypothalamic stimulation. At the highest doses tested, animals receiving this drug seemed to be physically rather incapacitated, but this does not prevent attack. Propranolol and fluvoxamine also affect hypothalamic aggression, the latter being a rather selective serotonin reuptake inhibitor.

Single–dose studies on phenytoin and carbamazepine suggest that antiepileptic drugs do not affect hypothalamic aggression at doses suppressing hypothalamic kindling. A pilot study with the nonspecific serotonin synthesis inhibitor parachlorophenylalanine (pCPA) seems to confirm the importance of 5–HT systems in hypothalamic attack.

GENERAL DISCUSSION

Behavioural characteristics of hypothalamic aggression

Use of the term 'hypothalamic aggression' suggests that the activated mechanisms are confined to anatomical boundaries of the hypothalamus but this is certainly not the case (Mos et al. 1982). One should recognize that responses have a spatial representation in the hypothalamus and can be induced by stimulation in this region. The data summarized in this chapter suggest that hypothalamic aggression involves only a limited subset of the properties attributed to the motivational system of aggression. Stimulation in the so–called "aggressive area" of the hypothalamus activates a brain mechanism which seems useful in many circumstances in which attack is required. The mechanism may be involved in territorial and maternal attack and in offence as well as defence. It may even be involved in predation. The topology of hypothalamic attack is similar to that used in many forms of violent interactions between animals. However, characteristic elements from the aggressive repertoire such as lateral threat are absent. Therefore, topology will not help us to allocate hypothalamic attack to one of the categories devised for spontaneous behaviour. The responses accompanying attack, also do not provide a reliable means of classifing hypothalamic attack. It is hard to reconcile these facts with the idea that hypothalamic stimulation induces the neural equivalent of a specific motivational state known from spontaneous aggression. It is interesting to note here that Hess in his first reports in the cat used the word "angriff" (attack) and not "aggression" (Hess 1928; Hess and Brugger 1943).

Hypothalamic fighting requires the processing of a lot of complicated sensory information. Therefore, such a mechanism should be able to coordinate motor patterns by taking into account internal and external sensory and postural cues essential to fighting. Parts of such mechanisms have been studied in detail in the cat. Stimulation of the hypothalamic attack sites in the cat creates areas around the lips and on the paws which, upon touch, elicit a biting or striking movement respectively (Bandler 1982; McDonnell and Flynn 1966a,b). These areas are more extensive at the side contralateral to the stimulated brain site and increase stepwise with increasing intensity of hypothalamic stimulation, suggesting that quite specific neural mechanisms are involved. In the rat, the attack–directing environmental stimuli during hypothalamic attack and during spontaneous fighting seem to be roughly similar. The main targets bitten are the back and the head of the opponent, while the thorax and abdomen are rarely bitten (Kruk et al. 1979; Mos et al. 1984, 1987). Other cues which generally determine the ultimate value of attack, such as the sex or the social position of the animals involved, are apparently without influence.

Our hypothesis is that the mechanism activated in the hypothalamus controls the immediate requirements for attack. Input to this mechanism includes information on e.g. orientation, posture, touch and movements of the stimulated animal as well as its opponent. Output is the set of attack patterns observed during stimulation. Normally this mechanism is subservient to other elements checking environmental, internal and experiential aspects which suppress, or activate the attack mechanism. The behavioural consequences of interactions between these components could be described as an agonistic behavioural system. This system may not, however, have a clear spatial representation in the brain. Indeed, it seems more likely that the apparent behavioural order in social interactions, perceived as an agonistic behavioural system, arises by the constant interplay of several of such mechanisms.

The hypothetical mechanisms controlling the attack mechanism need not necessarily be specific for behavioural categories such as offence or defence. They may process other aspects important to the organism such as pain, anxiety, experience, reproductive state, or sex of partner. These factors are known to be associated with aggression and other behaviours. The advantage of hypothalamic aggression is that it enables us to study the behavioural, pharmacological and anatomical properties of the attack mechanism in relative isolation. Such studies could serve as a starting point for the investigation of the interaction of brain mechanisms subservient to social or agonistic interactions. Also, since several of the mechanisms controlling attack seem to be absent following hypothalamic stimulation, a comparison with certain pathological aggression states in man could be worthwhile.

Pharmacological Properties of Hypothalamic Attack

Hypothalamic attack is insensitive to many drugs that have a profound influence on spontaneous aggression. Other hypothalamic responses such as stimulation escape (switch–off), locomotion and teeth–chattering are equally unresponsive to drug effects. Each of these responses has, in fact, its own rather selective pharmacological 'profile'. Of course, each drug tested here still has its own characteristic effects on the behaviour of the unstimulated rat. However, such a drug–induced behavioural state often is not reflected in changed thresholds, latencies or response topologies. When hypothalamic stimulation is applied the behaviour often seems similar in drug–treated and untreated animals. Relatively few drugs interfere with hypothalamic attack in a generally specific way, these include: TFMPP, fluprazine, propranolol and fluvoxamine. Naloxone and chlordiazepoxide have small effects on the topology of attack but not on the threshold or the latency. The magnitude of the effects of fluprazine (Van der Poel et al. 1982) and TFMPP suggests that serotonin receptors of the $5-HT_{1B}$ type play an important role in hypothalamic attack. The absence of effects of drugs like scopolamine, amphetamine, alcohol, oxazepam and 8–OH–DPAT on hypothalamic attack suggest that these compounds change spontaneous aggression by interacting with mechanisms outside the hypothalamic attack mechanism. The effect of such controlling mechanisms on hypothalamic attack may be either facilitatory or inhibitory perhaps accounting for the biphasic effects of some drugs.

The relative pharmacological selectivity of hypothalamic attack seems a little paradoxical: the hypothalamic area from which attack is induced, is at the cross–roads of many brainsystems. A considerable number of different neurotransmitters, neuromodulators and hormones reach relatively high densities in this area (Nieuwenhuys 1985) suggesting that much integrative processing may go on here. How is it possible then that hypothalamic attack is relatively insensitive to drugs? One possible explanation is that electrical stimulation overrules any chemical signal modulation by drugs in the vicinity of the electrode by direct activation of post–and presynaptic excitable mechanisms at the same time. Another possible explanation is that drugs affecting the "appraisal" stage of attacking work upstream of the attack mechanism and are therefore bypassed. The implication would be that drugs which do affect hypothalamic attack would affect downstream mechanisms located outside the vicinity of the electrodes and possibly outside the hypothalamus.

The pharmacology of hypothalamic aggressive responses in the rat and the cat are not entirely comparable (Fukada and Tsumagari 1983; Johansson et al. 1984; Katz and Thomas 1976; Marini et al. 1979; Mc Donnell et al. 1971;). Methodology and responses studied vary considerably making comparison with the present study difficult. Responses include directed attack, hissing, defence reaction and rage. In only one study (Katz and Thomas 1976) a non–aggressive response is used as a control for behavioural specificity of the drug effect. In the cat, effects of anxiolytics (Fukada and Tsumagari 1983), alcohol (Johansson et al. 1984), amphetamine (Marini et al. 1979) and pCPA have been demonstrated. Only the effects of pCPA and, to a certain extent, anxiolytics are comparable in rat and cat. Whether these differences reflect species differences, differences in response type or differences in methodology is hard to say. However, if hypothalamic attack mechanisms are primarily involved in the actual proximate organization of the fighting response, then one should not be surprised to find species differences in neural organization. Cats and rats have quite different attack topologies and use quite different sensory modalities in attack.

The use of ethopharmacological and brain mechanisms

In this kind of study, the usefulness of ethology lies in providing methodologies rather than theories. This is especially true if the aim is to look for brain mechanisms underlying the behavioural effects of drugs. However, ethology has also proposed a number of theoretical behavioural distinctions in aggression such as defence, offence, 'true' aggression and predation. It has often been implied that such distinctions must reflect the organization of brain mechanisms. That may be true in general terms but our results on actual brain mechanisms suggest a different type of organization i.e. a task–orientated proximate mechanism used in many behavioural functions requiring attack. This mechanism can be activated in several 'motivational states' requiring attack, and is controlled by other mechanisms determining the ultimate usefulness of fighting. The apparent pharmacological selectivity of the attack mechanism suggests that such a task–orientated brain mechanism may have a relatively simple organization. Therefore, it might be possible to determine where the drugs affecting this system actually work.

Attempts could also be made to determine whether some drug actions on spontaneous aggression represent effects on the central "attack" mechanism. Unfortunately, we have not yet fully developed the mathematical techniques to analyse the effects of TFMPP or other drugs on social interactions in rats (Haccou 1986) but simpler methods give an impression of the potential results of such a study. A reexamination of data of studies published earlier by Olivier et al. (1984) reveals that the aggressive repertoire of rats in a resident–intruder paradigm is affected quite differently by drugs which do and do not affect hypothalamic aggression. Table 2 shows which elements in the aggressive repertoire are significantly reduced by a number of drugs. Drugs inhibiting hypothalamic attack mechanisms do also reduce biting, the most violent aspect of territorial aggression. Drugs which do not affect hypothalamic attack, may inhibit biting, but also inhibit various other aspects of the aggressive repertoire in a rather less specific way, perhaps by affecting behavioural elements which are absent in hypothalamic attack. Chlordiazepoxide e.g. inhibits lateral threat but not biting. These results support the view that the serotonin–sensitive hypothalamic attack system is also involved in spontaneous aggression. It is also possible that drugs like scopolamine, amphetamine and chlordiazepoxide affect aggression at least in part by their action on mechanisms outside the hypothalamic attack system. Taken together, these results suggest that ethopharmacology is a useful approach for detecting behaviourally–selective drugs and adequately describing the effects of such drugs. In addition, ethopharmacology also provides useful concepts for the identification of the very mechanisms involved in such behaviourally specific drug effects. However, the approach is especially useful when accompanied by serious efforts to identify the neural mechanisms involved.

Table 2 Drug effects on Time spent on specific aggressive elements in a resident–intruder paradigm by the resident. Only significant decreases are shown. Page's Test: p<0.05:*; p<0.01:**; p<0.005:***; p<0.001:****. The significant effects shown here are all decreases in the time spent on specific behavioural elements. However, at the lowest doses tested, chlordiazepoxide increases the time spent on full aggressive posture and total time spent on aggression.

Aggressive Element	d–amphet- amine	scopol- amine	alcohol	chlordia- zepoxide	nalox- one	fluvox- amine	halo- peridol
Fighting	****	****	o	*	***	**	***
Biting	****	****	**	o	*	***	*
Full Aggr. Post.	****	****	o	*	**	o	o
Lateral Threat	****	****	*	***	*	o	****
Offensive Upr.	****	****	o	o	*	o	o
Chasing	***	***	o	o	o	*	*
Boxing	****	***	o	o	*	o	o
Total Aggression time	****	****	*	***	***	o	***
Doses tested	0.25,0.5 and 1.0 mg/kg IP	0.25,0.5 and 1.0 mg/kg IP	0.75,1.5 and 3.0 g/kg PO	2.5,5.0 and 10 mg/kg PO	0.1,1.0 and 10 mg/kg IP	5,10 and 20 mg/kg IP	0.125,0.25 and 0.5 mg/kg PO

Acknowledgements: Automation of the experimental set–up for electrical brain stimulation was supported by a generous gift from the Dr.Saal van Zwanenbergstichting. We thank Dr.B.Olivier (Duphar, Weesp) for his data on hypothalamic and spontaneous aggression.

REFERENCES

Adamec RE, Himes M (1978) The interaction of hunger, feeding and experience in alteration of topography of the rats predatory response to mice. Behav Biol 22: 230–243

Allikmets LK (1974) Cholinergic mechanisms in aggressive behaviour. Med Biol 52: 19–30

Bandler R (1982) Neural control of aggressive behaviour. TINS 5: 390–394

Dixon AK, Huber C, Kaesermann F (1984) Urinary odours as a source of indirect drug effects on the behaviour of male mice. In: Miczek KA, Kruk MR, Olivier B (eds) Ethopharmacological Aggression Research. Alan R Liss Inc, New York, pp 81–91

Eichelman B, Thoa NB (1973) The aggressive monoamines. Biol Psychiatry 6: 143–164

Flynn JP, Vanegas H, Foote W, Edwards S (1970) Neural mechanisms involved in a cat's attack on a rat. In: Whalen RE (ed) The neural control of behavior. Academic Press, New York, pp 135–173

Fukada T, Tsumagari T (1983) Effects of psychotropic drugs on the rage responses induced by electrical stimulation of the medial hypothalamus in cats. Jap J Pharmacol 33: 885–890

Haccou P (1986) Analysis of behaviour by means of continuous time Markov chain models and their generalizations. In: Colgan PW, Zayan R (1986) Quantitative Models in Ethology. Privat I.E.C. Toulouse, pp 81–96

Hess WR (1928) Stammganglien–reizversuche. Berichte der gesamten Physiologie 47: 554

Hess WR, Brugger M (1943) Das subkortikale Zentrum der affectiven Abwehrreactionen. Helv Physiol Acta 1: 33–52

Hunsperger RW, Bucher VM (1967) Affective behaviour produced by electrical stimulation in the forebrain and brainstem of the cat. In: Adey WR, Tokizane T (eds) Progr Brain Res 27, Elsevier, Amsterdam, pp 445–463

Johansson G, Huhtala A, Laakso ML (1984) Effects of Ethylalcohol on hypothalamic affective defense in the cat. Pharmacol Biochem Behav 20: 841–844

Katz RJ, Thomas E (1976) Effects of a novel anti–aggressive agent upon two types of brain stimulated emotional behavior. Psychopharmacologia 48: 79–82

Koolhaas JM (1978) Hypothalamically induced intraspecific aggressive behaviour in the rat. Exp Brain Res 32: 365–375

Kruk MR, Brain PF (1985) Mathematical methods and representations in ethological aggression research. Ethopharmacology Group, Leiden, The Netherlands

Kruk MR, Van der Laan CE, Mos J, Van der Poel AM, Meelis W, Olivier B (1984a) Comparison of aggressive behaviour induced by electrical stimulation in the hypothalamus of male and female rats. In: De Vries GJ, De Bruin JPC, Uylings HBM, Corner MA (eds) Progr Brain Res 61, Elsevier, Amsterdam, pp 303–314

Kruk MR, Van der Laan CE, Meelis W, Phillips RE, Mos J, Van der Poel AM (1984b) Brain–stimulation induced agonistic behaviour: a novel paradigm in ethopharmacological aggression research. In: Miczek KA, Kruk MR, Olivier B (eds) Ethopharmacological Aggression Research. Alan R Liss Inc, New York, pp 157–177

Kruk MR, Van der Poel AM (1980) Is there evidence for a neural correlate of an aggressive behavioural system in the hypothalamus of the rat? In: McConnel PS, Boer GJ, Romijn HJ, Van de Poll NE, Corner MA (eds) Adaptive capabilities of the nervous system. Progr Brain Res, 53, Elsevier, Amsterdam, pp 385–390

Kruk MR, Van der Poel AM, De Vos–Frerichs TP (1979) The induction of aggressive behaviour by electrical stimulation in the hypothalamus of male rats. Behaviour 70: 292–322

Kruk MR, Van der Poel AM, Meelis W, Hermans J, Mostert PG, Mos J, Lohman AHM (1983) Discriminant analysis of the localization of aggression–inducing electrode placements in the hypothalamus of male rats. Brain Res 260: 61–79

McDonnell MF, Fessok L, Brown SH (1971) Aggression and associated neural events in cats. Quart J Stud Alc 32: 748–763

McDonnell MF, Flynn JP (1966a) Sensory control of hypothalamic attack. Anim Behav 14: 399–405

McDonnell MF, Flynn JP (1966b) Control of sensory fields by stimulation of hypothalamus. Science 152: 1406–1408

Mackintosh JH, Chance MRA, Silverman AP (1977) The contribution of ethological techniques to the study of drug effects. In: Iversen LL, Iversen SD, Snyder SH (eds) Handbook of Psychopharmacology, Vol 7, Plenum Press, New York, London, pp 3–35

Marini JL, Walters JK, Sheard MH (1979) Effects of d- and l-amphetamine on hypothalamically–elicited movement and attack in the cat. Agressologie 20: 155–160

Miczek KA (1987) The psychopharmacology of aggression. In: Iversen LL, Iversen SD, Snyder SH (eds) Handbook of psychopharmacology, Vol 19: Behavioural pharmacology, Plenum Press, New York, pp 183–328

Miczek KA, DeBold JF, Thompson ML (1984) Pharmacological, hormonal, and behavioral manipulations in the analysis of aggressive behavior. In: Miczek KA, Kruk MR, Olivier B (eds) Ethopharmacological Aggression Research. Alan R Liss Inc, New York, pp 1–30

Mos J, Kruk MR, Van der Poel AM, Meelis W (1982) Aggressive behaviour induced by electrical stimulation in the central gray of male rats. Aggr Behav 8: 261–284

Mos J, Olivier B, Lammers JHCM, Van der Poel AM, Kruk MR, Zethof T (1987) Postpartum aggression in rats does not influence threshold currents for EBS–induced aggression. Brain Res 404: 263–266

Nieuwenhuys R (1985) Chemoarchitecture of the brain. Springer–Verlag, Berlin

Olivier B, Van Aken H, Jaarsma I, Van Oorschot R, Zethof T, Bradford LD (1984) Behavioural effects of psychoactive drugs on agonistic behaviour of male territorial rats (resident–intruder model). In: Miczek KA, Kruk MR, Olivier B (eds) Ethopharmacological Aggression Research. Alan R Liss Inc, New York, pp 137–156

Olivier B, Van Dalen D (1982) Social behaviour in rats and mice. An ethologically based model for differentiating psychoactive drugs. Aggr Behav 8: 163–168

Olivier B, Mos J, Van Oorschot R (1986) Maternal aggression in rats: lack of interaction between chlordiazepoxide and fluprazine. Psychopharmacology 88: 40–43

Page EB (1963) Ordered hypothesis for multiple treatments: a significance test for linear ranks. J Am Stat Ass 58: 216–230

Panksepp J (1971a) Drugs and stimulus–bound attack. Physiol Behav 6: 317–320

Panksepp J (1971b) Aggression elicited by electrical stimulation of the hypothalamus in albino rats. Physiol Behav 6: 321–329

Peroutka SJ (1985) Selective labeling of 5-HT$_{1A}$ and 5-HT$_{1B}$ binding sites in bovine brain. Brain Res 344: 167–171

Roberts WW (1969) Are hypothalamic motivational mechanisms functionally and anatomically specific? Brain Behav Evol 2: 317–342

Sills MA, Wolfe BB, Frazer A (1984) Determination of selective and nonselective compounds for the 5-HT$_{1A}$ and 5-HT$_{1B}$ receptor subtypes in rat frontal cortex. J Pharmacol Exp Ther 231: 480–487

Silverman AP (1966a) the social behaviour of laboratory rats and the action of chlorpromazine and other drugs. Behaviour 27: 1–38

Silverman AP (1966b) Barbiturates, lysergic acid diethylamine, and the social behaviour of laboratory rats. Psychopharmacologia 10: 155–171

Van der Poel AM, Mos J, Kruk MR, Olivier B (1984) A motivational analysis of ambivalent actions in the agonistic behaviour of rats in tests used to study effects of drugs on aggression. In: Miczek KA, Kruk MR, Olivier B (eds) Ethopharmacological Aggression Research. Alan R Liss Inc, New York, pp 115–135

Van der Poel AM, Olivier B, Mos J, Kruk MR, Meelis W, Van Aken JHM (1982) Anti-aggressive effect of a new phenylpiperazine compound (DU 27716) on hypothalamically induced behavioural activities. Pharmacol Biochem Behav 17: 147–153

Van der Poel AM, Remmelts M (1971) The effects of anticholinergics on the behaviour of the rat in a solitary and in a social situation. Arch Int Pharmacodyn 189: 394–396

Woodworth CH (1971) Attack elicited in rats by electrical stimulation of the lateral hypothalamus. Physiol Behav 6: 345–353

ETHOPHARMACOLOGY OF FLIGHT BEHAVIOUR

A.Keith Dixon and Hans-Peter Kaesermann. Sandoz Research Institute Berne Ltd., Postbox 2173, 3001 Berne, Switzerland.

INTRODUCTION

Fighting, or agonistic behaviour, is composed of two predominant modes of response, namely offensive activities associated with attacks, and flight responses or escape–oriented behaviour. Both modes occur during aggressive encounters between individuals of the same species. Studies of drug–induced changes in aggression particularly in rodents are very numerous. However, few studies have investigated flight behaviour even though it occurs in all situations conducive to fighting. Non–pharmacologically induced changes in offensive behaviour inevitably alter flight, either in terms of outright escape or as changes in ambivalent behaviour. It is, therefore, to be expected that drugs which alter offensive aggression will also alter patterns of flight behaviour. Flight is also a component of human non–verbal behaviour (Grant 1968, 1969), consequently the study of flight behaviour is relevant to man and animals alike. In this paper, some ethological concepts of flight behaviour are briefly outlined (see also Dixon 1986) so as to provide a backdrop against which some chosen examples of drug–induced changes in flight behaviour in animals are analysed.

The paucity of relevant studies means that in many cases, drug–induced changes in flight cannot always be discussed in strict adherence to the scheme described in the next section. However, this shortcoming is a testimony to the need for more detailed ethopharmacological studies on forms of flight behaviour, not only in rodents but also in higher species including man.

SOME ETHOLOGICAL ASPECTS OF FLIGHT BEHAVIOUR

Flight as an escape response

In this paper, flight behaviour refers to those activities of an individual which effect its escape or withdrawal from a source of danger or harm. In this sence, flight is used more specifically than the broader term "defence" which does not exclude retaliation (Adams 1980). The source of harm may be another individual of the same species, a predator or some inanimate environmental feature. Consequently, the usual inter– and intraspecific restrictions commonly applied to the study of aggressive behaviour may be less applicable to flight, especially since it is an emergency strategy. In its simplest form, unreserved flight is manifested as escape (fig. 1a, right) which if unchecked, leads to separation of individuals. In predator–prey encounters, successful escape is essential for the survival of the prey.

In social encounters, however, escape disrupts ongoing social activities. For example, border disputes between territorial species such as the house mouse, are rapidly terminated after one or both mice have fled back to inside their territorial boundaries (Mackintosh 1970). The act of escaping has consequences for the exchange of signals between the escapee and its opponent. Thus, with respect to visual cues, the act of escaping reduces the input of disturbing signals emitted from the source of threat i.e. the attacker is no longer in view.

This enables the animal's own level of arousal to subside. However, successful escape also removes an animal's own attack-evoking stimuli from the adversary's view thereby reducing further risk of attacks. Escape is, therefore, consummatory and reinforcing. Obviously, these arguments concerning the exchange of visual information also apply to other sensory modalities.

Consequences of blocked escape

Frequently, active escape is prevented either by physical barriers e.g. confinement due to caging, or social constraints e.g. when the escape-route is blocked by a more dominant animal. In this case, several strategies are possible depending upon the distance between the individuals. Since the animal cannot reduce the input of disturbing stimuli by escaping, it resorts to "cut-off" acts and postures (Chance 1962) which have an analogous function. The simplest cut-off is to avert the head away from the source of threat or cover the eyes or ears. In hamsters, such cut-offs can take the form of an animal actually turning its back towards its opponent. Cut-offs are also very common in humans and are especially apparent in psychotic patients (Dixon 1986; Grant 1968). Here too, they presumably serve to reduce the input of flight-evoking stimuli. In figure 1, a cut-off component can be seen in the averted head of the mouse exhibiting the defensive upright posture (fig. 1b, right) and in the closed eyes of the attacking mouse (fig. 1c, top).

Fig. 1   Some ethological elements of flight in mice.

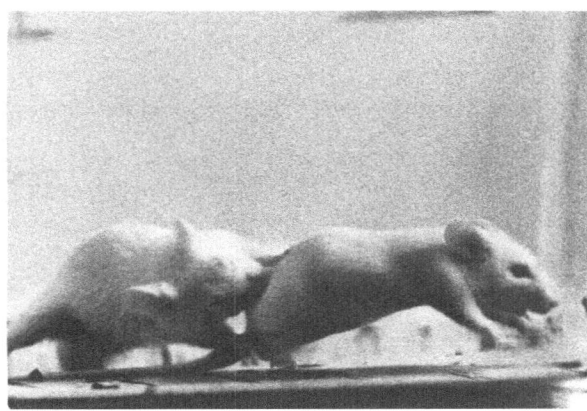

a) Bite, Left.
   Flee, Right.

b) Offensive
   Sideways, Left.
   Defensive
   Upright Right.

c) Attack, top (note closed eyes). Defensive Sideways, bottom.

d) Aggressive Groom, top. Crouch, bottom (blocked escape).

e) Defensive Sideways, Left. Offensive Upright, Right.

A second strategy of blocked escape is to adopt postures which have only a low output of aggression–releasing signals. These are often immobile postures e.g. crouching (fig. 1d, bottom). In rats, crouch often reduces the intensity of attacks in an opponent although it seems more effective when accompanied by 22 kHz ultrasonics (Lore et al. 1976). Grant (1963) showed that in rats, crouch is replaced by escape when encounters took place in large enclosures rather than cages, i.e. crouch is clearly a form of "blocked escape". It is evidence of this sort which enables flight function to be attributed to an immobile element. In other words, flight behaviour encompasses several responses which serve to avoid harm rather than seeking it, and which range from outright escape to cryptic immobility postures. Schizophrenics also have a crouch posture (Grant 1972) suggesting that here too, escape is blocked.

Ambivalence, submission and rank orders

Flight inevitably competes with approach–oriented activities and gives rise to groups of elements denoting ambivalence (Hinde 1970). These are either postures where one part of the body opposes the other, or are acts expressing alternating movements of approach or avoidance, e.g. the zig–zag responses described for sticklebacks (Tinbergen 1952). When flight is the predominant tendency, defensive ambivalence occurs. The defensive upright (fig. 1b) and sideways postures (fig. 1e, left) in mice and rats are typical ambivalent postures in which flight predominates (Grant 1963; Grant and Mackintosh 1963). By definition, therefore, such "defensive" elements cannot have a retaliatory function since the approach–component is confounded. As shown later, these are rather sensitive to anxiolytic drugs. Grant (1963) and Chance (1966) have provided evidence for a bifurcated flight–response in rats, one path leading to crouch (or escape) and the other to submission. Submissive postures serve to inhibit attacks and allow an animal to stay close to its attacker, again a 20–30 Hz ultrasonic call having aggression–inhibiting function may assist the process. Submissive postures also appear to contain two other features (fig. 2), i.e. the cut–off and the blocked escape components. By reducing the input and output of disturbing stimuli, they serve to reduce the individual's own flight tendency. By preserving social proximity submission also increases the chance of renewed social interactions.

Fig. 2   Submissive posture in the rat.
The full submissive posture in the (lower) rat inhibits attacks in the opponent. The posture contains both the blocked escape and cut–off components.

Not surprisingly, submission postures are typical of hierarchical or rank–ordered societies. By holding the balance of flight and aggression in check, submission assists group cohesion and helps an animal to maintain its position in the hierarchy

(Chance 1966; Dixon 1978b). Submission is also a prominent feature of primate societies including man. To date, little pharmacological work has been performed with respect to submission.

Interspecific forms of flight

In many prey–species, immobile postures are exhibited in order to reduce the chances of capture (Ratner 1976) best known of these being the "tonic–immobility" states of chickens (Gallup et al. 1971) and rabbits (Carli 1977; Klemm 1977). In many other species, freezing responses are shown both in social and in non–social situations as well as in interspecific encounters. As pointed out by Ratner (1976), freezing tends to occur at some distance from the source of threat. Experimental support for this contention has recently been provided by the studies of Blanchard et al (1986), showing that freezing in wild rats is reduced the nearer they are approached by a human observer (predator). Freezing is not identical to crouching, the latter being characterized as a hunched–back posture (figure 1d) which may involve covering of the sense organs whereas freeze can be any sudden arrest movement, the senses remaining unimpaired. Moreover, freeze postures do not involve cut–offs although like crouches, they exhibit crypticity. A cataleptic, immobile state induced in mice by pinching the scruff of the neck (Amir and Brown 1981) may also be an immobile response to capture by predators and this may be the reason why submissive postures in rats occasionally develop a cataleptic nature when fighting becomes intense. In any case, a strict distinction between an inter- and intra–specific function appears tenuous. Of interest is that immobile postures generally precede or follow escape attempts, irrespective of the context, showing that they are closely related forms of flight behaviour. The term "arrested flight' is adopted here to refer to collective changes in crouch and freeze during an encounter.

Another strategy, more frequent in prey escaping predators, is the display of Protean behaviours (Chance and Russel 1959; Driver and Humphries 1970) which consists of erratic, often bizarre movements and postures occurring in rapid succession, frequently interspersed with immobile postures. These patterns are unpredictable and induce conflict in an observer. Hence they delay capture by a predator and assist survival. Simple examples of Protean reactions in animals are shown in figure 3.

Sudden freezing, interspaced with fast irregular escape attempts, sometimes in the direction of the predator, often characterizes Protean behaviour. In humans, many patterns of behaviour appear erratic or confusing and appear to have Protean character. Of particular interest is the notion that certain types of seizures have been regarded as Protean (Chance 1957; Chance and Russel 1959).

PHARMACOLOGICAL STUDIES OF FLIGHT

To speak of an ethopharmacology of flight behaviour may be premature. This is because a) pharmacological studies of flight are relatively scarce, and b) when investigated, flight is rarely differentiated into its various forms, as discussed above. For these reasons, discussion is confined to a few selected examples which illustrate how drug–effect on flight can be assessed and interpreted. Most of the studies treated here describe the behaviour in terms of ethological categories of behaviour which, apart from a few exceptions, are generally based on those described for rodents by Grant (1963), Grant and Mackintosh (1963) and Lehman and Adams (1977) and our own work. However, reference is also made to other relevant sources.

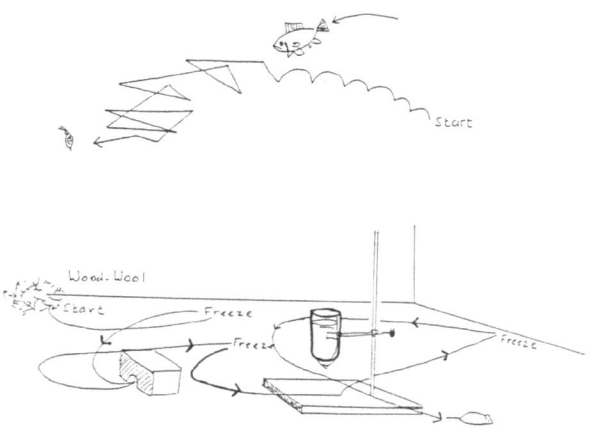

Fig. 3 Examples of protean behaviour in animals.
a) Escape of water–flea (Daphnia) from a fish. Simple erratic pathway.
b) Escape of a house–mouse from capture. Erratic pathway interjected with immobile postures.
Such Protean behaviour makes it difficult for a predator to anticipate where the prey animal will move next.

Effects of drugs on forms of Escape

Among the first detailed ethopharmacological studies of flight was the work of Silverman (1966a) who examined several compounds on the social behaviour of male rats. He showed that single doses of chlorpromazine (1, 2 and 4 mg/kg SC) increased flight behaviour whereby elements of blocked escape (crouch) were mainly unaffected. Submission was slightly increased after 1 and 2 mg/kg but active escape and defensive–ambivalent postures were not affected. Inevitably, aggression, social investigation and sexual elements were reduced, this being in accord with the reciprocal relationships existing between flight and approach–oriented behaviours.

In mice, chlorpromazine reduced the level of aggression displayed by individually housed males towards untreated intruders. Here too flight was increased, static elements being affected more than was active escape (Dixon 1978a). Silverman concluded that the main effects of chlorpromazine were to a) increase flight away from the other animal, and b) reduce responsiveness to those external stimuli which would divert the animal from its course of action. In a more recent study Dixon (1982a) showed that the antipsychotic drug clozapine, increased both defensive ambivalent and blocked escape elements in mice in a selective manner. Since defensive ambivalence as well as crouch and freeze postures were affected, these elements were referred to collectivily as "static flight". Since the mean durations of all other social elements apart from flight were unaffected, the increase in static flight was not due to motor impairment. It was concluded that clozapine somehow affected the perception of, or reaction to, flight–evoking stimuli. Of interest is that Schmidt (1983) has found that other dopamine antagonists, including clozapine and haloperidol, improve the placement of bites needed by ferrets to kill laboratory rats and has attributed this to a narrowing of the range of behavioural responses. Although this study does not directly concern flight behaviour, the conclusions of all of these three studies are similar and would indicate that neuroleptic drugs may somehow decrease distractibility, particularly those caused by signals incompatible with the tendency to escape. Whether such an action can account for the flight–evoking effects described above requires study. Moreover, these patterns of effects may change after chronic administration of such drugs but ethological studies have yet to be done. However, chlorpromazine, clozapine and other DA–antagonists are not the only drugs found to increase forms of escape in animals.

Dopaminergic drugs

In a review of the effects of drugs on agonistic behaviour Miczek and Krsiak (1979) concluded that whilst low doses of d–amphetamine generally increase aggression, most authors report an enhancement of flight in rats, mice and monkeys. Eight years later there seems to be no need to dispute this conclusion. It now seems evident that other DA–agonists, particularly L–DOPA and apomorphine, more often increase flight than offensive elements. Indeed much of the early literature describing "rage–like" reactions or aggressive responses after these drugs may well have been describing defensive ambivalent elements. A similar misinterpretation has been made with respect to foot–shock induced fighting in mice and rats which, as discussed by Blanchard and Blanchard (1977) and Brain (1981) involve defensive rather than offensive patterns of agonistic behaviour.

However, a number of studies have nevertheless implicated dopamine in aggressive behaviour. Krsiak et al. (1981) reported that apomorphine (0.2 mg/kg) increased attacks in both timid and aggressive mice. In rats, Hahn et al. (1982) and Pucilowski et al. (1986) report the occurrence of an aggressive response after a dose of 2.5 mg/kg IP apomorphine given to rats having 6–hydroxydopamine–induced lesions of the locus coeruleus but not when given to intact rats, thereby suggesting an involvement of noradrenaline in the response. However, judging from the descriptions of Randrup and Munkvad (1966, 1969), McKenzie (1971) and Kelly et al. (1980), giving apomorphine alone to intact rats, both in single and especially after repeated doses, leads to fighting which largely consists of defensive ambivalence and various forms of escapes.

It has been pointed out (Dixon 1978b; Miczek and Krsiak 1979) that apart from dosage, the social status of the drug recipient can determine the effects which a given drug may have on behaviour and this may account for some of the discrepancies in the above mentioned studies. If so, then apomorphine may simply exert opposite behavioural effects depending upon whether the animals are initially aggressive or defensive. In an unpublished study, apomorphine (0.5 and 1 mg/kg SC) was given to mice of different social status. Table 1 shows that irrespective of whether the treated mice were dominant or subordinate, apomorphine increased both the frequency and duration of static–flight elements at the expense of active escape responses. Moreover, all approach–oriented activities, including offensive behaviour were reduced, probably by competitive inhibition. However, whilst this study showed that apomorphine primarily elevated static flight, it was possible that the small cage size may have converted active escape responses to blocked escape. In addition, the effect of the drug may have been biased by the fact that the drugged mice encountered an opponent of opposite social status to themselves.

In a recent study (details to be published) apomorphine was given to pairs of equally matched aggressive male mice, encountering each other in a neutral cage large enough to enable active escapes to occur. Apomorphine (0.15, 0.3, 0.6 mg/kg SC), given 20 minutes before the 10 minute encounter to one of the partners, nevertheless, clearly elevated arrested flight and defensive ambivalence, these changes being sufficient to competitively decrease offensive, sexual and other approach–oriented behavioural categories (fig. 4). At this level of analysis, escape was slightly decreased. However, the detailed element analysis shown in fig. 5, reveals that a dissociation occurred between Retreat and Flee whereby the former was reduced and the latter enhanced as doses of apomorphine increased. Since Flee is the higher intensity escape element, its preferential increase over Retreat parallels the increase in Freeze at the expense of Crouch, Freeze occurring at higher intensities and at greater distance away from an opponent than does Crouch. This study confirms that the main effect of apomorphine under these conditions was to increase flight behaviour of all types.

Table 1 Effects of apomorphine sulphate on the agonistic behaviour of dominant and subordinate male mice

Category		Dominants (n=10)		Subordinates (n=8)	
		Doses in mg/kg SC			
		0.5	1.0	0.5	1.0
Non–social	F	– 42.7 *	– 52.4 *	– 4.8	+ 12.1 *
	D	+ 27.0	– 5.1	– 125.0 *	– 80.6 *
Soc.investigation + Sexual activity	F	– 42.4 *	– 41.4 *	+ 4.5 *	+ 3.0
	D	– 39.3 *	– 42.9 *	+ 19.0 *	– 4.6
Aggression	F	– 49.3 *	– 67.2 *	+ 0.5	+ 0.4
	D	– 32.5 *	– 41.4 *	+ 0.3	+ 1.3
Static flight	F	+ 26.1 *	+ 31.7 *	+ 58.3 *	+ 45.5 *
	D	+ 46.6 *	+ 90.5 *	+ 127.0 *	+ 132.3 *
Active escape	F	– 11.4 *	– 8.9 *	– 15.4 *	– 48.1 *
	D	– 3.0 *	– 1.4	– 21.5 *	– 48.4 *

10 week old adult male LAC mice were caged in pairs for 2 weeks during which time a clear dominant (aggressive/subordinate (flight oriented)) relationship was established. From each pair, a dominant or a subordinate mouse was removed to a "waiting cage", injected SC with either apomorphine sulphate or saline, and replaced back to their untreated partners 30 minutes later. Over the next 10 minutes, elements of behaviour shown by the mice were recorded by two observers using the method described by Dixon and Mackintosh (1971).
Results are expressed as the mean differences in frequency (F) and duration (D; in seconds), of elements shown by drugged mice relative to those shown by mice treated with the vehicle. In all cases, the untreated opponents of drugged dominants (N=10/dose) and subordinates (N=8/dose) were their original cagemates and were of opposite social status to the drugged animals. Tests were performed under reversed 12: 12 hr D:L lighting conditions between 8.00 and 12.00 hrs (under red light). Static Flight = Blocked Escape + Defensive Ambivalent Elements. Active Escape = Evade, Retreat and Flee. +/– = increase/decrease. Statistics: * = p<0.05; two tailed Mann – Whitney U test.

Nevertheless, when the various flight responses of apomorphine treated mice were expressed in relation to the aggression received from opponents (table 2) all elements of static flight were increased whereas active escape tended to decrease. This pattern held true irrespective of whether static flight was considered as a whole, or was subdivided into defensive ambivalence or arrested forms of flight. The fact that all but one of the apomorphine–treated mice lost their fights shows that apomorphine predisposed them to defeat.

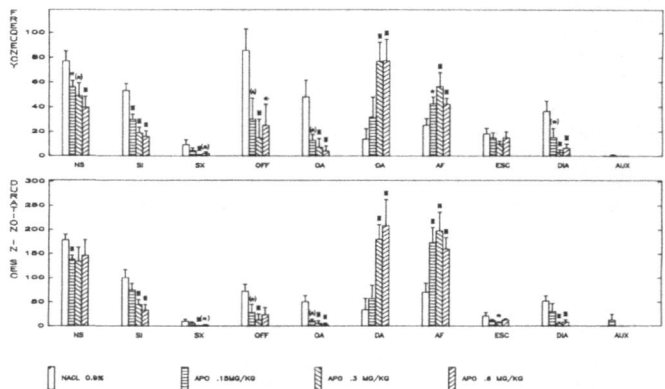

Fig. 4   Effects of apomorphine on the mean frequency (±SEM) and the mean
duration ((sec) ± SEM) of the behaviour of isolated male OF – 1 mice
against a strange intruder. The following behaviour elements were
measured; NS=Non Social, SI=Social Investigation, SX=Sex, OFF=Offence,
OA=Offensive Ambivalence, DA=Defensive Ambivalence, AF=Arrested
Flight, ESC=Escape. DIA=Distance Ambivalence (see Dixon 1986).
Method: Male OF-1 mice were housed individually for 3 weeks under
reversed lighting conditions (red light 8am–8pm, white light 8pm–8am) in
makrolon cages (26.6x15 cm), food and water being freely available. 2 days
before the study, all mice encountered in their home cage, group–housed,
foreign male intruders for 6 minutes. Based on the number of attacks the
isolates showed towards the intruder, pairs of isolates (N=7/group) equally
matched for aggressive scores, were designated to encounter each other in
a large neutral cage (makrolon 59x38.5x20 cm) chosen so as to allow the
mice to exhibit elements of active escape. Two days later, the test isolates
were given a SC injection of apomorphine sulphate or saline 0.9%, 20
minutes before they encountered their equally matched saline–treated
opponents in the neutral cage. Ethological elements of social and
non–social behaviour were measured during the 10 minutes encounters
using methods described elsewhere (Dixon 1982b). The category changes
depicted in fig. 4 show that mice treated with apomorphine exhibited
dose–related increases in flight categories but a corresponding decrease in
approach–categories. They also lost most of their fights against their
drug–free opponents. Doses of apomorphine were 0.15, 0.3 and 0.6 mg/kg
SC.

Fig. 5, also shows that the element Attend, in which the mouse orients its head
towards an opponent, was reduced by apomorphine. This may have been a
consequence of the overall decline in approach–oriented social activities. Sensory
deprivation studies in mice (Strasser and Dixon 1986) have shown that attend
ensures receipt of both auditory and visual cues, the latter being more relevant at
close quarters. It is, therefore, possible that the fall in attend reflects a cut–off
phenomenon associated with the general increase in flight. In mice, apomorphine
also induces two different types of approach, one being open–eyed and the other
being with the eyes closed and which are accompanied by frequent freezing
(Kaesermann, in prep.). These intriguing observations, which are reminiscent of

cut–off elements in our own behaviour, require closer study since they may throw light upon the ways flight–related events, including disturbances of social behaviour, affect the receipt and processing of sensory stimuli.

Both eye closure and increased defensive postures after apomorphine have also been reported for mice by Hodge and Butcher (1975). The same authors also found increases in defensive postures and blocked escape but decreases in aggression after L–DOPA, findings similar to those of Miczek and O'Donnell (1978) with other DA–agonists. In monkeys e.g. vervets, d–amphetamine appears to reduce social behaviour (Kjellberg and Randrup 1973; Schiorring 1981) including maternal care (Schiorring and Hecht 1979) and similar changes have been observed in marmosets

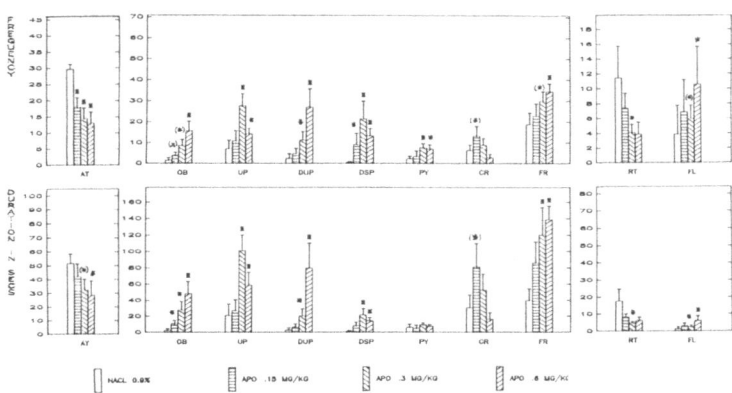

Fig. 5   Shows the mean frequencies and duration of individual flight elements underlying the category changes shown in fig. 4. These elements were measured from video films of the encounters. It can be seen that all flight elements i.e. those of Defensive Ambivalence (OB=Oblique, UP=Upright Posture, DUP=Defensive Upright Posture, DSP=Defensive Sideways Posture, PY=Parry), and Arrested Flight (CR=Crouch/Blocked Escape + FR=Freeze), increased in a dose–dependent manner, whereby Freeze increased at the expense of Crouch. Similarly Flee (FL) increased at the expense of Retreat (RT). The histogram also shows the reduction in Attend (AT) after apomorphine. See text for further comments.

after apomorphine (Scraggs et al. 1979). Stereotypies probably accounted for some of these effects (Kjellberg and Randrup 1972) but Miczek et al. (1981) concluded that the social withdrawal occurring in squirrel monkeys after d–amphetamine was not linked to the effects on stereotypies. Social status and rank was again shown to be an important determinant of d–amphetamine's effects on behaviour (Miczek and Gold 1983). Certainly, in the rodent studies discussed above, no stereotypies were observed, the decline in approach–oriented behaviours being secondary to the increases in flight. Whether this applies also to the social withdrawal syndromes in monkeys needs clarification since in the studies just cited, flight was not formally reported. Some evidence has been provided that in primates dopaminergic agonists can elevate certain forms of flight associated with submission, which are not confounded by stereotypies. Schlemmer et al. (1980) found that apomorphine (0.05 to 3 mg/kg IM) reduced social interactions but increased submissive gestures, i.e. lip–smacking, presenting, grimacing and related acts, in a group of 4–6 stumptail macaques living in a family group. In a later study (Schlemmer and Davis 1981) it was found that after chronic treatment for 12 days, both d–amphetamine (3.2 mg/kg) and apomorphine (1 mg/kg) increased submissive gestures in dominant individuals even though the aggression received by the monkeys did not increase in a corresponding manner. These effects were antagonized by haloperidol, pimozide and trifluperazine but not by the 5–HT antagonists cinanserine, methysergide, metergoline and cyproheptadine. This suggests that the effects on submission responses in this species were mediated through dopamine.

Table 2   Flight/Aggression relationships between opponents

| | | \multicolumn{4}{c}{Flight (apomorphine) /Aggression (vehicle) quotients} |
Interaction Quotients		Control	0.15	Apomorphine 0.3	0.6 mg/kg SC
SF/AGG	F	0.34	0.98*	0.94*	0.81*
	D	0.97	3.4	3.78*	3.62*
DA/OA	F	0.33	0.73	1.47	1.3
	D	0.13	1.65	5.0*	4.6*
AF/OFF	F	0.24	0.55*	0.5*	0.39
	D	0.87	2.96*	2.46*	2.0
FR/OA	F	0.45	0.52	0.57	0.58
	D	1.34	2.47	3.32*	3.02*
ESC/AGG	F	0.33	0.21	0.08	0.08
	D	0.36	0.17	0.07	0.12
ESC/OFF	F	0.7	0.4	0.13	0.13
	D	0.77	0.47	0.11	0.18

Table 2 shows Interaction Quotients which were obtained by dividing the drugged animals' Flight scores by the Aggression scores of their vehicle–treated opponents. The left column depicts different types of Flight/Aggression quotients derived for static as well as dynamic forms of flight. For example, the value of 0.34 obtained for SF/AGG (frequency) shows that under vehicle–vehicle conditions, 3 aggressive acts were required to evoke one static element of flight in the test animals. However, the quotients of 0.98, 0.94 and 0.81 show that on almost all occasions, apomorphine treated mice responded to an aggressive act of their partners with an element of static flight. In contrast, the quotients calculated for ESC/OFF shows that compared to control conditions, apomorphine lowered the amount of active escape shown as a response to aggression. Values above and below 1 show that an animals' flight is greater or less than the amount of aggression it received.
OFF = Offence, OA = Offensive Ambivalence, AGG = Aggression (OFF+OA), AF = Arrested Flight (Crouch+Freeze), FR = Freeze, ESC = Escape, DA = Defensive Ambivalence, SF = Static Flight (AF+DA).

Taken together, both the data from the primate and the rodent studies suggest that the capacity of dopaminergic drugs to increase the arrested components of flight behaviour, clearly disrupts approach oriented activities and impairs the capacity to switch to dynamic acts. The ensuing rigidification of the behaviour induced in mice by apomorphine, is analogous to Schmidt's (1984) interpretation of the reduced fighting between ferrets induced by L–DOPA and apomorphine i.e. that DA agonists disrupt long chain behavioural responses but increase short ones. It is however, somewhat at odds with his suggestion that DA–agonists widen the range of possible responses open to an animal. Clearly, it is necessary to examine a number of factors, e.g. species differences, doses, and the effects the situational context has on responsiveness, in order to clarify the exact conditions governing these effects. Nevertheless, the experiments with apomorphine described above confirm the impression gained from the literature that dopaminergic drugs reduce offensive patterns of behaviour but increase flight. In addition, they clearly attest to the reciprocal relationship existing between flight and approach–oriented social activities.

Cannabis, opiates and peptides

Flight–evoking effects in the mouse as well as an "immobility" posture have been described for cannabis (Cutler et al. 1975a) as well as for delta-9-tetrahydrocannabinol (Δ^9-THC) (Cutler and Mackintosh 1975). Proof that flight was increased directly, was obtained by the rebound in aggression which occurred after withdrawal from a 2 week diet containing cannabis extract (Cutler et al. 1975b). Increases in static flight responses and immobility in mice treated after one or after 4 daily doses of Δ^9-THC (20 mg/kg) have also been reported by Sieber et al. (1980a) who also observed a decline in social investigation. The immobility in this study appeared to correlate with "freeze". Somewhat puzzling is the disappearance of major behavioural effects after four daily doses of Δ^9-THC in a second study with mice (Sieber et al. 1980b), a lack the authors attributed to territorial occupancy or to the level of social stimuli to which the mice were exposed. In rats, immobile responses were also increased by cannabis (Cutler and Mackintosh 1975) but in contrast to mice, flight behaviour was not increased. Although requiring clarification, the immobility seen in these experiments would not seem to be a form of blocked escape like the types discussed so far and may reflect muscle impairment. Cutler et al. (1975a) noted differences in the dose–reponse curves for flight and immobility which suggested the two behaviours are separable. Why cannabis elevated flight in mice but not in rats is unclear. Whilst species differences do exist (Mechoulam 1973) many of the reported descriptions of "aggression" in rats after cannabis intake, resemble defensive ambivalence. As noted by Miczek and Krsiak (1979) these substances certainly reduce aggression but are only weakly active against defensive behaviour. This would fit with the drug's observed augmenting action on flight responses but clearly the confounding effects of accompanying side effects as well as species specificity need investigation.

Flight elevating effects after morphine, methadone, β–endorphin and D–ala D–leu enkephalin are marked (Puglisi–Allegra et al. 1984), observations which appear to be in accord with the recent discovery that opioids are involved in defensive behaviour (Miczek et al. 1982; Rodgers and Hendrie 1983). It still remains to be determined which forms of escape are specifically related to the opiates. Such information is necessary since the role of opiates in flight behaviour could have important consequences for certain human disorders, e.g. addiction (Hendrie 1985). Following abrupt termination of chronic morphine treatment by challenging with an opiate antagonist such as naloxone, rodents frequently exhibit a persistent jumping response which has been used as an index of withdrawal reactions analogous to the human state (Wiley and Downs 1979). Jumping is also one of several forms of escape in rats and mice, thus indicating that opiate withdrawal is associated with flight. In mice, the marked avoidance of brightly illuminated areas following abrupt withdrawal from alcohol ingestion (Costall et al. 1986) is a related phenomenon. These observations suggest that flight is a predominant component of withdrawal states in animals and presumably underlies the marked feelings of fear and anxiety associated with withdrawal from addictive drugs in man.

Effects on arrested flight and defensive ambivalence behaviour

Krsiak (1975) reported that a proportion of individually-housed male mice became withdrawn and avoided non-aggressive intruders. He coined the term "timidity" to describe a number of elements which were associated with flight. These were the arrested flight postures freeze and crouch, the ambivalent element defensive–upright posture, and elements of escape. He showed that diazepam and chlordiazepoxide reduced these elements at doses which did not cause ataxia. Inspection of his results suggests that these drugs exerted their main effects on the defensive ambivalent postures and on the immobile elements rather than escape. In later studies, Krsiak et al. (1981, 1984) extended the list of compounds decreasing flight to include barbiturates and gaba–ergic drugs. Of interest was the observation that serotonin agonists, which are known to decrease aggression (Valzelli 1981), increased flight behaviour in aggressive as well as timid mice. To some extent, these results were presaged by Hoffmeister and Wuttke (1969) who demonstrated that in mice, chlordiazepoxide preferentially reduced shock–induced fighting postures which they showed to be the defensive ambivalent elements described by Grant and Mackintosh (1963). The offensive elements shown both in the mouse and the cat were only reduced at ataxic doses. Chlorpromazine was equally effective against both forms of fighting. Although many studies have confirmed these effects on defensive behaviour, several reports also indicate that offensive patterns can be increased after benzodiazepine treatment (Fox and Snyder 1969; Krsiak 1979; Miczek 1974) and barbiturates (Silverman 1966b). Although these effects are seen at low doses, and to some extent depend upon the type of recipient, Dixon (1982a,b) showed that the two effects i.e. the reduction in flight and the pro–aggressive response could be separated. Use was made of the fact that intruder mice, when subjected to attack from an isolated male, show marked increases in flight behaviour including defensive–ambivalence and the arrested flight responses freeze and crouch. This increase occurs at the expense of approach–oriented behaviours which are reduced. When given to the intruders, diazepam, cloxazolam and several other anxiolytic drugs reduced static flight elements and increased approach oriented social activities (Dixon 1982a). Further differentiation of the flight responses showed that the main inhibitory action of these drugs was on defensive–ambivalence. These changes in behaviour can be seen in Table 3, which summarizes the effects of both benzodiazepine and atypical anxiolytic drugs e.g. ipsapirone, together with other psychotropic agents on social behaviour of mice in a flight–evoking situation. Some of these drugs, i.e. diazepam, chlordiazepoxide, cloxazolam and tizanidine even increased aggression.

Since aggression and flight are reciprocally related, two actions were postulated to explain the results (see fig. 6), viz. a direct flight reducing effect which would disinhibit the release of the approach–oriented activities and/or b) a primary effect on the approach activities such that flight was reduced indirectly. It was concluded that the results were best explained by a direct flight reducing effect. Thus, a primary social-promoting action exerted on mice in a flight–evoking situation would lead to conflict between the two components and result in more defensive ambivalence. Since this was not the case, a direct-flight reducing action of diazepam and related drugs was likely. In fact, a similar conclusion was reached by Silverman (1966b) for amylobarbital which in rats, increased aggression and markedly reduced arrested flight. When the rats were tested in opaque boxes which normally increased flight even further and thereby lowered social activities, the increase in aggression after amylobarbital was even greater. Poole (1976) obtained similar findings on the basis of studies with chlordiazepoxide in hamsters. Taken together, these studies all point to a

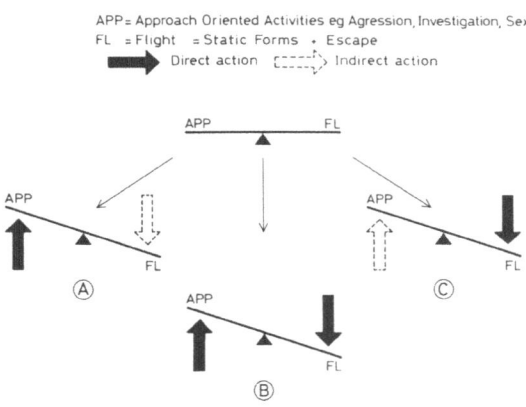

Fig. 6 The reciprocal relationship existing between flight (FL) and various approach–oriented activities (APP) is represented as a see–saw (top). According to this scheme, the actions of diazepam, other benzodiazepines and also the barbiturates on social behaviour may occur in three ways:

A) In this case, diazepam exerts a direct promoting action upon approach–activities and reduces flight indirectly. In a flight provoking situation, such an effect would give rise to ambivalence but this does not happen after diazepam.

B) Here, diazepam exerts both a direct promoting action upon approach–activities and a direct inhibitory effect upon flight. Evidence for indirect effects of diazepam on APP via urinary odours has been found.

C) This scheme infers that diazepam reduces flight directly and hence releases approach–oriented activities from active inhibition. This appears the most likely mode of action of diazepam and related drugs.

Evidence from ethological studies suggest that case C fits the results best and could explain certain clinical events as "paradoxical aggression" in humans taking anxiolytic drugs. See text for further details (Dixon 1982a).

disinhibition of aggression as a result of diazepam and other putative anxiolytics exerting a direct attenuation of flight behaviour. As proposed by Dixon (1982a) this effect would explain the so–called paradoxical aggression occasionally observed in humans imbibing benzodiazepines (DiMascio 1973). However, whilst this interpretation remains valid, it has also been shown that at least in mice, diazepam can evoke aggression indirectly via changes in urinary odours of the drug recipient (Dixon 1982b, 1984). Flight reduction, therefore, is probably only part of a complex set of events which predispose an individual to aggressive acts after taking these drugs.

Krsiak et al. (1981, 1984) and Poshivalov (1981) report that besides the benzodiazepines and barbiturates, many GABA–ergic drugs reduce defensive behaviour including defensive–ambivalence, in rodents. Table 3 shows that THIP (4,5,6,7, tetrahydroiso–oxazolo (5,4–c) pyridine 3–ol), a GABA agonist, also reduces

Table 3 Summary of drug–induced changes in social behaviour of intruder–mice after single oral doses (flight–evoking situation)

	NS	SIM	AGG	DA	AF	ESC	mg/kg PO
Diazepam	↓↑	↑↑	↑	↓↓	(↑)	(↑)	0.1 – 1
Chlcrdiazepoxide	↓	↑↑	↑	↓↓		(↑)	1 – 9
Cloxazolam	↓	↑↑	↑	↓↓		–	0.1 – 1
Alprazolam	↓	↓	–	↓↓	↑↑	(↑)	0.3 – 1
Ipsapirone	↑	↑↑	–	↓	↓↓	–	1 –10
Tizanidine	↓↓	↑↑	↑	↓		–	0.3 – 3
BI 27–062	↑	↑	–	↓↓	–	–	0.1 – 1
THIP	↑	–	(↓)	↓↓	↓↓	(↓)	0.1 – 1
Amitriptyline	↓	↑↑	–	↑	–	–	1 – 9
Apomorphine	↓	–	–	↑↑	↑↑	↓↓	0.5 – 1 SC
LSD	(↑)	↓↓	–	(↑)	(↑)	(↑)	0.5 – 8γ/kg
Clozapine	↓↓	↓↓	(↓)	↑↑	↑↑	–	0.3 – 3

( )=weak, ↑/↓=moderate, ↑↑/↓↓=strong, –=unchanged. Based on changes in frequency and duration of elements (p<0.05 MW U–test). NS=Non Social, SIM=Social Investigation+Sex, AGG=Aggression, DA=Defensive Ambivalence, AF=Arrested Flight, ESC=Escape.

these behaviours. This suggests that GABA exerts an attenuating influence on flight behaviour. Indeed THIP injected into the brain ventricles of rats actually promotes offensive behaviour whereas GABA–antagonists e.g. bicuculline methiodide, reduce offence but promote defensive patterns of behaviour (DePaulis and Vergnes 1984). GABA–antagonists also increase defensive–ambivalence responses to foot–shock in rats (Rodgers and DePaulis 1982). These findings are in line with experiments showing that GABA–antagonists evoke flight when injected into the periaqueductal grey regions of the rat brain (DiScala et al. 1984) and support the proposal of Brandao et al. (1986) made on the basis of similar experiments, that at the level of the mesencephalic central grey area and medial hypothalamus, GABA exerts a tonic inhibition of the "flight motivating system". The recent report of a decrease in social activities but of an increase in avoidance behaviour after the β–carboline derivative FG 7142 in rats (Beck and Cooper 1986) further supports the involvement of the GABA–benzodiazepine interaction in the control of flight behaviour. However, more detailed ethological studies with benzodiazepine–antagonists are needed.

Effect on a non–social form of ambivalence

Recently, evidence was obtained showing that a non–social, albeit not operant, form of conflict is susceptible to the inhibitory action of putative anxiolytic drugs. Grant and Mackintosh (1963) first described a posture in mice which involved the stretching of the body when confronting a conspecific. They classified this behaviour as an ambivalent element reflecting an approach–avoidance tendency. Van der Poel (1967, 1979) provided experimental support for this idea based on the relation between the incidence of the posture in a passive avoidance situation and exposure to shocks. Kaesermann (1986) found that mice when placed singly onto an elevated platform, which provided a novel environment, initially displayed this posture. He renamed it Stretched Attend Posture or SAP to distinguish it from its occurrence in a social context. Foreign mouse odours were lightly rubbed onto the platform so as to increase its aversiveness. When naive mice were repeatedly exposed to the platform for several days, SAP's decreased. It was found that in naive mice, single doses of diazepam, clobazam, and phenobarbital but not imipramine and chlorpromazine reduced SAP's as if the mice were habituated to the situation. The conclusion was that the SAP represented a non–social form of conflict suitable for detecting putative anxiolytic drugs. Further work has supported this contention (Kaesermann and Dixon 1986). Table 4 shows that in addition to the above mentioned drugs, alprazolam, the atypical anxiolytic buspirone and 8–OH–DPAT (8–hydroxydipropylaminotetralin), both putative 5–HT$_{1A}$ agonists, as well as the 5–HT$_2$ antagonist pizotifen, all reduced SAP's in this test. Thus, in addition to GABA–involvement, these results suggest that serotonin is also involved in the control of defensive–ambivalence responses.

It is of interest that the pattern of effects obtained in the SAP test closely parallels the pattern obtained with these drugs on defensive ambivalence in social encounters between mice (Table 3). This underlines the defensive character of the SAP. Indeed, the findings that d–amphetamine, caffeine and apomorphine increase

Table 4 Drug–induced changes in stretched attend posture (SAP), a non–social ambivalent element

	threshold dose in mg/kg PO	
Drug	Decrease	Increase
Alprazolam	0.1	
Clobazam	0.1	
Chlordiazepoxide	10.0	
Diazepam	1.0	
Buspirone	3.0	
Phenobarbital	30.0	
Pizotifen	3.0	
8–OH–DPAT	0.1	
d–Amphetamine		10.0
Caffeine		3.0
Apomorphine (SC)		3.0

Methods: see Kaeserman HP (1986) Psychopharmacology 89: 31–37

the incidence of SAP's is also in accord with the bias of this element towards flight. It should be noted that the drugs which decrease SAP's occasionally increase immobility. Since immobility competes for the available time to perform SAP's, their effects on the latter could be regarded as indirect. However, some drugs reduced SAP's in doses which did not increase immobility, e.g. phenobarbital and clobazam. Additionally, untreated mice which had habituated to the situation, showed progressively more immobility, as the SAP's declined. Though not ruling out an indirect action on SAP, the evidence obtained so far clearly favours a direct inhibitory action of putative anxiolytic drugs on SAP–postures and, along with the data from social situations suggests that these drugs have a propensity for attenuating flight–oriented ambivalent responses.

INTERSPECIFIC ESCAPE RESPONSES

Effects of drugs on immobility

Apart from social situations, immobile postures, particularly freezing, frequently occur in non–social situations but also in the context of predator–prey interactions. Ratner (1976), Woodruff (1977) and others have proposed that immobile responses are defense strategies against capture by predators.

The freezing response shown by wild rats in the presence of a human adversary is a further example of this (Blanchard et al. 1986). As already pointed out, immobile postures frequently occur under conditions of blocked escape, particularly in a social context. However, most of the pharmacological work on immobile responses has concerned changes in tonic immobility reactions in chickens and rabbits in an inter-specific context. In rabbits, Klemm (1977) reported that large doses of chlorpromazine enhanced tonic-immobility. Kelly and Whishaw (1977) confirmed this and also found that atropine prolonged tonic-immobility whereas escerine and d-amphetamine curtailed the response. These authors showed that an inverse relationship existed between body temperature and the duration of tonic-immobility, chlorpromazine and atropine causing a fall, and escerine and d-amphetamine causing a rise in core temperature. Of interest is that atropine has also been shown to induce an immobile response in rats confined to a low narrow alleyway from which escape was possible (Schallert et al. 1980). The changes in body temperature which were recorded in these studies raise an important issue with respect to the potential mechanisms by which immobile responses are modified by drugs. Huddling and crouching, apart from having communicative value presumably also serve to conserve body warmth (Alberts 1978a,b). Since many drugs alter body temperature e.g. reserpine (Askew 1963), chlorpromazine (Kollias and Bullard 1964) and apomorphine (Fuxe and Sjoqvist 1972) this may be a potential determining factor in the way these drugs modify immobility in general and blocked escape in particular. However, it may be premature to separate drug-effects on body temperature from an action on flight tendencies per se. In some species, e.g. the desert lizard (Hertz et al. 1982) changes in body temperature determines whether the animal attacks or flees. Clearly, the links between changes in behaviour and the effects drugs have on body temperature require further investigation.

In chickens, where field evidence has attested to the survival value of immobility responses (Ratner 1967, 1976) considerable evidence exists to show that drugs increasing synaptic concentrations of 5-HT, prolong tonic immobility. Thus LSD, methysergide, pargyline, iproniazid (Maser et al. 1975), tryptophan (Gallup et al. 1977) but not tryptamine, prolonged this response whereas a tryptophan-free diet attenuated the response. Morphine, a compound which also induces catalepsy in rats (Stille 1971) also prolonged immobility but the effect was antagonized by p-chlorophenylalanine, an inhibitor of serotonin synthesis. Wallnau and Gallup (1978) have concluded, on the basis of this evidence, that drugs which inhibit firing of the midbrain raphe nucleus promote tonic immobility and those that increase raphe firing reduce the immobility. Indeed inhibition of 5-HT synthesis by p-chlorophenylalanine dit not affect tonic immobility. Whilst the drug depleted brain 5-HT, it did not affect firing of the midbrain raphe nucleus. Here too, it would be of interest to know how these drugs affect body temperature.

Effects on falling convulsions - a possible Protean reaction
    When held by the tail and jerked in a falling manner, mice exhibit a tonic seizure-like response which has been termed "falling convulsions" since it is close to the posture falling mice adopt (Chance 1953). Figures 7a to 7c show the different appearance of mice displaying this behaviour. The intensity of the seizure is indicated both by the position of the ears and the grimace of the face and can last up to several seconds if mice are slightly twirled by the tail. This "falling convulsion" occurs in a varying proportion of most strains of mice and has been regarded as a polyethic character (Chance 1957). In rodents, spontaneous convulsions can occur after handling or when under attack (e.g. Mus musculus), and/or when forcibly confined in a new surrounding (e.g. Meriones; Chance and Russel 1959; Kaplan 1975), i.e. in situations which evoke escape. Convulsions are composed of highly varied running movements, tonic postures, immobility and jerking movements alternating rapidly in a startling manner. As Chance noted,

Fig. 7   Falling convulsions in a hanging mouse

a) Falling convulsion showing open
mouthed grimace, erect ears,
partial eye closure, tonically
extended limbs with splayed digits.
The back is slightly arched, rigid
and is accompanied by a fine tremor.
The mouse may urinate spontaneously.

b) This posterior view of a falling
convulsion clearly shows the
widely splayed hind limbs and the
tonic extension of the forepaws
held almost perpendicularly
from the trunk.

c) Posture of a mouse after a
falling convulsion. The
grimace is absent, the ears
are relaxed back along the
head, eyes are wide open, limbs
are relaxed, the back no longer
arched and tremor is absent.
The animal may begin to struggle.

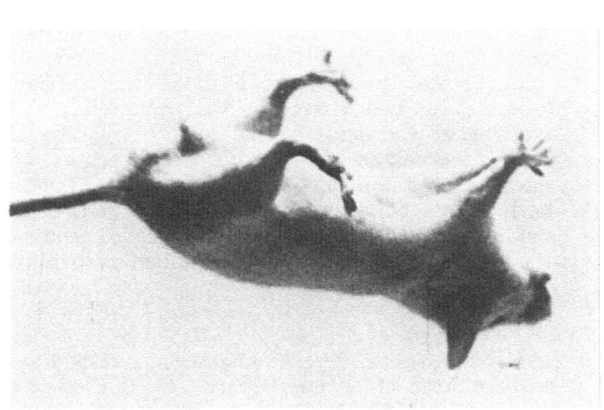

Table 5 Effects of psychoactive drugs on falling convulsions in male mice

Compound	Dose mg/kg SC	seizure counts*	Compound	Dose mg/kg SC	seizure counts*
Apomorphine	1	+ 45*	Mianserin	1	+ 62*
	3	+136***		3	+ 72.5***
	3	+123***		6	+ 44*
d–Amphetamine	1	+ 1.5	Oxotremorine	0.1	+ 63***
	3	– 6		0.5	+389***
	12	+ 41**			
Atropine	3	+ 27**	Chlordiazepoxide	1	– 2.5
				3	– 25**
Chlorpromazine	0.5	+ 27	Diazepam	1	– 6
	1	+ 46*		3	– 28**
	3	+ 50**			
	6	+ 38*			
Clozapine	1	+ 68.5***	Alcohol (g/kg PO)	0.5	– 10.5
	6	+ 60***			
Imipramine	1	+ 8	Pentylenetetrazol	3	– 4
	3	– 9.5		10	+ 20
	20	– 14*		20	+ 39*
				30	+ 84.5***

*Difference between counts of placebo and treated groups assessed 30 and 40 minutes after drug. Method: Groups of 10 OF–1 male mice, 20 to 25 g bodyweight, under normal lighting conditions, were kept in large makrolon cages (59x38.5x20 cm) overnight in the laboratory. Food and water were freely available. Between 8.00 and 12.00 hr, the next day, mice were given either the drug or 0.9% saline via the subcutaneous route. 30 minutes later each mouse was placed on the palm of the left hand, grasped with the right hand by the tail, raised into the air, and then jerked so as to simulate a fall, whereby the left hand was withdrawn. This left the mouse hanging by its tail, held in the right hand. This procedure generally produced falling convulsions usually lasting about 3 to 7 seconds, in about 40% of untreated mice. Since timing with a clock was difficult, seizure–durations were assessed by two observers who counted at circa half second intervals starting at the onset of the seizure (fig. 7a) until the animal relaxed (fig. 7c). The mean of the two observer's counts were then recorded for each mouse tested. Seizures were assessed at 30 and 40 minutes after treatment both counts being combined for each mouse. Differences between counts obtained for the drug and vehicle treated groups of mice were assessed using the Mann–Whitney U–test (one tailed probabilities). * $p<0.005$; ** $p<0.01$; *** $p<0.001$.

these features possess a Protean character and he has suggested that convulsions are a protean form of escape (Chance 1957; Chance and Russel 1959). Certainly convulsions are very startling and inevitably evoke uncertainty in an observer, the latter being a criterion of protean reactions (Driver and Humphries 1970).

We have noticed that a number of drugs appears to facilitate the release of "falling convulsions" in mice. Accordingly, a procedure was developed which enabled the duration of the seizures to be measured in drugged and placebo-treated mice both when they are held by their tails and after they have been gently twirled for ca 2 revolutions. By counting how long the seizures lasted (i.e. when the rigidity, the tonically splayed limbs and erect ears relaxed, and when the tremor and grimace disappeared), both in placebo and drug-treated mice, the effects of various compounds were assessed. Table 5 shows that apomorphine, oxotremorine, atropine, mianserin, clozapine and chlorpromazine facilitated release of falling convulsions. Pentylenetetrazol, a convulsant, also increased falling convulsions, whereas chlordiazepoxide and diazepam reduced them. d-Amphetamine only increased falling convulsions at a relatively high dose, imipramine weakly inhibited them and a single dose of alcohol had no effect.

If falling convulsions are indeed a Protean form of escape, then it should be possible to find an association between falling convulsive activity and changes in flight behaviour. Unfortunately, few of the many studies of drug effects on agonistic behaviour have measured changes in flight. Consequently, since a reciprocal relationship exists between aggression and flight in social encounters (Dixon 1982a; Grant 1963; Mackintosh 1981) it was predicted that those compounds which reduce aggression should increase flight, either actively or at least predispose towards it. Moreover, those compounds which predispose towards flight should increase falling convulsions. As can be seen from Table 6, this prediction appears to fit surprisingly well.

With the exception of mianserin and possibly amphetamine most of the compounds which have been shown to increase flight, or at least reduce aggression, appear to increase falling convulsions in mice. Even though direct evidence is lacking, there is, as already discussed, sufficient circumstantial evidence to suggest that withdrawal from chronic alcohol induces an escape syndrome. It also induces strong falling convulsions (Crabbe et al. 1981). Whilst no ethological studies of pentylenetetrazol on social behaviour were found, the results of File and Lister (1984) in the social "anxiety" test suggest that this compound increases flight. How far this fits with the compound's reported "anxiogenic" action remains a matter for conjecture.

It should be emphasised that the above scheme is hypothetical and some of the inferred effects on flight require confirmation. It also depends on a comparison between flight in social situations and a form of escape which has largely been attributed to interspecific contexts, although this division is still debateable. Yet, the hypothesis can be tested and could prove of heuristic value. Taken together, the available evidence suggests that a) falling convulsions may be a Protean form of escape, and b) that drugs which in social encounters predispose to flight, may also facilitate the release of protean behaviours, one form of which are convulsions. The results also suggest that Protean behaviours are not strictly confined to interspecific situations but may be of general value.

HORMONAL ASPECTS OF FLIGHT

Since flight is evoked by exposure to stressful, potentially threatening situations, it is not surprising that the hormones of the pituitary-thyroid and/or adrenal axes have been implicated in the development of defensive behaviour. Flight behaviour is generally the most predominant mode of response in socially subordinate or defeated individuals and this appears to be linked to an altered hormonal state. The majority of studies show that subordinate animals have high levels of circulating glucocorticosteroids, corticosterone and ACTH being higher, whilst testosterone and related androgen levels are generally lower than in dominant individuals. Valenti et al. (1981) have also shown that socially subordinate

Table 6 Effect of psychoactive drugs on falling convulsions and agonistic behaviour of male mice

Compound	falling convulsions expected	observed	flight: AGGR		Literature source
Apomorphine	yes	yes	↑	↓	Miczek and O'Donnell 1978; Dixon 1986
Amphetamine	yes	(yes)	↑	↓	Silverman 1966 Miczek 1978
Chlorpromazine	yes	yes	↑	↓	Silverman 1966
Clozapine	yes	yes	↑	↓	Dixon 1978a; 1982a
Alcohol acute	no	no	↑	↑	Chance et al. 1973 Winslow and Miczek 1985
Alcohol chronic/withdrawal	?	yes	?	?	Goldstein 1972
Atropine	yes	yes	↑	↓	DaVanzo et al. 1966
Imipramine	no	no	↓	↑	Dixon 1978a Spiegel and Dixon 1982
Amitryptyline	no	no	↓	↑	Dixon unpublished
Mianserin	no	yes	↓	↑	Dixon unpublished
Chlordiazepoxide	no	no	↓	↑	Krsiak 1975 Zwirner et al. 1975
Diazepam	no	no	↓	↑	Dixon 1982a
Pentylenetetrazol	yes	yes	"flight?"		File and Lister 1984

see text for further details	↓ ↑	confirmed
	↓ ↑	needs confirmation

mice have substantially lower plasma levels of thyroid hormones than dominants, a finding compatible with own unpublished observations, that flight is negatively associated with thyroid hormone levels (Dixon et al. 1980). Since these hormones, e.g. thyroid (Atterwil et al. 1984; Gross et al. 1980; Whybrow and Prange 1981), steroids (Biegon and McEwen 1982; Gordon et al. 1980a, 1980b; Greengrass and Tonge 1974a,b) and castration (Bernard and Paolino 1974; Engel et al. 1979) have profound actions on neurotransmitters in the brain, it seems reasonable to suspect them of being involved in the expression of agonistic behaviour. Brain (1972), Leshner and his colleagues (Leshner et al. 1973; Leshner and Pollitch 1979; Nock and Leshner 1976; Pollitch and Leshner 1977) and others, e.g. Simon and Gandelman (1978) have provided evidence suggesting that ACTH and corticosterone predisposes male mice towards defeat, whereby ACTH seems to reduce aggression (at the pituitary level) whereas corticosterone (at the adrenal level) controls submissiveness (Leshner and Pollitch 1979). As proposed by Nock and Leshner (1976) ACTH mediates the effects of defeat on readiness to attack but corticosteroids are directly involved in the submissive syndrome. Injections of ACTH and vasopressin, another pituitary hormone, into male mice immediately after defeat increases the extent to which these mice acquire and retain the defensive upright posture upon subsequent encounters with an aggressive opponent (Roche and Leshner 1979; Siegfried et al. 1982, 1984). Siegfried et al. regard this effect as being mediated by opiates which protect the mice from pain and (presumably) also facilitates memory processes. In fact, the outcome of social encounters between mice have recently been associated with changes in the endogenous opiate system. Mice subject to attacks by aggressive opponents develop analgesia (Miczek et al. 1982; Rodgers and Hendrie 1983) whereas the winners of such encounters develop hyperalgesia (Hendrie 1985; Rodgers and Hendrie 1983, 1984). The analgesia of the defeated mouse which occurs after the mouse has adopted a defensive upright posture is naloxone sensitive and differs from the immediate phase of analgesia (Hendrie 1985). Of interest, is the observation that active escape also appears after this time (Hendrie 1986). Whilst the exact mechanisms underlying the appearance of flight behaviour are clearly highly complex, the bulk of the studies just discussed suggest that for mice, the hormones of the pituitary adrenal axis, including the opiate peptides appear to play a permissive role with respect to the maintainance of flight behaviour. However the role of the thyroid, and other hormonal systems e.g. progesterone in females, needs to be examined more closely.

FLIGHT AND BEHAVIOURAL SWITCHING

In this paper, evidence has been presented showing that dopaminergic drugs promote flight thus implicating the neurotransmitter dopamine in the expression of this behaviour. However, dopamine appears to mediate other forms of behaviour including aggression and sex which are reciprocally related to flight. What then, are the factors which determine whether changes in the dopaminergic system (or in other neurotransmitter systems) lead to flight, aggression or some other activity?

The relative balance of neurotransmitters in the brain must play an important role in this respect. Serotonin has long been regarded as exerting an inhibitory influence on behaviour and suppresses positively rewarded behaviours (Stein and Wise 1974) and some social activities (Soubrie 1986). For example, many studies show that serotonergic drugs generally counteract the actions of noradrenaline and dopamine, as e.g. in sexual behaviour in both rodents and primates (Ahlenius et al. 1980; Everitt et al. 1974, 1975; Everitt 1979; Gessa and Tagliamonte 1974; Gradwell et al. 1975), although it depended upon the gender of the drug recipient and possibly the type of serotonin receptors affected (Ahlenius et al. 1981; Mendelson and Gorzalka 1985). This reciprocity between serotonin and catecholamines also appears to apply to aggressive behaviour e.g. Hodge and Butcher (1974, 1975), Soubrie (1986), Valzelli (1981), whereby serotonin exerts mainly an inhibitory influence. Social dominance in agonistic encounters is also associated with changes in serotonergic function in rodents (Kostowski et al. 1984) and primates (Raleigh et al. 1984, 1985) and an association between suicide, aggressive acts and brain 5–HT metabolism in humans is emerging (e.g. Brown et al. 1982; Edman et al. 1986; Korpi et al. 1986). Since flight is generally reciprocally related to aggression and sexual behaviour, serotonin must also affect the extent to which dopaminergic or noradrenergic events in the brain are translated into flight.

As already discussed, serotonin appears to predispose animals towards certain immobile forms of flight and these would attenuate approach–oriented activities. Cazala and Garrigues (1983) have also suggested that in mice serotonin modulates flight via the lateral hypothalamus and mesencephalic central grey area, whereas dopamine appeared to be involved at the latter sites. However, GABA too is intimately involved in these brain areas and, according to Brandao et al. (1986), exerts here an inhibitory action on a "flight system". Whilst it is premature to ascribe to serotonin a flight–permissive function analogous to the "serotonergic punishment system" of Stein and Wise (1974), the relative balance of serotonergic and GABAergic inputs at such brain sites could influence the extent to which dopaminergic and noradrenergic neuronal systems, which themselves are mutually interactive (Antelman and Caggiula 1977; Plaznik and Kostowski 1983), lead to escape as opposed to approach–oriented social behaviour. The hypothalamus and mesencephalic grey are probably only part of the neurological system governing flight and the superior colliculus and surrounding areas also appear important, at least in rats (Dean and Redgrave 1984; Sahibzada et al. 1986). The precise brain areas subserving flight clearly need elucidation. Changes in flight behaviour are probably affected by other neurotransmitters besides the catecholamines, serotonin and GABA. Nevertheless, it was pointed out in the previous section, that hormones exert profound effects upon neurotransmitter dynamics and so alter behaviour. In view of this and the well known behavioural responses of fish, birds and mammals to seasonal changes in endocrine status, hormones would appear to qualify as suitable candidates involved in the behavioural switching process. As has been discussed, corticosterone, ACTH and other pituitary peptides are apparently involved in the development of flight–related activities. Similarly, estrogens and androgens are involved in the development of sexual or aggressive behaviour. Therefore, assuming, for example, that the balance of neurotransmitters in the brain favoured dopamine, the presence or absence of such endocrine hormones, could determine whether or not flight, as opposed to some other dopamine–mediated behaviour, appears.

CONCLUSIONS

This article has only touched upon selected issues concerning drug effects on flight behaviour and was not intended to be a comprehensive survey. By focussing primarily on flight, other aspects of behaviour have inevitably been ignored. Yet it must be emphasised that no one component of behaviour can be understood without reference to the dynamic structure of behaviour as a whole (Chance 1966; Dixon 1982a). Indeed this is one of the cornerstones of the ethopharmacological approach.

Flight is clearly an intrinsic component of social, non social and interspecific behaviour. Consequently, it should be taken into consideration even if the main thrust of a study is to investigate aggression, sexual activity, exploration, or some other aspect of behaviour. As we have seen, flight takes various forms and drugs may affect these differently. Nonetheless, drugs increasing (or decreasing) flight need not necessarily have a common behavioural mechanism of action. Thus, a certain overlap between the effects of some neuroleptics and apomorphine on flight responses was evident from the literature. However, whilst the behavioural elements affected are not always identical, most of the studies reviewed examined only single doses of drugs. Obviously, chronic treatment and the time of testing needs closer scrutiny. Also, other factors as social status, experience and the endocrinological state of the drug recipient as well as the environmental context will all contribute to a drug's effect.

For reasons already stated, this discussion has focussed on aspects of flight behaviour primarily observed in rodents. Whilst ethograms of primates exist (Altmann 1967; Angst 1974) and sophisticated analyses of primate social interactions have been performed (Van den Bercken and Cools 1980a,b) very little relevant drug studies have been attempted, though some findings pertaining to dopaminergic drugs and stimulants (e.g. Schlemmer and Davis 1981) have been mentioned. Consequently, not all aspects of flight occurring in rodents can be unreservedly applied across species. Nevertheless, flight does occur in all species. Despite the morphological differences in actual elements, the way flight covaries with other behaviours i.e. offence or sex, appears to be common to man as well as animals. Certainly, studies of human non-verbal behaviour indicate that just as in animals, high levels of flight are associated with impairment of contact-oriented behaviour, particularly in the mentally ill (Dixon 1986; Grant 1972; Spiegel and Dixon 1982). The study of flight behaviour in both animals and man is therefore essential. The ethopharmacological approach is particularly suitable for this task. Thus, it applies the same concepts and methodology to the study of animals and humans. Moreover, the units of behaviour studied, i.e. acts, postures, gestures etc. are also common to both. Since new types of psychotropic drugs affecting flight as well as other social responses are now being developed on the basis of ethopharmacological criteria, future clinical trials will have to take these new ethological aspects into account.

REFERENCES

Adams DB (1980) Motivational systems of agonistic behaviour in muroid rodents: a comparative review and neural model. Aggr Behav 6: 295-346

Alberts JR (1978a) Huddling by rat pups: multisensory control of contact behavior. J Comp Physiol Psychol 92: 220-230

Alberts JR (1978b) Huddling by rat pups: group behavioral mechanisms of temperature regulation and energy conservation. J Comp Physiol Psychol 92: 231-245

Altmann J (1974) Observational study of behaviour: sampling methods. Behaviour 49: 227-267

Ahlenius S, Larsson K, Svensson L (1980) Further evidence for an inhibitory role of central 5-HT in male sexual behaviour. Psychopharmacology 68: 217-220

Ahlenius S, Larsson K, Svensson L, Hjorth S, Carlsson A, Lindberg PO, Wilstrom H, Sanchez D, Arvidsson LE, Hacksell U, Nilsson JLG (1981) Effects of a new type of 5-HT receptor agonist on male rat sexual behaviour. Pharmacol Biochem Behav 15: 785-792

Amir S, Brown ZW, Amit Z, Ornstein K (1981) Body pinch induces long lasting immobility in mice: behavioural characterization and the effect of naloxone. Life Sci 10: 1189-1194

Angst W (1974) Das Ausdrucksverhalten des Javaneraffen Macaca Fascicularis Raffles. Fortschritte der Verhaltens forschung, Heft 15. Beiheft zur Z.Tierpsychol. Verlag Paul Parey, Berlin, Hamburg.

Antelman SM, Caggiula AR (1977) Norepinephrine-dopamine interactions and behavior. Science 195: 646-653

Askew BM (1963) A simple screening procedure for imipramine-like antidepressant drugs. Life Sci 10: 723-730

Atterwil CK, Bunn SJ, Atkinson DJ, Smith SC, Heal DJ (1984) Effects of thyroid status on presynaptic alpha-2 adrenoceptor function and beta adrenoceptor binding in the rat brian. J Neural Transmission 59: 43-55

Beck CGM, Cooper SJ (1986) The effect of the β–carboline FG 7142 on the behaviour of male rats in a living cage: An ethological analysis of social and nonsocial behaviour. Psychopharmacology 89: 203–207

Berken van den JHL, Cools AR (1980a) Information statistical analysis of social interaction and communication: an analysis of variance approach. Anim Behav 28: 172–188

Bercken van den JHL, Cools AR (1980b) Information statistical analysis of factors determining ongoing behaviour and social interaction in Java monkeys (Macaca fascicularis). Anim Behav 28: 189–200

Bernard BK, Paolino RM (1974) Time–dependent changes in brain biogenic amine dynamics following castration in male rats. J Neurochem 22: 951–956

Biegon A, McEwen BS (1982) Modulation by estradiol of serotonin receptors in brain. J Neurosci 2: 199–205

Blanchard RJ, Blanchard DC, Takahashi LK (1977) Reflexive fighting in the albino rat: aggressive or defensive behavior? Aggr Behav 3: 145–155

Blanchard RJ, Flannelly KJ, Blanchard DC (1986) Defensive behaviors of laboratory and wild Rattus norvegicus. J Comp Psychol 100: 101–107

Brain PF (1972) Mammalian Behaviour and the adrenal cortex–A review. Behav Biol 7: 453–477

Brain PF (1981) Differentiating types of attack and defense in rodents. In: Brain PF, Benton D (eds) Multidisciplinary approaches to aggression research. Elsevier, North Holland Biomedical Press, Amsterdam, pp 53–78

Brandao ML, DiScala G, Bouchet MJ, Schmitt P (1986) Escape behaviour produced by blockade of glutamic acid decarboxylase (GAD) in mesencephalic central gray or medial hypothalamus. Pharmacol Biochem Behav 24: 497–501

Brown GL, Ebert MH, Goyer PF, Jimerson DC, Klein WJ, Bunney WE, Goodwin FK (1982) Aggression, suicide, and serotonin: Relationships to CSF amine metabolites. Am J Psychiatry 139: 741–746

Cazala P, Garrigues AM (1983) Effects of apomorphine, clonidine or 5–methoxy–N,N–dimethyltryptamine on approach and escape components of lateral hypothalamic and mesencephalic central gray stimulation in two inbred strains of mice. Pharmacol Biochem Behav 18: 87–93

Carli G (1977) Animal hypothesis in the rabbit. Psychol Rec 27 (Suppl): 123–143

Chance MRA (1953) The posture of a falling mouse. Br J Anim Behav 1: 118–119

Chance MRA (1957) Role of convulsions in behaviour. Behav Sci 2: 30–45

Chance MRA (1962) An interpretation of some agonistic postures: the role of "cutt–off" acts and postures. Proc Zool Soc Lond 8: 71–89

Chance MRA (1966) Resolution of social conflict in animals and man. In: de Reuck AVS, Knight J (eds) Ciba Foundation Symposium on Conflict in Society. Churchill JA, London, pp 16–35

Chance MRA, Russel WMS (1959) Protean displays: a form of allaesthetic behaviour. Proc Zool Soc Lond 132: 65–70

Chance MRA, Mackintosh JH, Dixon AK (1973) The effects of ethylalcohol on social encounters between mice. J Alcoholism 8: 90–93

Costall B, Kelly EM, Naylor RJ (1986) Alcohol treatment and withdrawal in a mouse model of anxiety. Neuropharmacology (in press)

Crabbe JC, Young ER, Janowsky J, Rigter H (1981) Pyrazole exacerbates handling–induced convulsions in mice. Neuropharmacology 20: 605–609

Cutler MG, Mackintosh JH, Chance MRA (1975a) Effects of cannabis resin on social behaviour in the laboratory mouse. Psychopharmacologia 41: 271–276

Cutler MG, Mackintosh JH, Chance MRA (1975b) Behavioural changes in laboratory mice during cannabis feeding and withdrawal. Psychopharmacologia 44: 173–177

Cutler MG, Mackintosh JH (1975) Effects of delta–9–tetrahydro–cannabinol on social behaviour in the laboratory mouse and rat. Psychopharmacologia 44: 287–289

DaVanzo JP, Daugherty M, Ruckart R, Kang L (1966) Pharmacological and biochemical studies in isolation–induced fighting mice. Psychopharmacologia 9: 210–215

Dean P, Redgrave P (1984) The superior colliculus and visual neglect in rats and hamsters. III Functional implications. Brain Res Rev 8: 155–163

DePaulis A, Vergnes M (1984) Involvement of central gabaergic receptors in the control over offensive and defensive behaviours in the rat. In: Miczek KA, Kruk MR, Olivier B (eds) Ethopharmacological Aggression Research. Alan Liss Inc, New York, pp 249–264

DiMascio A (1973) The effects of benzodiazepines on aggression: Reduced or increased? Psychopharmacologia 30: 95–102

DiScala G, Schmitt P, Karli P (1984) Flight induced by infusion of bicuculline methiodide into periventricular structures. Brain Res 309: 199–208

Dixon AK (1978a) Changes in the social behaviour of mice after antidepressants and neuroleptics. Experientia 34: 923

Dixon AK (1978b) Rodent social behaviour in relation to biomedical research. In: Weihe WH (ed) Das Tier im Experiment. Huber Verlag, Bern, pp 128–146

Dixon AK (1982a) Ethopharmacology: a new way to analyse drug effects on behaviour. Triangle 21: 95–105

Dixon AK (1982b) A possible olfactory component in the effects of diazepam on social behaviour of mice. Psychopharmacology 77: 246–252

Dixon AK (1984) Urinary odours as a source of indirect drug effects on the behaviour of male mice. In: Miczek KA, Kruk MR, Olivier B (eds) Ethopharmacological Aggression Research. Alan Liss Inc, New York, pp 81–91

Dixon AK (1986) Ethological aspects of psychiatry. Schw Arch Neurol Psychiat 137: 151–163

Dixon AK, Huber C, Rotach K (1980) The effects of imipramine HCL on thyroid function and on the social behaviour of dominant and subordinate mice. Int Soc Psychoneuroendocrinology, Florence, Abstract No. 19 Fondazione, Giovanni, Lorenzini

Driver PM, Humphries DA (1970) Protean displays as indices of conflict. Nature 226: 967–968

Edman G, Asberg M, Levander S, Schalling D (1986) Skin conductance habituation and cerebrospinal fluid 5-hydroxyindole acetic acid in suicidal patients. Arch Gen Psychiatry 43: 586–592

Engel J, Ahlenius S, Almgren O, Carlson A, Larsson K, Soedersten P (1979) Effects of gonadectomy and hormone replacement on brain monoamines synthesis in male rats. Pharmacol Biochem Behav 10: 149–154

Everitt BJ (1979) Monoamines and sexual behaviour in non–human primates. Sex Horm Behav 62: 329–358

Everitt BJ, Fuxe K, Hökfelt T (1974) Inhibitory role of dopamine and 5-hydroxytryptamine in the sexual behaviour of female rats. Eur J Pharmacology 29: 187–191

Everitt BJ, Fuxe K, Hökfelt T, Jonsson G (1975) Role of monoamines in the control by hormones of sexual receptivity in the female rat. J Comp Physiol Psychol 89: 556–572

File SE, Lister RG (1984) Do the reductions in social interaction produced by picrotoxin and pentylenetetrazol indicate anxiogenic actions? Neuropharmacology 23: 793–796

Flanningan KP, Whishaw IQ (1977) The effects of some pharmacological agents on the duration of immobility shown by rabbits placed in various postures. Bull Psychonom Soc 10: 499–502

Fox KA, Snyder RL (1969) Effect of sustained low doses of diazepam on aggression and mortality in grouped male mice. J Comp Physiol Psychol 69: 663–666

Fuxe K, Sjoqvist F (1972) Hypothermic effect of apomorphine in the mouse. J Pharm Pharmacol 24: 702–705

Gallup GG, Nash RF, Donegan NH, McClure MK (1971) The immobility response: A predator–induced reaction in chickens. Psychol Rec 21: 513–519

Gallup GG, Wallnau LB, Boren JL, Gagliardi GJ, Maser JD, Edson PH (1977) Tryptophan and tonic immobility in chickens: Effects of dietary and systemic manipulations. J Comp Physiol Psychol 91: 642–648

Gessa GL, Tagliamonte A (1974) Possible role of brain serotonin and dopamine in controlling male sexual behaviour. Adv Biochem Psychopharmacol 11: Raven Press, New York, pp 217–228

Goldstein DB (1972) An animal model for testing effects of drugs on alcohol withdrawal reactions. J Pharmacol Exp Ther 183: 14–22

Gordon JH, Gorski RA, Borison RL, Diamond BI (1980a) Postsynaptic efficacy of dopamine: possible suppression by estrogen. Pharmacol Biochem Behav 12: 515–518

Gordon JH, Borison RL, Diamond BI (1980b) Modulation of dopamine receptor sensitivity by estrogen. Biol Psychiatry 15: 389–395

Gradwell PB, Everitt BJ, Herbert J (1975) 5–Hydroxytryptamine in the central nervous system and sexual receptivity of female rhesus monkeys. Brain Res 88: 281–293

Grant EC (1963) An analysis of the social behaviour of the male laboratory rat. Behaviour 21: 260–281

Grant EC (1968) An ethological description of non–verbal behaviour during interviews. Br J Med Psychol 41: 172–184

Grant EC (1969) Human facial expression. Man 4: 525–536

Grant EC (1972) Non–verbal communication in the mentally ill. In: Hinde RA (ed) Non–Verbal Communication. Cambridge University Press, Cambridge, pp 349–358

Grant EC, Mackintosh EC (1963) A comparison of the social postures of some laboratory rodents. Behaviour 21: 246–259

Greengrass PM, Tonge SR (1974a) Suggestions on the pharmacological actions of ethinylestradiol and progesterone on the control of monoamine metabolism in three regions from the brains of gonadectomised male and female mice and the possible clinical significance. Arch Int Pharmacodyn 211: 291–304

Greengrass PM, Tonge SR (1974b) Further studies on monoamine metabolism in three regions of mouse brain during pregnancy: monoamine metabolite concentrations and the effect of injected hormones. Arch Int Pharmacodyn 212: 48–59

Gross G, Brodde OE, Shuuman HS (1980) Decreased number of β receptors in cerebral cortex of hypothyroid rats. Eur J Pharmacol 61: 191–194

Hahn RA, Hynes MD, Fuller RW (1982) Apomorphine–induced aggression in rats chronically treated with oral clonidine: modulation by central serotonergic mechanisms. J Pharmacol Exp Ther 220: 389–393

Hendrie CA (1985) Opiate dependence and withdrawal: a new synthesis. Pharmacol Biochem Behav 23: 863–870

Heritage AS, Stumpf WE, Sar M, Grant LD (1980) Brainstem neurones are target sites for sex steroid hormones. Science 207: 1377–1379

Hertz PE, Huey RB, Nevo E (1982) Fight versus flight: body temperature influences defensive responses of lizards. Anim Behav 30: 676–679

Hinde RA (1970) Animal Behaviour. McGraw–Hill Inc, New York

Hodge GK, Butcher LL (1974) 5–hydroxytryptamine correlates of isolation induced aggression in mice. Eur J Pharmacol 28: 326–337

Hodge GK, Butcher LL (1975) Catecholamine correlates of isolation induced aggression in mice. Eur J Pharmacol 31: 81–93

Hoffmeister F, Wuttke W (1969) On the actions of psychotropic drugs on the attack and aggressive–defence behaviour of mice and rats. In: Garattini S, Sigg EB (eds) Aggressive Behaviour. Excerpta Medica, Amsterdam, pp 273–280

Kaplan H (1975) What triggers seizures in the gerbil Meriones unguiculatus? Life Sci 17: 693–698

Kaesermann HP (1986) Stretched attend posture, a non–social form of ambivalence, is sensitive to a conflict–reducing drug action. Psychopharmacology 89: 31–37

Kaesermann HP, Dixon AK (1986) Further validation of the SAP test, a simple behavioural conflict test in mice. Psychopharmacology 89: S54

Kelly PF, Whishaw IQ (1977) The effects of some pharmacological agents on the curation of immobility shown by rabbits placed in various postures. Bull Psychonom Soc 10: 499–502

Kelly M, Lynch M, Leonard BE (1980) Induction of two distinct behavioural responses by chronic treatment with apomorphine. J Neurosci Res 5: 35–42

Kjellberg B, Randrup A (1972) Stereotypy with selective stimulation of certain items of behaviour observed in amphetamine treated monkeys (Cercopithecus). Pharmacopsychiatrie 5: 1–12

Kjellberg B, Randrup A (1973) Disruption of social behaviour of vervet monkeys (Cercopithecus) by low doses of amphetamines. Pharmacopsychiatry Neuro–Psychopharmacol 6: 287–293

Klemm WR (1977) Identity of sensory and motor system that are critical to the immobility reflex (animal hypnosis) Psychol Rec 1: 145–159

Kollias J, Bullard RW (1964) The influence of chlorpromazine on physical and chemical mechanisms of temperature regulation in the rat. J Pharmacol Exp Therap 145: 373–381

Korpi ER, Kleinman JE, Goodman SI, Phillips I, DeLisi LE, Linnoila M, Wyatt JR (1986) Serotonin and 5–hydroxyindole acetic acid in brains of suicide victims. Arch Gen Psychiatry 43: 594–600

Kostowski W, Plewako M, Bidzinski A (1984) Brain serotonergic neurones: Their role in a form of dominance–subordination behaviour in rats. Physiol Behav 33: 365–371

Krsiak M (1975) Timid singly housed mice: Their value in prediction of psychotropic activity of drugs. Br J Pharmacol 55: 141–150

Krsiak M (1979) Effects of drugs on behaviour of aggressive mice. Br J Pharmacol 65: 525–533

Krsiak M, Sulcova A, Tomasikova Z, Dlohozkova N, Kosar E, Masek K (1981) Drug effects on attack, defence and escape in mice. In: Miczek KA (ed) The psychopharmacology of aggression and social behavior. Pharmacol Biochem Behav 14 (suppl 1): 47–52

Krsiak M, Sulcova A, Donat P, Tomasikova Z, Dlohozkova N, Kosar E, Masek K (1984) Can social and agonistic interactions be used to detect anxiolytic activity of drugs? In: Miczek KA, Kruk MR, Olivier B (eds) Ethopharmacological Aggression Research. Alan R Liss Inc, New York, pp 93–114

Lehman MN, Adams DB (1977) A statistical and motivational analysis of the social behaviours of the male laboratory rat. Behaviour 61: 238–275

Leshner AI, Walker WA, Johnson AE, Kelling SJ, Kreisler SJ, Svare BB (1973) Pituitary adrenocortical activity and intermale aggressiveness in isolated mice. Physiol Behav 11: 705–711

Leshner AI, Pollitch JA (1979) Hormonal control of submissiveness in mice: Irrelevance of the androgens and relevance of the pituitary–adrenal hormones. Physiol Behav 22: 531–534

Lore R, Flannelly K, Farina PH (1976) Ultrasounds produced by rats accompany decreases in intraspecific fighting. Aggr Behav 2: 175–181

Mackintosh JH (1970) Territory formation by laboratory mice. Anim Behav 18: 177–183

Mackintosh JH (1981) Behaviour of the house mouse. Symp Zool Soc Lond 47: 337–365

Maser JD, Gallup GG, Hicks LE (1975) Tonic immobility: Possible involvement of monoamines. J Comp Physiol Psychol 89: 319–328

McKenzie GM (1971) Apomorphine–induced aggression in rats. Brain Res 34: 323–330

Mechoulam R (1973) Marijuana, chemistry, pharmacology and clinical effects. Academic Press, London

Mendelson SD, Gorzalka BB (1985) A facilitatory role for serotonin in the sexual behaviour of the female rat. Pharmacol Biochem Behav 22: 1025–1033

Miczek KA (1974) Intraspecies aggression in rats: Effects of d–amphetamine and chlordiazepoxide. Psychopharmacologia 39: 275–301

Miczek KA, Krsiak M (1979) Drug effects on agonistic behavior. In: Thompson T, Dews PB (eds) Advances in Behavioral Pharmacology. Academic Press, New York, pp 87–162

Miczek KA, Gold LH (1983) Ethological analysis of amphetamine action on social behavior in squirrel monkeys (Saimiri sciureus). In: Miczek KA (ed) Ethopharmacology: Primates models of neuropsychiatric disorders, Alan R Liss Inc, New York, pp 137–155

Miczek KA, Woolley J, Schlisserman S, Yoshimura H (1981) Analysis of amphetamine effects on agonistic and affiliative behavior in squirrel monkeys (Saimiri sciureus). In: Miczek KA (ed) The psychopharmacology of aggression and social behaviour. Pharmacol Biochem Behav 14 (suppl 1): 103–107

Miczek KA, O'Donnell JM (1978) Intruder–evoked aggression in isolated and non–isolated mice: effects of psychomotor stimulants and L–DOPA. Psychopharmacology 57: 47–55

Miczek KA, Thompson ML, Shuster L (1982) Opioid–like analgesia in defeated mice. Science 215: 1520–1522

Nock BL, Leshner AI (1976) Hormonal mediation of the effects of defeat on agonistic responding in mice. Physiol Behav 17: 111–119

Plaznik A, Kostowski W (1983) The interrelationships between brain noradrenergic and dopaminergic neuronal systems in regulating animal behaviour: possible clinical implications. Psychopharmacol Bull 19: 5–11

Pollitch JA, Leshner AI (1977) Relationship between plasma corticosterone levels and the tendency to avoid attack in mice. Physiol Behav 19: 781–785

Poole TB (1973) Some studies on the influence of chlordiazepoxide on the social interaction of the golden hamster. Br J Pharmacol 48: 538–545

Poshivalov VP (1981) Pharmaco–ethological analysis of social behaviour of isolated mice. In: Miczek KA (ed) The psychopharmacology of aggression and social behavior. Pharmacol Biochem Behav 14 (suppl 1): 53–59

Pucilowski O, Kozak W, Valzelli L (1986) Effect of 6–OHDA injected into the locus coeruleus on apomorphine–induced aggression. Pharmacol Biochem Behav 24: 773–775

Puglisi–Allegra S, Mele A, Cabib S (1984) Involvement of endogenous opioid systems in social behaviour of individually housed mice. In: Miczek KA, Kruk MR, Olivier B (eds) Ethopharmacological Aggression Research. Alan R Liss Inc, New York, pp 209–225

Raleigh MJ, McGuire MT, Brammer G, Yuwiler A (1984) Social and environmental influences on blood serotonin concentrations in monkeys. Arch Gen Psychiatry 41: 405–410

Raleigh MJ, Brammer G, McGuire MT, Yuwiler A (1985) Dominant status facilitates the behavioural effects of serotonergic agonists. Brain Res 348: 274–282

Randrup A, Munkvad I (1966) DOPA and other naturally occurring substances as causes of stereotypy and rage in rats. Acta Psychiat Scand 42: 193–199

Randrup A, Munkvad I (1969) Relation to brain catecholamines of aggressiveness and other forms of behavioural excitation. In: Garattini S, Sigg EB (eds) Aggressive Behaviour. Excerpta Medica, Amsterdam, pp 228–235

Ratner SC (1967) Comparative aspects of hypnosis. In: Gordon JE (ed) Handbook of Clinical and Experimental hypnosis. MacMillan, New York, pp 550–587

Ratner SC (1976) Animal's defenses: fighting in predator–prey relations. In: Tlinir T, Alloway TM, Kramis L (eds) Advances in the study of communication and affect: II non–verbal communication of aggression. Plenum Press, New York, pp 175–190

Roche KE, Leshner AI (1979) ACTH and vasopressin treatments immediately after a defeat increases future submissiveness in male mice. Science 204: 1343–1344

Rodgers RJ, Hendrie CA (1983) Social conflict activates status–dependent endogenous analgesic or hyperalgesic mechanisms in male mice: effects of naloxone on nociception and behaviour. Physiol Behav 30: 775–780

Rodgers RJ, Hendrie CA (1984) On the role of endogenous opioid mechanisms in offense, defense and nociception. In: Miczek KA, Kruk MR, Olivier B (eds) Ethopharmacological Aggression Research. Alan R Liss Inc, New York, pp 27–41

Rodgers RJ, Depaulis A (1982) GABAergic influence on defensive fighting in rats. Pharmacol Biochem Behav 17: 451–456

Sahibzada N, Dean P, Redgrave P (1986) Movements resembling orientation or avoidance elicited by electrical stimulation of the superior colliculus in rats. J Neurosci 6: 723–733

Schallert T, De Ryck M, Teitelbaum PH (1980) Atropine stereotypy as a behavioral trap: a movement subsystem and electroencephalographic analysis. J Comp Physiol Psychol 94: 1–14

Schiorring E, Hecht A (1979) Behavioural effects of low, acute doses of d–amphetamine on the dyadic interaction between mother and infant vervet monkeys (Cercopithecus aethiops) during the first six postnatal months. Psychopharmacology 64: 219–224

Schiorring E (1981) Psychopathology induced by "speed drugs" In: Miczek KA (ed) The psychopharmacology of aggression and social behavior. Pharmacol Biochem Behav 14 (suppl 1): 109–122

Schlemmer FR, Davis JM (1981) Evidence for dopamine–mediation of submissive gestures in the stumptail macaque monkey. In: Miczek KA (ed) The psychopharmacology of aggression and social behavior. Pharmacol Biochem Behav 14 (suppl 1): 95–102

Schlemmer RF, Narasimhachari N, Davis JM (1980) Dose dependent behavioural changes induced by apomorphine in selected members of a primate social colony. J Pharm Pharmacol 32: 285–289

Schmidt WJ (1983) Involvement of dopaminergic neurotransmission in the control of goal–directed movements. Psychopharmacology 80: 360–364

Schmidt WJ (1984) L–DOPA and apomorphine disrupt long- but not short behavioural chains. Physiol Behav 33: 671–680

Scraggs PR, Baker HF, Ridley RM (1979) Interaction of apomorphine and haloperidol: effects on locomotor and other behaviour in the marmoset. Psychopharmacology 66: 41–43

Sieber B, Frischknecht HR, Waser PG (1980a) Behavioural effects of hashish in mice. I: social interactions and nest building of males. Psychopharmacology 70: 149–154

Sieber B, Frischknecht HR, Waser PG (1980b) Behavioral effects of hashish in mice. II: social interactions between two residents and an intruder male. Psychopharmacology 70: 273–278

Siegfried B, Frischknecht HR, Waser PG (1982) A new learning model for submissive behaviour in mice: Effects of naloxone. Aggr Behav 8: 112–115

Siegfried B, Frischknecht HR, Waser PG (1984) Vasopressin impairs or enhances retention of learned submissiveness behaviour in mice depending on the time of application. Behav Brain Res 11: 259–269

Silverman P (1966a) Social behaviour of laboratory rats and the action of chlorpromazine and other drugs. Behaviour 27: 1–38

Silverman P (1966b) Barbiturates, lysergic acid diethylamine, and the social behaviour of laboratory rats. Psychopharmacologia 10: 155–171

Simon NG, Gandelman R (1978) Influence of corticosterone on the development and display of androgen–dependent aggressive behavior in mice. Physiol Behav 20: 391–396

Soubrie PH (1986) Reconciling the role of central serotonin neurons in human and animal behavior. Behav Brain Sci 9: 319–364

Spiegel R, Dixon AK (1982) Psychotropic drug experiments in normal subjects, their relation to animal studies and clinical trials. In: Spiegelstein MY, Levy A (eds) Behavioural Models and the analysis of drug action. Proc 27th OHOLO Conf, Israel, Elsevier, Amsterdam, pp 39–55

Stein L, Wise CD (1974) Serotonin and behavioral inhibition. In: Adv Biochem Psychopharmacology 11. Raven Press, New York, pp 281–291

Stille G (1971) Zur Pharmakologie Katatonigener Stoffe. Arzneimittel Forsch 21: 30–42

Strasser ST, Dixon AK (1986) Effects of visual and acoustic deprivation on agonistic behaviour of the albino mouse (M.musculus L). Physiol Behav 36: 773–778

Tinbergen N (1952) The curious behaviour of the stickleback. Sci Amer 187: 22–26

Valenti G, Vescovi PP, Volpi R, Mainardi D, Mainardi M, Cavaggioni A (1981) Thyroid hormone pattern and aggressiveness. La Ricerca Clin Lab 11: 117–122

Valzelli L (1981) Psychopharmacology of aggression and violence. Raven Press, New York, pp 98–103

Van der Poel AM (1967) Ethological study of the behaviour of the albino rat in a "passive–avoidance" test. Acta Physiol Pharmacol Neerl 14: 503–504

Van der Poel AM (1979) A note on "stretched attention", a behavioural element indicative of an approach–avoidance conflict in rats. Anim Behav 27: 446–450

Wallnau LB, Gallup GG (1978) Serotonergic and electric shock effects on tonic immobility: Evidence for independent systems. Physiol Behav 21: 869–872

Whybrow PW, Prange AJ (1981) A hypothesis of thyroid–catecholamine receptor interaction. Arch Gen Psychiatry 38: 106–113

Wiley JN, Downs DA (1979) Naloxone–precipitated jumping in mice pretreated with acute injections of opioids. Life Sci 25: 797–801

Winslow JT, Miczek KA (1985) Social status as determinant of alcohol effects on aggressive behavior in squirrel monkeys (Saimiri sciureus). Psychopharmacology 85: 953–958

Woodruff ML (1977) Limbic modulation of contact defensive immobility (animal hypnosis) Psychol Rec 1: 161–175

Zwirner O, Porsolt RD, Loew DM (1975) Intergroup aggression in mice: a new method for testing the effects of centrally active drugs. Psychopharmacology 45: 133–138

SITUATIONAL–DEPENDENCE AND DIFFERENTIAL MEDIATION OF ANALGESIC REACTIONS TO CONSPECIFIC ATTACK IN MICE

R. John Rodgers and Jill I. Randall. Pharmacoethology Laboratory, School of Psychology, University of Bradford, Bradford BD7 1DP, U.K.

INTRODUCTION

Within the past decade, a growing appreciation and application of ethological perspectives has led to substantial advances in our understanding of the physiology of aggression and defeat. Such progress stems from the important theoretical contribution of Paul Scott (1958) with his concept of 'agonistic behaviour' and the elegant descriptive studies of Ewan Grant and John Mackintosh (1963) on social patterns in common laboratory rodents. The former directly drew attention to aggression as part of a much larger behavioural repertoire subserving adaptation to situations involving intraspecific conflict, whilst the latter documented the specific acts and postures assocated with such interactions in a variety of species. Early examples of the extension of this knowledge to the study of drug effects on agonistic behaviour can be found in the work of Chance and Silverman (1964), Krsiak and Steinberg (1969), Miczek (1974) and Poshivalov (1974). The impact of the pharmacoethological (or is it ethopharmacological?) approach in contemporary laboratory studies on fighting behaviour in animals is self–evident in the literature.

Whilst pharmacoethological methods have largely been employed in the study of drug effects on behaviour during agonistic encounters, they may also be successfully applied to the analysis of the physiological and behavioural results of engaging in such interaction.

This point is clearly alluded to by Eleftheriou and Scott (1971) in 'The Physiology of Aggression and Defeat'; these authors emphasized the importance of studying not only the antecedents of fighting but also its behavioural, neuroendocrine and neurochemical consequences. Indeed, evidence has accumulated suggesting that these consequences are markedly different for victors and the vanquished (Bronson and Eleftheriou 1965; Daruna 1978; Leshner 1980; Raab et al. 1986; Scott and Marston 1953; Welch and Welch 1969). The aim of the present chapter is to review work from our laboratory and elsewhere which has clearly established that exposure to conspecific attack has profound antinociceptive consequences which vary in form and mediation according to specific circumstances.

THE CONCEPT OF ENVIRONMENTALLY–INDUCED ANALGESIA

In their Gate Control Theory of pain, Melzack and Wall (1965) postulated the existence of powerful internal inhibitory controls over pain responding. Only four years later, Reynolds (1969) reported that focal electrical stimulation of rat midbrain yielded very potent analgesia, a finding obviously confirming the Melzack and Wall proposal. Whilst research on the mechanisms and clinical applications of 'stimulation–produced analgesia' has continued in its own right (Terman and Lieteskind 1986), the very existence of an intrinsic analgesia system prompted questions regarding the types of stimuli which might normally lead to its activation. The first clue to this problem came with three independent reports in 1976, all of which indicated that environmental 'stressors' (eg. footshock) could elicit strong analgesia in laboratory rats (for review: Rodgers and Randall 1986c). Since then, a quite 'bewildering' array of external factors have been found capable of activating intrinsic analgesia mechanisms (Table 1).

Of course, from the outset, studies have also been directed at elucidating the mechanisms underlying environmentally–induced analgesia. In this context, a logical starting point was with the then recently–discovered endogenous opioid systems but, although several forms of analgesia were indeed found to be opioid–mediated, many others were not. In fact, it is currently thought that multiple intrinsic analgesia mechanisms exist which can be dissociated on a number of dimensions (Terman and Liebeskind 1986; Watkins and Mayer 1982). Thus, in addition to the obvious biological importance of nociception in signalling injury, the pervasive existence of environmentally–activated antinociception would imply quite profound adaptive significance.

Table 1   Factors Eliciting Analgesia in Rodents

Body pinch	Insulin
Centrifugal rotation	Irradiation
Classical conditioning	Novelty
Copulation (Males)	Predation
2–deoxy–D–glucose	Presence of natural predator
Electric shock	Restraint
Food deprivation	Social isolation
Forced swimming	Stress odours
Heat	Tail pinch
Hypertonic saline	Vaginal stimulation

(adapted from Rodgers and Randall 1986c)

ECOLOGY OF PAIN INHIBITION

Ten years ago, Liebeskind et al. (1976) argued that endogenous analgesia mechanisms may have evolved to suppress pain in situations in which responding to noxious input might otherwise disrupt effective behavioural action. Certainly, the ethological literature is replete with examples of prey species which seem capable of inhibiting response to injury (e.g. death feigning) until an opportunity to escape arises (Ratner 1967). A similar phenomenon has recently been reported under laboratory conditions, in which rats were rendered analgesic by exposure to the presence of a natural predator (Lester and Fanselow 1985).

Despite these revealing observations on predator–prey interactions, intraspecific conflict is probably the most frequent source of noxious stimulation for many organisms. It was this line of reasoning which, some six years ago, led us to begin an analysis of the effects of fighting on nociceptive responding in rats and mice. The results of our early work have recently been reviewed (Rodgers and Hendrie 1984) and, as such, will only briefly be summarized to provide perspective for our more recent investigations.

Our initial experiments on intermale encounters in rats gave very disappointing results. Despite a variety of housing manipulations, we could obtain no evidence that fighting elicits analgesia in this species. However, this 'absence of evidence' could not be taken as 'evidence for absence' since videotape analysis of encounters revealed a consistent pattern of low–intensity interactions. In other words, these early studies did not constitute a valid test of the general hypothesis.

The importance of stimulus intensity was confirmed in two subsequent studies. Firstly, exposure to defensive attack from lactating conspecifics was found to induce a partially naloxone–sensitive analgesia in male intruder rats, suggesting activation of both opioid and non–opioid intrinsic analgesia mechanisms by this experience (Rodgers et al. 1983). Secondly, exposure to persistent attack from isolated male resident conspecifics was found to elicit a fully naloxone–sensitive analgesia in male intruder–mice (Rodgers and Hendrie 1983).

Although it was tempting to directly attribute both of these positive findings to the increased intensity of encounters, correlational analyses (degree of analgesia versus the latency/frequency/duration of particular behaviours) suggested the critical importance of 'psychological' factors. Thus, the degree of analgesia in intruder rats was found to be a positive function of the latency to female attack and the duration of partner/cage investigation by the intruder. Thus, somewhat counter–intuitively, the longer it took for residents to attack and the longer intruders spent exploring the resident/resident's cage, the greater the degree of analgesic response. Obviously, if physical stimulation per se is the most important factor, then the degree of intruder analgesia ought to have negatively correlated with resident attack latency and positively correlated with the number of attacks received; no such relationships were evident. Rather, the profile to emerge implicated factors such as fear, unpredictability or uncontrollability in the generation of the analgesic reaction.

Although the degree of analgesia evident in intruder mice was positively correlated with the number of resident attacks, several findings militated against such a simple relationship. Firstly, although bitten during encounters, resident mice actually exhibited hyperalgesia. Although we did not follow up this finding, it would be interesting to determine whether such enhanced pain reactivity in residents is a function of physical stimulation or whether, like intruder analgesia, it seems to relate to psychological factors. Secondly, the degree of intruder analgesia correlated highly with the duration of intruder immobility, a reaction observed when escape and defensive strategies have been tried but fail to terminate resident attack. Finally, in separate studies, we consistently failed to find evidence of analgesia following unresolved interactions between pairs of isolated male mice, despite the very high intensity of fighting. Again, the critical factor(s) seemed to be largely psychological in nature, and we argued that perceived loss of control as a function of a failure in behavioural coping strategies may be most important. On this view, the physical stimulation of attack is relevant only in so far as it leads to this end–point. Importantly, this early tentative proposal was totally in accord with ideas emerging from the study of the opioid analgesic response to footshock in rats (Maier et al. 1982).

In general agreement with the above findings, Miczek et al. (1982) reported that intruder mice (B6AF$_1$ strain) display naloxone–sensitive and morphine cross–tolerant analgesia in response to high frequency attack. Subsequently, they showed that such analgesia is (a) most evident in those mouse strains which evidence a good analgesic response to morphine, (b) associated with changes in whole–brain beta–endorphin content and brain–stem opiate receptor binding, (c) blocked by naloxone microinjections into arcuate and PAG sites, (d) unaffected by peripherally–acting opiate antagonists or manipulations of pituitary–adrenal function and (e) associated with profound long–term changes in the functioning of endogenous opioid systems (for review Miczek et al. 1986). These elegant studies, together with complementary findings from other laboratories (Hendrie 1985; Kavaliers and Hirst 1985; Siegfried et al. 1984; Teskey and Kavaliers 1984, 1985; Teskey et al. 1984) provide fairly conclusive evidence for the involvement of central opioid mechanisms in the analgesic response of intruder mice to intense attack.

TWO FORMS OF ANALGESIA IN INTRUDER MICE

Despite the above consensus, opinions have varied with respect to the critical precipitant(s) of intruder analgesia. For example, although agreeing that analgesia cannot readily be attributed to the influence of physical stimuli per se, Miczek et al. (1982) and Rodgers and Hendrie (1983) proposed related, though somewhat different, suggestions as to the nature of the critical 'psychological' factor(s). The former argued that opioid analgesia is related to the special biological significance of defeat whilst the latter suggested (as reviewed above) that loss of control may be the important factor. Unfortunately, for both proposals, early experimental designs (fixed attack criterion or fixed encounter duration, respectively) did not allow for a clear separation of the contribution of defeat per se from extended exposure to attack. In other words, intruders in both types of experiment would invariably have been defeated prior to termination of encounters. That defeat experience itself may not be the critical precipitant of opioid analgesia was suggested by two findings: (a) no correlation was found between parameters of defeat (upright submissive posture) and degree of analgesia in BKW strain mice (Rodgers and Hendrie 1983) and (b) intruders of the CF1 strain were found to exhibit naloxone–sensitive analgesia irrespective of the display of defeat (Teskey et al. 1984). This uncertainty regarding the psychological 'trigger' of opioid analgesia, stimulated several new series of studies with DBA/2 mice – a strain that is very sensitive to the analgesic effects of morphine.

Analgesic Reaction to Extended Attack

In view of potential strain differences in mediation, we began by studying the analgesic reaction (reported, but not characterized, by Miczek and Thompson (1984) and Siegfried et al. (1984)) in DBA/2 intruders. Group–housed DBA/2 mice were individually exposed to a moderate attack criterion (35 attack bites) in dyadic confrontations with singly–housed experienced BKW residents (Rodgers and Randall 1985 a, b). Under these test conditions, intruders evidenced a strong analgesic reaction of 40–60 minutes duration which was (i) blocked and reversed by a low dose (3 mg/kg) of naloxone, (ii) insensitive to peripherally–acting doses of methyl naloxone and (iii) bidirectionally cross–tolerant with morphine (5 mg/kg). These data paralleled the findings of Miczek et al. (1982) in B6AF$_1$ mice in confirming endogenous opioid mediation of the analgesic reaction of intruder mice to extended conspecific attack. In order to determine the potential contribution of defeat per se to the development of this analgesia, we adopted a rather different test criterion in our next series of studies.

Analgesic Reaction to Defeat Experience

Videotape analysis of the above studies confirmed our earlier suspicion that intruders display unambiguous defeat behaviour (upright submissive postures) very early on in encounters. In the present context, clear defeat was shown within 60 seconds of introduction to the resident's home cage and in response to fewer than ten attack bites. We therefore conducted a series of experiments in which 'defeat' was employed as the behavioural end–point for encounters (Rodgers and Randall 1986a). Under these altered conditions, intruders indeed showed post–encounter analgesia, but this was a very different reaction to that observed in response to extended attack. Firstly, it had a duration of < 10 minutes (versus 40–60 minutes); secondly, it was insensitive to naloxone (up to 10 mg/kg) and, thirdly, it did not display cross–tolerance either to or from morphine (5 mg/kg). Thus, defeat experience per se is associated not with opioid analgesia but with an acute non–opioid analgesia. This finding is obviously at variance with the 'defeat' hypothesis of opioid analgesia but is not inconsistent with the 'uncontrollability' hypothesis. The latter argues that defeat is important in the generation of opioid analgesia only in so far as it is ineffective in terminating sustained attack. As such, it would be predicted that opioid–typical analgesia should develop when encounters are allowed to proceed beyond the display of defeat i.e. intruders must learn that this behavioural strategy is ineffective. Consistent with prediction, exposure to as few as ten attack bites beyond defeat was found to be sufficient to induce an enduring analgesic reaction (Rodgers and Randall 1985c). These data directly parallel work on footshock analgesia which indicates that extended exposure to aversive stimulation is necessary for organisms to learn that their behaviour is ineffective in coping with environmental contingencies (Maier et al. 1982).

Acute Analgesia as a Preparatory Defense Response

In the studies reviewed in the preceding section, display of defeat was merely the behavioural end–point for encounters. As such, any factor(s) up to and including the end–point could been responsible for the initiation of analgesia. In other words, display of defeat may not have been a critical prerequisite for the onset of non–opioid analgesia. To further define the most important stimulus feature of defeat experience, we exposed a large number of intruder mice to 30–second encounters with aggressive residents (Rodgers and Randall 1986b). This exposure period was chosen as the most likely to yield sub–groups of intruders as defined by specific behavioural criteria. All encounters were videotaped and individual intruders assigned to three groups (not bitten; bitten but not defeated; defeated and exposed to further attack) on the basis of tape analysis. Compared to controls (no encounter), all categories of intruder (including those which were not attacked) displayed significant analgesia following brief interaction with residents; however, only those intruders exposed to attack beyond defeat showed opioid–typical enduring analgesia. Thus, physical stimulation would not appear to be necessary for the activation of the non–opioid analgesia system.

Additional studies (Rodgers and Randall 1986b), in which environmental novelty, social novelty and olfactory stimuli were independently examined, revealed that 'mere' exposure to the soiled home cage of an isolated male conspecific is sufficient to induce acute analgesia in DBA/2 intruder mice. Furthermore, variations in the duration of exposure to the soiled substrate indicated that (a) scent alone can result in a degree of analgesia similar to that induced by defeat experience, (b) 'defeat' appears to result in a temporal advancement of peak analgesia since 60 seconds scent exposure is necessary to achieve the same strength of analgesia induced by a 30–second defeat experience, (c) the analgesic reaction to scent completely dissipates if intruders are allowed to remain in the resident's cage for more than two minutes, and (d) like 'defeat' analgesia, the analgesic reaction to scent is insensitive to the opiate antagonist naloxone. The most parsimonious interpretation of these findings is that acute non–opioid analgesia represents a biologically–adaptive preparatory defense response to the detection of conspecific territorial scent–marking. This reaction can apparently be temporally exacerbated by the 'realization' of the perceived threat (i.e. initial attack) whilst, in the absence of the latter, it rapidly dissipates.

NEUROHUMORAL MEDIATION OF NON–OPIOID ANALGESIA?

To describe a form of environmentally–induced analgesia as 'non–opioid' is akin to the archaic definition of instinct as that which is not learned! In other words, it is a largely negative definition. However, a review of the general literature on other forms of environmentally–induced analgesia would suggest that, as with the labelling of a given behaviour as 'instinctive', the labelling of a form of analgesia as 'non–opioid', is associated with a relative dearth of subsequent research (Rodgers and Randall 1986c). At best, our knowledge of the substrates of non–opioid forms of analgesia is nowhere near as extensive as our understanding of the mechanisms of opioid analgesia phenomena. In this context, the question arises as to the particular neurohumoral substrates underlying the analgesic reaction to defeat experience/scent of a territorial conspecific.

Previous research (e.g. Jones and Nowell 1973; Sandnabba 1985; Sawyer 1978) has clearly shown that substrates soiled by dominant/isolated male mice are actively avoided by male conspecific intruders. Similarly, there is a wealth of data confirming that exposure to attack (or in some instance, the mere threat of attack) is a potent stimulus leading to pituitary–adrenal activation and active avoidance behaviour (e.g. Brain 1980; Bronson and Eleftheriou 1965; Leshner 1980). As fear/anxiety would be an undoubted correlate of exposure to such biologically–relevant aversive stimuli, we have recently undertaken an analysis of the potential involvement of benzodiazepine receptor mechanisms in the acute analgesic reaction to defeat experience.

Although benzodiazepines (BZPs) were originally developed some 25 years ago, it is only within the past few years that we have begun to understand the mechanisms whereby they exert their multiple effects. These advances have been the direct result of (a) the identification of specific high–affinity binding sites for BZPs in CNS, (b) the clarification of an intimate functional relationship between these binding sites and the GABA receptor complex and (c) the development of a wide range of pharmacological tools with which the functional significance of these sites may be studied (for review, Haefely et al. 1983). The ligands now available have been classified into three broad categories on the basis of intrinsic activity: agonists (e.g. chlordiazepoxide, midazolam), antagonists (e.g. Ro 15–1788) and inverse agonists (e.g. CGS8216, FG7142). As the latter category is novel, it ought to be understood that these drugs bind to BZP sites with high affinity but exert actions opposite to those of classic benzodiazepines.

Benzodiazepine Ligands and Basal Nociception

In attempting to define the neurohumoral substrates of 'defeat' analgesia, it is essential to assess possible effects on basal nociception of the various drugs to be used. Without such advance knowledge, interpretation of data would be severely compromised. Unfortunately, the existing literature concerning BZP agonist effects on nociceptive responding is highly contradictory whilst virtually nothing is known about potential effects of BZP antagonists and inverse agonists. In this context, we have recently (Rodgers and Randall 1986d) screened a range of BZP ligands for effects on tail–flick responding in mice (i.e. our present analgesia assay). The results are summarized in Table 2. Briefly, three agonists (chlordiazepoxide, midazolam, diazepam) and an antagonist (Ro 15–1788) were devoid of antinociceptive effects. However, the three inverse agonists employed (CGS8216, FG7142, DMCM) each induced significant analgesia. It was confirmed that this action relates to a direct effect at BZP recognition sites by drug

Table 2. Effect of benzodiazepine ligands on basal tail–flick responding in mice. Drugs were administered (IP) 20–30 minutes pretest.

Ligand	Dose (IP)	Basal Nociception
Chlordiazepoxide	5–30 mg/kg	No effect
Midazolam	0.625–10 mg/kg	No effect
Diazepam	0.5–2 mg/kg	No effect
Ro 15–1788	5–80 mg/kg	No effect
CGS8216	10–20 mg/kg	Analgesia
FG7142	5–20 mg/kg	Analgesia
DMCM	0.06–2 mg/kg	Analgesia

interaction studies indicating that the analgesic reactions induced by CGS8216 and FG7142 were completely inhibited by combined treatment with either Ro 15–1788 or chlordiazepoxide. In summary, substances which act as inverse agonists at BZP recognition sites consistently produce analgesia as assessed by the tail–flick test in mice. In contrast, substances which act as agonists or antagonists at these sites do not alter basal nociceptive responding. With this data–base, we have recently begun an analysis of the influence of BZP ligands on analgesia induced by defeat experience.

Benzodiazepine Ligands and 'Defeat' Analgesia

In view of the 'stressful' nature of many of the situations which engender, analgesia, it is very surprising that minimal research attention has been paid to potential effects of benzodiazepines. The literature that does exist presents a confused picture. Neither opioid nor non–opioid forms of footshock analgesia are affected by chlordiazepoxide or diazepam (Chance et al. 1979; Hayes et al. 1978; Kinsheck et al. 1984). Although the non–opioid analgesic response to forced cold–water swim is potentiated by diazepam (Leitner and Kelly 1984), a possible hypothermic synergism between drug and 'stressor' may underlie this effect. More recently, however, Willer and Ernst (1986) have reported that human stress analgesia (induced by anticipation of pain) is attenuated by diazepam treatment. Nevertheless the lack of research in this area obviates any concrete statement regarding the influence of benzodiazepines on environmentally–induced analgesia.

One of the characteristic features of BZP inverse agonists, apart from analgesic efficacy, is their ability to induce a state of anxiety in animals and man (for review: Pellow and File 1984). As defeat experience also induces analgesia and may be expected to provoke fear/anxiety in intruders, it might be predicted that 'defeat' analgesia should be blocked by treatments (Ro 15–1788; chlordiazepoxide) which antagonise the analgesic and anxiogenic effects of BZP inverse agonists. Consistent with this prediction we have found that the BZP antagonist Ro 15–1788 completely blocks acute 'defeat' analgesia (fig. 1). It should be noted that, compared to controls, neither latency to defeat nor number of attacks to defeat were significantly altered by the antagonist. Complete antagonism of 'defeat' analgesia by Ro 15–1788 would be consistent with the proposal (Corda and Biggio 1986) that mild 'stressors' (e.g. handling) evoke the release of an endogenous 'inverse agonist–like' ligand for BZP recognition sites. However, despite the obvious attractions of this interpretation, we have been unable to block 'defeat' analgesia with either chlordiazepoxide (up to 30 mg/kg) or midazolam (up to 5 mg/kg). If stimuli associated with defeat experience cause the release of an endogenous 'anxiogenic' ligand which produces analgesia via interaction with BZP sites, then current theory demands that the analgesia ought to be blocked by both antagonist and agonist treatment. This is certainly true for the analgesic, anxiogenic and anorectic effects of inverse agonists such as FG7142 and DMCM (Cooper 1985; Pellow and File 1984; present observations). Very recently, however, we have shown that, at a non–sedative dose (2 mg/kg), diazepam completely blocks the antinociceptive consequences of 'defeat' (fig. 2). Although the reason for this discrepancy in findings with chlordiazepoxide, midazolam and diazepam are at present unclear, it may relate to the much greater receptor affinity of the last–mentioned compound (Jensen et al. 1986). Importantly, despite their ability to block acute 'defeat' analgesia, neither Ro 15–1788 nor diazepam significantly alters the enduring (opioid) analgesic reaction to intense attack (unpublished observations). This finding implies that the two forms of analgesia associated with murine social conflict are totally independent reactions. That is, prevention of the initial acute analgesia by Ro 15–1788 or diazepam does not prevent the development of the enduring (opioid) reaction seen in response to sustained attack.

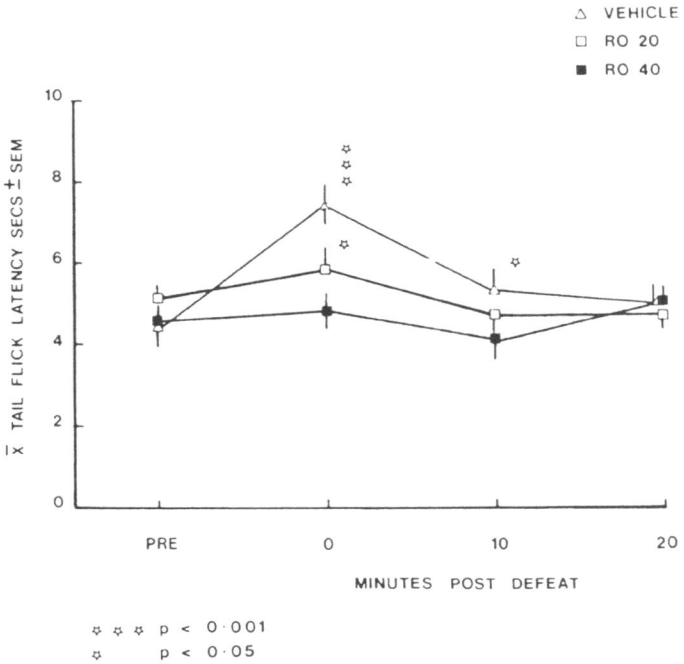

Fig. 1 Effect of Ro 15–1788 (20–40 mg/kg, IP, 30 min pretest) on acute analgesic reaction to defeat experience in DBA/2 mice. Pre = baseline. Statistical values relate to comparisons with baseline.

SUMMARY AND CONCLUSIONS

Over the past five years, research from this laboratory and elsewhere has unequivocally demonstrated the ecological validity of the concept of environmentally–induced analgesia. It is now evident that social conflict in male mice is associated with at least two forms of analgesic reaction. Thus, upon exposure to the scent–marked territory of an aggressive male conspecific, intruders display an acute analgesia which, in the absence of additional aversive/noxious input, rapidly dissipates. However, if the olfactory 'threat' is accompanied by attack from the resident culminating in the display of defeat by the intruder, this acute analgesic reaction is temporally exacerbated. If intruder strategies (escape, defense, submission) prove ineffective in reducing aversive stimulation, a more enduring analgesic reaction develops. Whilst pharmacological/biochemical studies have clearly shown the latter reaction to be opioid–mediated, the neurohormoral substrates underpinning the initial acute analgesia would appear to involve benzodiazepine receptor mechanisms. However, our present inability to demonstrate efficacy of certain BZP agonists in preventing the reaction would serve to caution against premature conclusions.

88

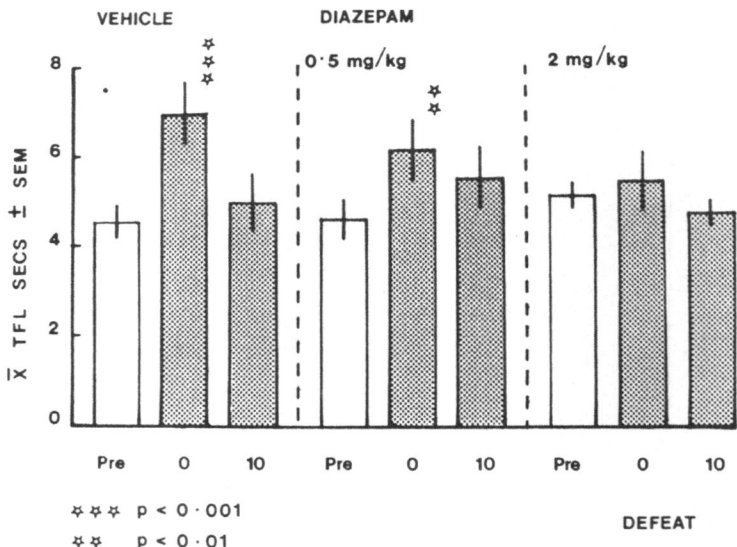

Fig. 2   Effect of diazepam (0.5–2 mg/kg, IP, 30 min pretest) on acute analgesic
reaction to defeat experience in DBA/2 mice. Details as for fig. 1.

Returning to an ecological perspective, the potential contribution of intrinsic
analgesia mechanisms to <u>longer–term behavioural adaptation</u> (e.g. social learning
and memory) should not be underestimated. Siegfried et al. (1985) have recently
begun to address this important issue. Two strains of mice (C57BL/6, DBA/2)
differing in their reactions to stress, were assessed for behavioural change 24 hours
after exposure to varying degrees of conspecific attack. The 24h retest consisted
of interactions with either aggressive or non–aggressive opponents. In response to
aggressive encounters, C57 mice showed significantly less escape but more crouch
and defensive upright postures compared to day 1. In contrast, DBA/2 mice
displayed no change in defensive pattern from day 1 to day 2. A different profile of
results was observed in response to non–aggressive patterns. Thus, C57 mice
showed conditioned 'recall' of escape and submissive postures as a positive function
of the degree of attack to which they had been exposed on day 1; in direct
contrast, retention of learned behaviour by DBA/2 mice decreased as a function of
day 1 attack intensity.

Importantly, the C57 strain is relatively insensitive to the analgesic effects of
morphine and high frequency attack whilst, as reviewed extensively in this chapter,
the DBA/2 strain is exquisitely sensitive to the analgesic effects of both
manipulations. Although further studies are obviously necessary, a link between the
activation of intrinsic analgesia mechanisms and subsequent behavioural adaptation
is certainly suggested by these novel and exciting data.

Finally, the question may be asked as to the potential significance of present
findings for the human condition. At first glance, it may appear that studies on the
consequences of fighting behaviour in mice have no relevance whatsoever for our
understanding of human physiology and behaviour. However, setting aside specific

details of the animal 'models', several more general principles regarding intrinsic analgesia mechanisms appear to be emerging. Firstly, the studies reviewed in this chapter have provided ecological validity for the concept of environmentally–induced analgesia, in that such phenomena can readily be observed in animals engaging in biologically–relevant activity. Thus, analgesic reactions to footshock and other artificial forms of stimulation are not mere artefacts of esoteric laboratory procedures. Secondly, research on the antinociceptive consequences of social conflict in mice has confirmed the existence of more than one intrinsic analgesia system. Not only have opioid and non–opioid (possibly benzodiazepine receptor–mediated) systems been identified, but the general conditions under which each may be activated have also been described. Thus, a major psychological correlate of non–opioid analgesia would appear to be acute anxiety/fear, whilst opioid analgesia seems to result from a loss of control over environmental events. These general principles lead to obvious predictions for testing in human subjects. Indeed, both naloxone–sensitive and diazepam–sensitive forms of human 'stress analgesia' have quite recently been reported (Willer and Albe–Fessard 1980; Willer and Ernst 1986). Further studies are obviously required in this area. Thirdly, clinical advances invariably arise as a result of progress in pure (or basic) science. Thus, by contributing to a more complete and sophisticated understanding of mechanisms of pain and pain inhibition, animal studies provide the vital foundation from which more rational therapeutic approaches may be developed for the relief of human suffering and distress. The introduction of direct brain stimulation techniques for the relief of intractable pain conditions is but one excellent example of the intimate inter–relationship between pure and applied research in this field. Furthermore, the identification of multiple intrinsic analgesia systems offers the possibility of developing yet more important advances in pain therapy and management. In this context, and given the problems inherent in opioid manipulations (i.e. tolerance and dependence), it is surprising that relatively little research attention has thus far been paid to the substrates underlying the so–called 'non–opioid' forms of environmentally–induced analgesia. Hopefully, this situation will be rectified in the not–too–distant future.

Acknowledgements: This work was supported by the Medical Research Council. The authors gratefully acknowledge the kind gifts of drugs from Endo Laboratories Inc. (New York), Ciba–Geigy Corporation (New Jersey), Hoffman–La–Roche (Switzerland), Roche Products Ltd (UK) and Schering AG (Germany).

REFERENCES

Brain PF (1980) Adaptive aspects of hormonal correlates of attack and defence in laboratory mice: a study in ethobiology. In: McConnell PS, Boer GJ, Romijn HJ, Van de Poll NE, Corner MA (eds) Adaptive Capabilities of the Nervous System, Progress in Brain Research, Vol. 53, Elsevier, Amsterdam, pp 391–413

Bronson FH, Eleftheriou BE (1965) Adrenal response to fighting in mice: separation of physical and psychological causes. Science 147: 627–628

Chance MRA, Silverman AP (1964) The structure of social behaviour and drug action. In: Steinberg H, de Reuck AVS (eds) Animal Behaviour and Drug Action, Ciba Foundation Symposium, Churchill, London, pp 65–79

Chance WT, White A, Krynock GM, Rosecrans JA (1979) Autoanalgesia: acquisition, blockade and relationship to opiate binding. Eur J Pharmacol 58: 461–468

Cooper SJ (1985) Bidirectional control of palatable food consumption through a common benzodiazepine receptor: theory and evidence. Brain Res Bull 15: 397–410

Corda MG, Biggio G (1986) Stress and GABAergic transmission: biochemical and behavioural studies. Adv Biochem Psych Pharm 41: 121–136

Daruna JH (1978) Patterns of brain monoamine activity and aggressive behaviour. Neurosci Biobehav Rev 2: 101–113

Eleftheriou BE, Scott JP (1971) The Physiology of Aggression and Defeat. Plenum, New York

Grant EC, Mackintosh JH (1963) A comparison of the social postures of some common laboratory rodents. Behaviour 21: 246–259

Haefely W, Polc P, Pieri L, Schaffner R, Laurent J–P (1983) Neuropharmacology of benzodiazepines: synaptic mechanisms and neural basis of action. In: Costa E (ed) The Benzodiazepines: From Molecular Biology to Clinical Practice. Raven Press, New York, pp 137–146

Hayes RL, Bennett GJ, Newlon PG, Mayer DF (1978) Behavioural and physiological studies of non-narcotic analgesia in the rat elicited by certain environmental stimuli. Brain Res 155: 69–90

Hendrie CA (1985) On the adaptive significance of the opioid withdrawal syndrome. In: LeMoli F (ed) Multidisciplinary Approaches to Conflict and Appeasement in Animals and Man. University of Parma Publications, Parma, p 50

Jensen LH, Petersen EN, Honore T, Drejer J (1986). Bidirectional effects of benzodiazepine receptor ligands, seizure activity and interaction studies. In: Biggio G, Spano PF, Toffano G, Gessa GL (eds) Modulation of Central and peripheral transmitter function, Liviana Press, Padova, pp 401–404

Jones RB, Nowell NW (1973) Aversive and aggressive-promoting properties of urine from dominant and subordinate male albino mice. Anim Learn Behav 3: 207–210

Kavaliers M, Hirst M (1985) FMRFamide, a putative endogenous opiate antagonist: evidence from suppression of defeat-induced analgesia and feeding in mice. Neuropeptides 6: 485–494

Kinsheck IB, Watkins LR, Mayer DJ (1984) Fear is not critical to classically-conditioned analgesia: the effects of periaqueductal gray lesions and administration of chlordiazepoxide. Brain Res 298: 33–44

Krsiak M, Steinberg H (1969) Psychopharmacological aspects of aggression: a review of the literature and some new experiments. J Psychosomat Res 13: 243–252

Leitner DA, Kelly DD (1984) Potentiation of cold swim stress analgesia in rats by diazepam. Pharmacol Biochem Behav 21: 813–816

Leshner AI (1980) The interaction of experience and neuroendocrine factors in determining behavioural adaptations to aggression. In: McConnell PS, Boer GJ, Romijn HJ, Van de Poll NE, Corner MA (eds) Adaptive Capabilities of the Nervous System, Progress in Brain Research, Vol 53, Elsevier, Amsterdam, pp 427–438

Lester LS, Fanselow MS (1985) Exposure to a cat produces opioid analgesia in rats. Behav Neurosci 99: 756–759

Liebeskind JC, Giesler GJ, Urca G (1976) Evidence pertaining to an endogenous mechanism of pain inhibition in the central system. In: Zotterman Y (ed) Sensory Functions of the Skin in Primates, Vol. 1, Pergamon Press, London, pp 561–573

Maier SF, Drugan RC, Grau JW (1982) Controllability, coping behaviour and stress–induced analgesia in the rat. Pain 12: 47–56

Melzack R, Wall PD (1965) Pain mechanisms: a new synthesis. Science 150: 971–979

Miczek KA (1974) Intraspecies aggression in rats: effects of d–amphetamine and chlordiazepoxide. Psychopharmacologia 39: 275–301

Miczek KA, Thompson ML (1984) Analgesia resulting from defeat in a social confrontation: the role of endogenous opioids in brain. In: Bandler R (ed) Modulation of Sensorimotor Activity During Altered Behavioral States, Alan R Liss, New York, pp 431–456

Miczek KA, Thompson ML, Shuster L (1982) Opioid–like analgesia in defeated mice. Science 215: 1520–1522

Miczek KA, Thompson ML, Shuster L (1986) Analgesia following defeat in an aggressive encounter: development of tolerance and changes in opioid receptors. Ann NY Acad Science 467: 14–29

Pellow S, File SE (1984) Multiple sites of action for anxiogenic drugs: behavioural, electrophysiological and biochemical correlations. Psychopharmacology 83: 304–315

Poshivalov V (1974) Pharmacological and psychophysiological analysis of aggressive behaviour under conditions of zoosocial interactions. In: Valdman VA (ed) Psychopharmacology of Behaviour, First Pavlov Medical Institute, Leningrad, pp 60–82

Raab A, Dantzer R, Michaud B, Mormede P, Taghzouti K, Simon H, Le Moal M (1986) Behavioural, physiological and immunological consequences of social status and aggression in chronically coexisting resident–intruder dyads of male rats. Physiol Behav 36: 223–228

Ratner SC (1967) Comparative aspects of hypnosis. In: Gordon JE (ed) Handbook of Clinical and Experimental Hypnosis, Macmillan, New York, pp 550–587

Reynolds DV (1969) Surgery in the rat during electrical analgesia induced by focal brain stimulation. Science 164: 444–445

Rodgers RJ, Hendrie CA (1983) Social conflict activates status–dependent endogenous analgesic and hyperalgesic mechanisms in male mice: effects of naloxone on nociception and behaviour. Physiol Behav 30: 775–780

Rodgers RJ, Hendrie CA (1984) On the role of endogenous opioid mechanisms in offense, defense and nociception. In: Miczek KA, Kruk MR, Olivier B (eds) Ethopharmacological Aggression Research, Alan R Liss, New York, pp 27–41

Rodgers RJ, Hendrie CA, Waters AJ (1983) Naloxone partially antagonizes post–encounter analgesia and enhances defensive responding in male rats exposed to attack from lactating conspecifics. Physiol Behav 30: 781–786

Rodgers RJ, Randall JI (1985a) Strain differences in behaviourally–induced antinociception and morphine analgesia in male mice. Br J Pharmacol 84: 105P

Rodgers RJ, Randall JI (1985b) Social conflict analgesia: studies on naloxone antagonism and morphine cross–tolerance in male DBA/2 mice. Pharmacol Biochem Behav 23: 883–888

Rodgers RJ, Randall JI (1985c) Relationship between opioid and non–opioid forms of social conflict analgesia. In: LeMoli F (ed) Multidisciplinary Approaches to Conflict and Appeasement in Animals and Man, University of Parma Publications, Parma, p 95

Rodgers RJ, Randall JI (1986a) Acute non–opioid analgesia in defeated male mice. Physiol Behav 36: 947–950

Rodgers RJ, Randall JI (1986b) Resident's scent: a critical factor in acute analgesic reaction to defeat experience in male mice. Physiol Behav 37: 317–322

Rodgers RJ, Randall JI (1986c) Environmentally–induced analgesia: situational factors, mechanisms and significance. In: Rodgers RJ, Cooper SJ (eds) Endorphins, Opiates and Behavioural Processes, Wiley, Chichester, In Press

Rodgers RJ, Randall JI (1986d) Benzodiazepine ligands, basal nociception and 'defeat' analgesia in male mice. Psychopharmacology, In Press

Sandnabba NK (1985) Differences in the capacity of male odours to affect investigatory behaviour and different urinary marking patterns in two strains of mice, selectively bred for high and low aggressiveness. Behav Proc 11: 257–267

Sawyer FT (1978) Aversive odors of male mice: experiential and castration effects, and the predictability of the outcome of agonistic encounters. Aggr Behav 4: 263–275

Scott JP (1958) Aggression. University of Chicago Press, Chicago

Scott JP, Marston MV (1953) Nonadaptive behaviour resulting from a series of defeats in fighting mice. J Abnorm Soc Psychol 48: 417–428

Siegfried B, Frischknecht H-R, Waser PG (1984) Defeat, learned submissiveness and analgesia in mice: effect of genotype. Behav Neural Biol 42: 91–97

Siegfried B, Kulling P, Waser PG, Frischknecht H-R (1985) Defeat–induced defense reactions in mice. In: LeMoli F (ed) Multidisciplinary Approaches to Conflict and Appeasement in Animals and Man, University of Parma Publications, Parma, p 73

Terman GW, Liebeskind JC (1986) Relation of stress–induced analgesia to stimulation–produced analgesia. Ann NY Acad Sci 467: 300–308

Teskey GC, Kavaliers M (1984) Opioids and aggression: influences of opioids agonists and antagonists on aggressive behaviour in mice. Soc Neurosci Abstr 10: 1132

Teskey GC, Kavaliers M (1985) Prolyl–leucyl–glycinamide reduces aggression and blocks defeat–induced opioid analgesia in mice. Peptides 6: 165–167

Teskey GC, Kavaliers M, Hirst M (1984) Social conflict activates opioids analgesic and ingestive behaviours in male mice. Life Sci 35: 303–315

Watkins LR, Mayer DJ (1982) Organization of endogenous opiate and non–opiate pain control systems. Science 216: 1185–1192

Welch BL, Welch AS (1969) Fighting: preferential lowering of norepinephrine and dopamine in the brain stem, concomitant with a depletion of epinephrine from the adrenal medulla. Commun Behav Biol 3: 125–130

Willer JC, Albe–Fessard D (1980) Electrophysiological evidence for a release of endogenous opiates in stress–induced 'analgesia' in man. Brain Res 198: 419–426

Willer JC, Ernst M (1986) Diazepam reduces stress–induced analgesia in humans. Brain Res 363: 398–402

STUDIES CONTRASTING DRUG EFFECTS ON REPRODUCTION INDUCED AGONISTIC BEHAVIOUR IN MALE AND FEMALE MICE

H. Yoshimura. Department of Pharmacology, Ehime University School of Medicine, Ehime 791-02, Japan.

Numerous research papers have followed the introduction of the concept of the "animal model" into psychiatric medicine. Experimental animals may be used to investigate the psychopathology of a behavioural disorder, if one realizes the significance and limitations of the findings. This principle holds true for research on aggression.

Ever since the classic finding that benzodiazepine anxiolytics have a "taming" effect on savage zoo animals without producing excessive sedation and sleep, various experimental models of aggression have been developed and used for screening of new psychotropic drugs. There has been, however, a tendency to overemphasize the suppressive effects of a drug on attack behaviour, whilst directing little attention to the social interactions between the attacking and the attacked animal.

Recently, new approaches based on ethological considerations have been introduced for the behavioural analysis of drug action, under the general heading of "ethopharmacology". Such procedures make an interdisciplinary approach possible for the study of both species-typical and atypical, disordered behaviour (Miczek 1983) In this context, agonistic behaviour, which is a system of behavioural patterns, proven adaptive in conflict situations, is the focus rather than aggression per se.

Several excellent reviews and books concerning the effects of drugs on agonistic behaviour in animals ranging from rodents to primates have recently been published (e.g. Miczek 1986; Miczek and Barry 1976; Miczek and Krsiak 1979; Olivier et al. 1984). This paper will focus only on the characteristics of agonistic behaviour in male and female mice and the actions of drugs on such activities. The significance of displacement activity during agonistic confrontations in ethopharmacological studies of aggression will also be considered.

PRESENT STATUS OF ANIMAL STUDIES

Over the last 25 years, a variety of experimentally-induced aggressive behaviours in laboratory animals have been developed for evaluating the psychotropic action of drugs. Most of the testing procedures require, however, potentially aversive experimental manipulations such as electrical foot-shock (Tedeschi et al. 1959), prolonged isolation (Ginsberg and Allee 1942), or physical provocation (e.g. the dangling method: Scott 1966 or the warm-up technique: DaVanzo et al. 1966). The results obtained in pharmacological or behavioural studies, in many cases, appear dependent upon the type of experimental manipulation employed. Moreover, the topography of fighting episodes under these conditions differs from the species-specific patterns of aggressive behaviour seen ferally.

To minimize the complex interactions between aversive situations and drug effects, a resident-intruder paradigm which does not employ aversive experimental manipulation has been introduced into the preclinical evaluation of psychotropic drugs (Miczek and Krsiak 1979; Miczek and O'Donnell 1978; Olivier 1981; Olivier and Van Dalen 1982). Resident male mice which have cohabited with a female for several weeks will reliably show species–specific agonistic behaviour when confronting an unfamiliar intruder male in the resident's home cage. The ethological literature has indicated, in a variety of animal species, that the appearance of a "stranger" is the strongest stimulus which evokes an aggressive response (Wilson 1975). It is also possible to analyse the behavioural topography of agonistic and non–agonistic episodes quantitatively with the aid of video monitoring systems and the computerized event–recorder. Using ethological techniques, Miczek and co–workers have performed detailed analyses of drug effects on various elements of agonistic behaviour in both rodents and primates (Miczek 1974; Miczek and Barry 1976; Miczek and O'Donnell 1978; Miczek et al. 1981; Miczek and Yoshimura 1982).

In preclinical psychopharmacology, little attention has been paid, however, to female aggression in laboratory animals. There are at least three possible reasons for this: 1) aggressive behaviour in females is infrequent, 2) females are insensitive to the usual experimental manipulations for inducing male aggression, and 3) evaluation of drug effects is complicated by the endocrine changes that occur in the oestrus cycle. Investigators have focussed on male aggression rather than on the female, even though lactating animals of a variety of species are known to vigorously attack intruders (Hafez 1962; Scott 1966). Systematic behavioural studies on maternal aggression showed that female Rockland–Swiss mice displayed intense aggressiveness toward intruders during early (between 0 and 13 days after parturition) lactation (Gandelman 1972). St.John and Corning (1973) reported significant differences in the incidence of female attackers between strains of mice a feature recently confirmed (Ogawa and Makino 1981). Since nonparturient virgin females rarely attack intruders (Gandelman 1972; Noirot et al. 1975), hormonal changes during gestation and lactation were considered important likely variables (for reviews: see Svare 1981; Svare and Mann 1983) but nipple growth and suckling stimulation by pups seem to play a proximal role in the initiation of such behaviour (Svare and Gandelman 1976a,b; Svare et al. 1980).

There is, however, little information concerning drug effects on maternal aggression, compared to the voluminous psychopharmacological studies on intermale behaviour. An early study by Brain and Al–Maliki (1979) found that lithium chloride did not alter maternal aggression in TO strain mice. Most recently, Olivier et al. (1985) found that oral administration of chlordiazepoxide (5 and 10 mg/kg) increased maternal aggression in rats, and that fluprazine (5, 10 and 20 mg/kg, IP) suppressed maternal aggression in a dose–dependent manner. With regard to neuroregulatory mechanisms of maternal aggression, the serotonin depletor parachlorophenylalanine significantly suppressed maternal aggression in mice (Ieni and Thurmond 1985; Svare 1983). Because a stable level of maternal aggression is confined to a restricted postpartum period (Svare et al. 1981) determination of acute effect of a drug has to be performed within a one week period.

TOPOGRAPHY OF AGONISTIC BEHAVIOUR IN MALE AND FEMALE MICE

The pattern of species–specific agonistic behaviour shown when a resident male mouse which has been cohabiting with a female partner confronts an unfamiliar male intruder, is illustrated in fig. 1. Immediately before the introduction of the intruder, the female mouse and, her pups (if present), were removed, and after completion of the test the removed mice were returned to their home cage (Ogawa and Yoshimura 1983; Yoshimura 1985).

To evaluate the effects of drugs on male agonistic behaviour, the following acts and postures, based on the definitions by Grant and Mackintosh (1963), were

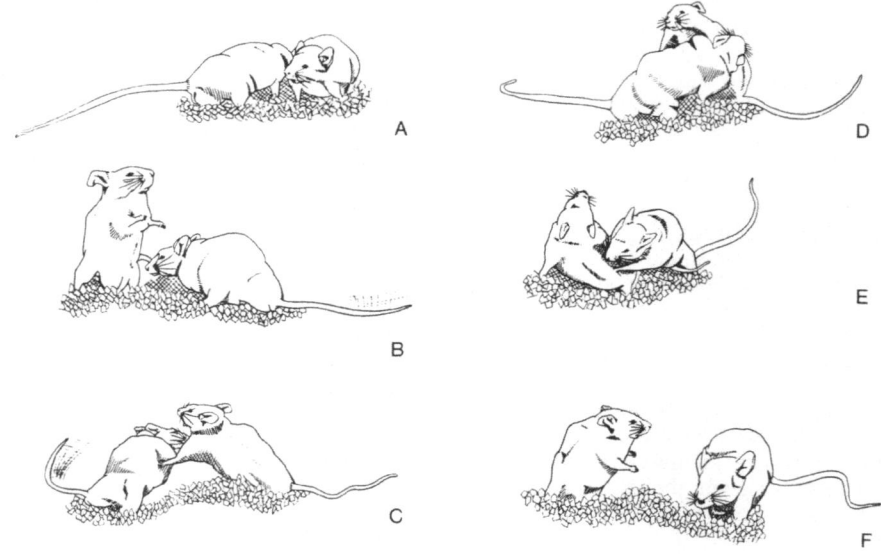

Fig. 1   Agonistic acts and postures shown by resident and intruder mice. (A)
Anogenital contact by the resident mouse (left). (B) Mincing with tail
rattling by the resident mouse (right) and defensive upright posture by the
intruder mouse (left). (C) and (D) Offensive sideways posture by the
resident mouse (left). (E) Attack bite by the resident mouse (right). (F)
Defensive upright posture by the intruder mouse (left).

determined: resident's behaviour (offensive sideways posture, tail rattle, attack
bite, pursuit, locomotion, rearing, and self-grooming); intruder's behaviour
(defensive upright posture, escape, locomotion, rearing, and self-grooming).
Ethograms of agonistic and nonagonistic behaviours in mice have been described
repeatedly (e.g. Krsiak 1975; Miczek and O'Donnell 1978; Poshivalov 1974; Scott
1966).

The incidence of male aggression gradually increases as a function of the length
of his cohabitation with the female (fig. 2). Only ICR strain albino mice were used
and experience over several years with this paradigm shows that 5 weeks of
cohabitation results in virtually all resident male mice displaying intense attack
biting on male intruders. It must be noted that not all strains do show a similar
attack profile and also the nature of the attacked animal seems an important
variable. The variability caused by the opponent's behaviour could be reduced by
using the following types of opponent: nonaggressive and sexually naive young
males (Brain and Nowell 1970), olfactory bulbectomized males (Denenberg et al.
1973; Ropartz 1968) and castrated males (Brain and Poole 1974; Edwards et al.
1972). As neither bulbectomized nor castrated opponents are suitable for
determining direct and indirect effects of drugs on agonistic behaviour, sexually
naive group-housed males were used as intruders.

In the behavioural analysis of drug actions, the experimental design employed is
also   an   important   factor.   In   general,   a   within-group   comparison
(repeated-measurement design) is preferred over a between-group comparison
(independent sample design). The "round-robin" technique has often been used to

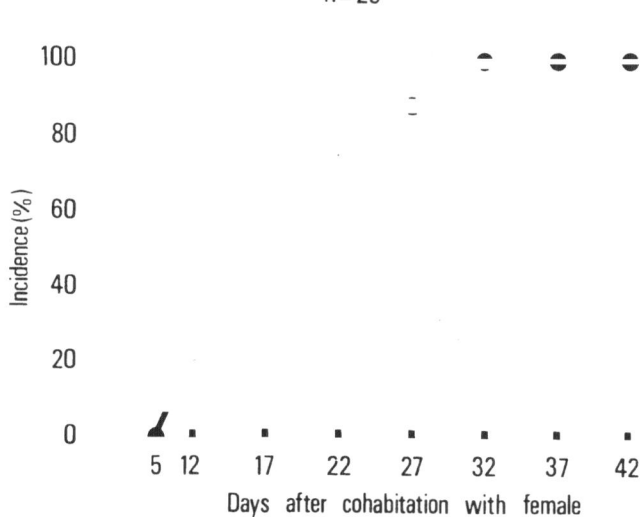

Fig. 2 Incidence of attack bites in male mice as a function of the day after cohabitation with females.

determine agonistic behaviour: each subject is exposed once to every stimulus animal in rotation, until all possible combinations have been completed (e.g. Brain and Nowell 1970; Scott and Fredericson 1951). The major difficulty with this technique seems to be in the statistical analysis of drug-induced behavioural changes. Previously, Yoshimura and Miczek (unpublished) investigated whether the level of agonistic behaviour alters as a function of repeated confrontations between resident and intruder mice. Confrontations produced variable and sometimes extremely intense sequences of attack and flight behaviour, but after three to four initial confrontations both opponents maintained a stable level of agonistic behaviour during 10 consecutive tests (each test was separated by 3 to 4 days from the next). Consequently, in pharmacological studies, a fixed combination of one resident and one intruder mouse was used throughout the experimental period and a treatment-by-subject design was employed, which permits analysis of repeated measures on the same individuals.

Maternal aggression in mice is characterized by immediate attack bites and threat is actually rare, (e.g. Al-Maliki et al. 1980; Svare 1981). Fig. 3 shows the typical pattern of maternal aggression by a lactating female on an unfamiliar male intruder. Agonistic confrontations were performed on postpartum day 5 in the female's home cage, and pups were not removed during the testing period.

Young, sexually naive male mice (6 to 7 weeks of age) were used as intruders. Although we have employed the ICR strain albino mice as subjects for investigating maternal aggression, it should be noticed that strain differences are more explicit in female aggression than in male aggression. For example, lactating female ICR mice, unlike other strains such as Rockland-Swiss (e.g. Svare and Gandelman 1973), did not attack female intruders. Moreover, female ICR mice cohabiting with the mating partner throughout pregnancy, parturition, and lactation periods rarely showed attack behaviour toward a strange male intruder, even when the cohabiting male was removed from the home cage during the confrontation. Green (1978) also observed this behavioural characteristic of ICR mice, whereas Brain and Al-Maliki

98

Fig. 3   Agonistic acts and postures shown by a parturient female mouse (left). A young (6–7 weeks of age) male mouse was employed as intruder.

(1979) and Al–Maliki et al. (1980) have observed similar phenomena in another strain of mouse.

Maternal and intermale aggression in mice differ with respect to the sequence and the frequency of occurrence of components of agonistic behaviour. In aggressive male mice, attack bite is usually preceded by offensive sideways posture and frequently followed by rapid pursuing the intruder. However, as shown in fig. 4, aggressive female mice infrequently displayed these elements. Little tail rattling was evident in females. Although no significant difference in the frequency of attack bite was found between male and female mice, it is of great interest that behavioural elements reflecting a conflict between approach and avoidance showed a sex difference. Moreover, a difference in the latency from the introduction of an intruder to the first attack bite on that animal between male and female mice has been repeatedly reported (e.g. Svare 1981): most of the aggressive females attack the opponent within 10 sec, but aggressive males have longer latencies with larger

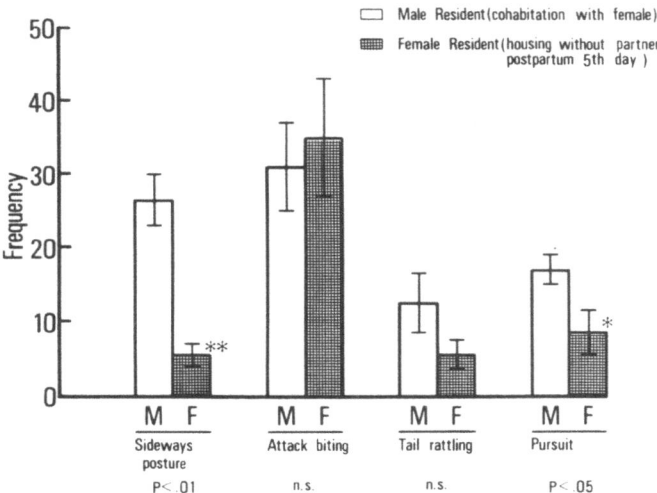

Fig. 4   Comparison of the frequency of aggressive elements occurring in intermale
and maternal aggression in mice.

variation. There was a significant difference in the incidence of attack bite by
females that were housed together with the mating partner and solitary housed
counterparts (fig. 5). The percentage of attacking females in both the paired and
isolated condition was 15% or less during pregnancy but, immediately after
parturition, the isolated females became aggressive and displayed intensive attack
bites on postpartum day 3. In contrast, paired females rarely showed aggressive
behaviour even during the early lactation period. These results clearly indicate that
housing conditions during pregnancy and early lactation play an essential role in the
manifestation of maternal aggression in mice.

To confirm the significance of housing conditions, three series of experiments
were conducted (Yoshimura and Ogawa 1984b). First, maternal aggression in: 1)
females which cohabited with their mating partner throughout the experimental
period, 2) virgin females that were housed individually for 30 days without mating,
and 3) females that were individually housed after the mating period of 4 days were
compared. As virgin females did not manifest maternal aggression 30 days after
isolation, pregnancy and parturition appear necessary to induce female aggression.
As mentioned previously significant differences in both the incidence and the
frequency of attack bites between categories 1 and 3 were evident. Thus, individual
housing after the female has become pregnant also seems important.

Second, it was determined whether the nature of the mating partner affects the
manifestation of maternal aggression. For this, partner males were castrated 5
days after initial cohabitation with females. Interestingly, females that were
housed with the castrated partners showed a remarkable increase in both the
incidence and the frequency of attack bites as compared to the group housed with
sham–operated mating partners: 90% of females housed with a castrated male
exhibited attack bites. This finding clearly indicates that the nature of the mating
partner has an important influence of maternal aggression. As it is well known that
individual recognition in mice is mainly mediated by olfactory cues, especially by
pheromones from the preputial glands, in the third experiment partner males
underwent preputialectomy after the mating period. However, there was no
significant difference in the incidence of maternal aggression between females

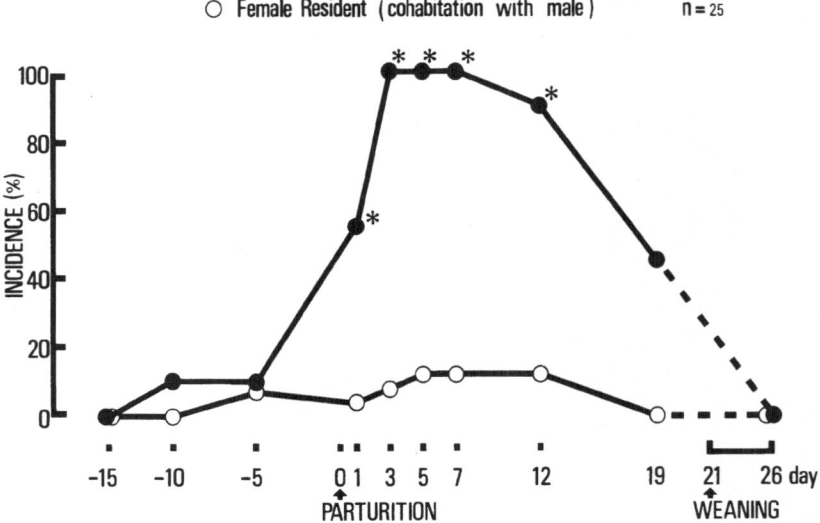

● Female Resident (housing without mating partner) n = 25
○ Female Resident (cohabitation with male) n = 25

Fig. 5   Incidence of maternal aggression in mice during pregnancy and lactation.

housed with preputialectomized or females housed with sham–operated males. The behavioural characteristics of the partner may be of a greater importance.

ETHOPHARMACOLOGY OF MALE AGONISTIC BEHAVIOUR

Benzodiazepine anxiolytics such as chlordiazepoxide and diazepam are widely prescribed psychotropic drugs not only in psychiatric hospitals but also in general practice. However, the effectiveness of these anxiolytics on aggressiveness in patients or normal volunteers is incompletely understood. Contrary to the effectiveness of these drugs in the management of anxiety and tension, some physicians have noticed that anxiolytics may facilite rather than suppress aggression under certain conditions (Bond and Lader 1979; DiMascio 1973; Lion et al. 1975). In laboratory animals, also there is increasing evidence that benzodiazepine anxiolytics may increase or induce aggressive behaviour (Fox and Snyder 1969; Krsiak 1979; Miczek 1974; Miczek and O'Donnell 1980), depending upon the experimental situation. Although such behavioural changes have been considered "paradoxical reactions" or "disinhibitory actions", it is unclear whether this augmentation is specific to benzodiazepine derivatives, and whether this phenomenon relates to the anxiolytic actions of the drug. To answer these questions the effects of benzodiazepine anxiolytics on agonistic behaviour were compared with those of nonbenzodiazepine anxiolytics such as suriclone. Moreover, effects of other psychotropic drugs such as antidepressants and antipsychotics were also examined.

The intensity of agonistic behaviour seems dependent upon the confronted opponent's nature. Drug–induced changes in one animal may change behaviour in the nondrugged opponent (Miczek 1974; Miczek and Barry 1977; Miczek and O'Donnell 1978; Yoshimura and Ogawa 1983, 1984a). The finding that the effect of a drug on agonistic interactions depends on whether aggressive or defensive animals are the drug recipients provides a clue to differentiate anxiolytics from

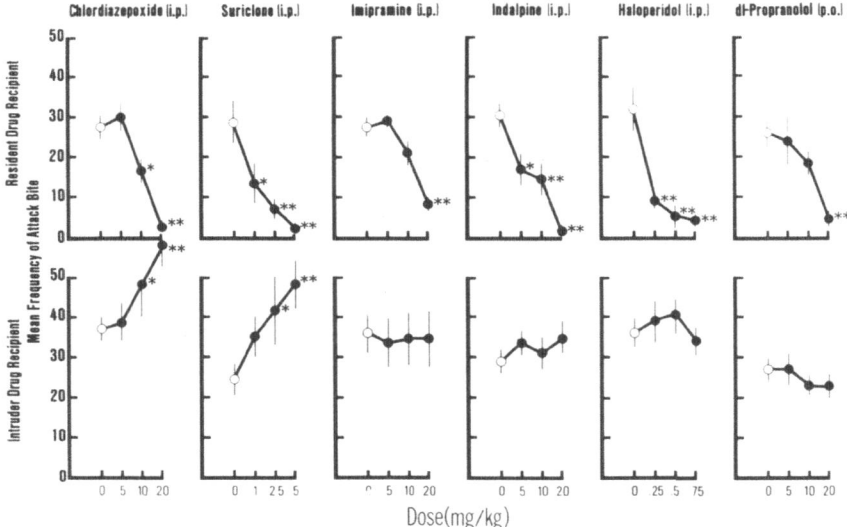

Fig. 6 Effects of psychotropic drugs on the frequency of attack bites in male mice. Top panels indicate the resident's attack bites when the resident mice were the drug recipients, and bottom panels show the resident's attack bites when the intruder mice were the drug recipients.

other psychotropic drugs. As shown in Fig. 6, drugs which have an anxiolytic action caused intruder mice to be more frequently attacked by nontreated resident mice, while antidepressants or antipsychotics had no such effect.

It is especially noteworthy that a nonbenzodiazepine anxiolytic like suriclone increased the frequency of the resident's attack bites when the intruder was drugged. As reported previously (Yoshimura and Ogawa 1983), ethyl alcohol (0.5 to 2.0 g/kg, PO) also showed similar indirect effect. The question that arises is why only drugs which possess anxiolytic actions show indirect effects on the resident's aggression. Perhaps the frequency of the intruder's defensive postures is decreased after administration of anxiolytics (Yoshimura and Ogawa 1984a)? It is likely that fear or anxiety in intruder mice, produced in such confrontations, are suppressed by anxiolytics. The fact that offensive sideways posture and tail rattles shown by residents are decreased when intruders are drugged, allowing the resident to attack more easily, supports this suggestion. On the other hand, antidepressants and antipsychotics have only direct suppressive effects on the resident's attack behaviour.

Recently, a series of experiments on the effects of adrenergic beta-blockers on agonistic behaviour (in press) were carried out administering the following drugs orally at four dose levels (0, 5, 10 and 20 mg/kg) to either resident or intruder mice: dl-propranolol, d-propranolol, l-propranolol, and practolol. When the resident was treated with either dl- or l-propranolol, its aggressive episodes (offensive sideways posture, attack bite, tail rattle) were suppressed in a dose-dependent manner, whereas d-propranolol and practolol were ineffective. When the intruder was treated with beta-blockers, all failed to alter agonistic behaviour. Thus, it is likely that effects of dl- and l-propranolol on agonistic behaviour resemble the action of antidepressants or antipsychotics which only directly suppress the resident's attack behaviour; beta–blockers have little or no anxiolytic action. Furthermore, we confirmed that other beta–blockers such as oxprenolol and carteolol also showed only direct suppressive effects on male aggression (Yoshimura and Ogawa 1985).

ETHOPHARMACOLOGY OF MATERNAL AGGRESSION

As shown in Fig. 7, a stable and reliable level of maternal aggression in ICR mice was obtained between 5 and 9 days after parturition. The frequency of attack bites was highest on postpartum day 3, but there was wide individual variation. Based on this observation, behavioural testing for evaluation of acute effects of drugs was performed on postpartum days 5 and 7: drugs were administered only on the 7th day and the frequency of each behavioural item was then compared with the corresponding control levels which had been determined on postpartum day 5. All female mice were individually housed after a mating period of 4 days.

The acute effect of anxiolytics on maternal aggression is shown in Fig. 8. The benzodiazepine anxiolytic chlordiazepoxide (10 mg/kg, IP) significantly increased the frequency of attack bites by females. By contrast, the nonbenzodiazepine anxiolytics, suriclone (1.5 and 3 mg/kg, IP) and CL–218,872 (3 and 6 mg/kg, IP), suppressed maternal aggression in a dose–dependent manner. Augmentation of aggressive behaviour in female mice following administration of chlordiazepoxide seems not necessarily due to an anxiolytic action alone. Olivier et al. (1985) reported a similar phenomenon in rats using this drug (5 and 10 mg/kg, PO).

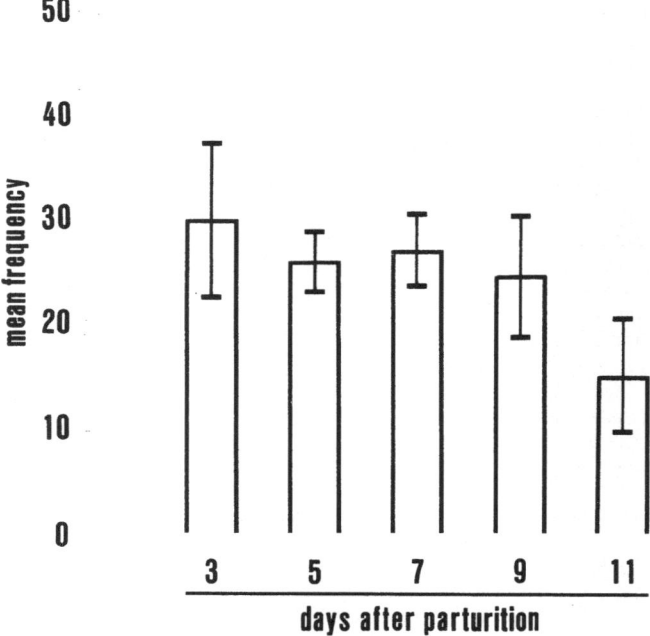

Fig. 7   Frequency of attack bites shown by female mice as a function of the day after parturition.

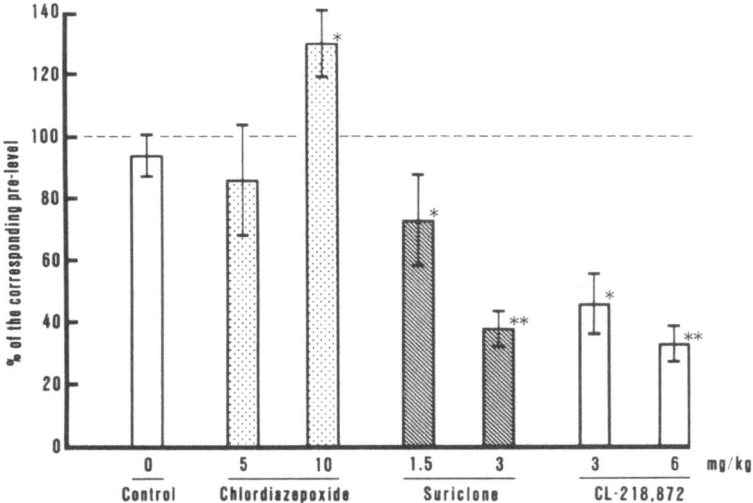

Fig. 8 Acute effects of anxiolytic drugs on maternal aggression in mice.

Further experiments with other types of benzodiazepine anxiolytics are necessary to elucidate whether or not augmentation of maternal aggression is specifically observed only after chlordiazepoxide.

Maternal aggression in mice was also altered by administration of antidepressants. The tricyclic antidepressants imipramine (5 and 10 mg/kg, IP) and clomipramine (5 and 10 mg/kg, IP), decreased the frequency of attack bites in a dose–dependent manner with little change in motor activity (Ogawa et al. 1985). Indalpine (5 and 10 mg/kg, IP), a new nontricyclic antidepressant, also suppressed maternal aggression. These findings are interesting as antidepressants are effective in the management of postpartum behavioural disorders in women.

As psychological factors during pregnancy and early lactation seem to play an essential role in the manifestation of maternal aggression, the effects of chronic administration of psychotropic drugs were studied. Drug treatment was started immediately after the removal of the partner male (5th day of cohabitation), and terminated on postpartum day 3. Behavioural testing was performed on postpartum day 5 without drug injection. As compared to the vehicle–treated group (Fig. 9), imipramine (5 and 10 mg/kg/day, IP) and suriclone (1.5 and 3 mg/kg/day, IP) significantly suppressed maternal aggression, whereas chlordiazepoxide (5 and 10 mg/kg/day, IP) and haloperidol (0.1 and 0.2 mg/kg/day, IP) tended to facilitate maternal aggression. All drugs employed did not alter the female's motor performance.

DISPLACEMENT ACTIVITY DURING AGONISTIC INTERACTIONS IN FEMALE MICE

Several "displacement activities" can be observed during an agonistic confrontations; (see Grant and Mackintosh 1963) e.g., when a female resident encounters a strange male intruder in her home cage, she displays not only intense attack but also digging activity. Such females suddenly stops their movement towards the intruder and start digging. Digging females are indifferent to the intruder and pups even if the former is close to the latter. No one appears to have

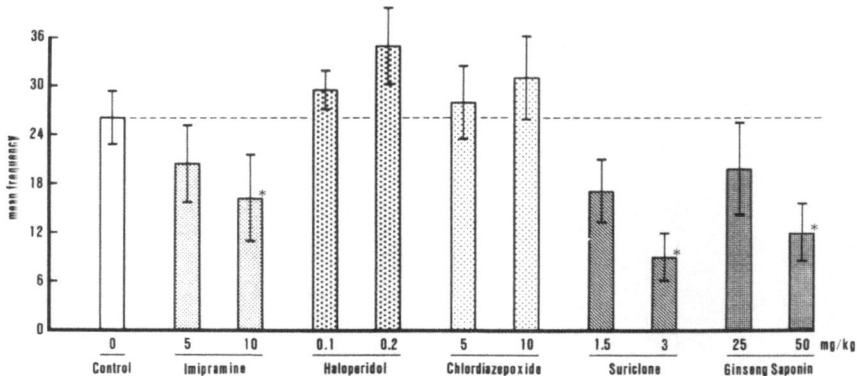

Fig. 9   Chronic effects of drugs on maternal aggression in mice.

focussed on this displacement activity in the analysis of maternal aggression and an attempt was made to determine whether maternal aggression and digging could be correlated. A significant negative correlation between the frequency of digging and the frequency of attack bites ($y=-0.65x + 29$, $r=0.72$, $t=5.49$, $df=28$, $p<0.001$) was found. Most of the females that frequently showed intense attack rarely exhibited displacement digging, whereas nonaggressive females showed high levels of such activity. To confirm this reciprocal relationship, the effect of housing conditions on digging activity was re-examined, using video tape as described previously. It was found that females housed with their mate showed more digging activity than the individually housed females. Moreover, even in the isolated virgin females that never attacked intruders, half the mice displayed digging activity. If a reciprocal relationship exists between digging and attack bites in females, the possibility arises that suppression of attack bites by a drug may accelerate digging activity. Interestingly, chronic treatment with either imipramine or suriclone, effective in suppressing maternal aggression, tended to increase the frequency of digging, while haloperidol and chlordiazepoxide tended to decrease this activity. This reciprocal relationship may be accounted for by assuming a behavioural shift between aggression and digging.

These findings suggest that the inner state of a parturient female mouse when the intruder is introduced into the female's home cage, can be expressed by at least two different types of behaviour, namely agonistic behaviour and displacement activity. It appears that experimental manipulations such as housing condition or chronic treatment with a drug act differentially on the female's inner state.

GENERAL DISCUSSION AND PERSPECTIVES

It is well-documented that the intensity of attack is altered by the nature of the opponent, but in pharmacological studies the interrelation between attack and defeat has only recently been determined. As previously stated, it is important to note that effects of psychotropic drugs on agonistic behaviour depends on whether attacking or attacked animals were drugged. The finding here that drugs with anxiolytic actions increase the resident's attack when the intruder is the drug recipient, suggest that this kind of approach is a useful tool to differentiate the actions of psychotropic drugs.

On the other hand, Miczek et al. (1982) found the experiences in an agonistic confrontation may alter the perception of painful stimuli. This important finding stimulated an attempt to investigate whether the pharmacological action of a drug is modified by the experience of an agonistic confrontation. A significant difference in the hypnotic effect of pentobarbital was evident between attacking and defeated mice (Yoshimura et al. 1985). This study seems to be of obvious interest for future research on the psychopathology of agonistic behaviour.

The mechanism responsible for the manifestation of postpartum aggression is unknown at the present time. Because of the high incidence of attack behaviour in females that were housed without a mate and because females housed with a castrated partner also became aggressive, it is likely that social factors after mating may play an essential role.

Although it has been generally accepted that maternal aggression toward an intruder serves to protect the young, it is possible that the manifestation of postpartum aggression is brought about by a more complex motivational state. Another possibility is that aggression can be considered as "pathological aggression" because isolation of such females leads to enhanced aggression. The finding that there is a reciprocal relationship between attack bites and digging, may provide a clue to solve this problem.

In addition to behavioural and pharmacological studies, knowledge of neurochemical changes in the brain seems to be necessary for understanding behaviour. As compared with neurochemical studies in interspecies (predatory) aggression such as mouse–killing behaviour by rats (e.g. Yoshimura 1981; Yoshimura et al. 1974a,b; Yoshimura and Ueki 1977, 1981), little information concerning the neuro–regulatory mechanisms mediating agonistic behaviour is available at present. By combining the ethopharmacological analysis of the behaviour with precise measurements of neurochemical changes in the brain, we could understand more precisely the "psychopathology of aggression".

Acknowledgements: This study was supported by a Grant–in–Aid for Scientific Research (C 59570983 and C 61571124) from the Ministry of Education, Science and Culture, Japan, a Grant from the Pharmacopsychiatry Research Foundation, and a Grant from Funds for Research on Red Ginseng. The assistance of Kohki Watanabe, Ichizo Matsuzaki, Nobuyuki Takaoka, Shigehiro Ohdo, and Misuzu Kurokawa is gratefully acknowledged.

REFERENCES

Al–Maliki S, Brain PF, Childs G, Benton D (1980) Factors influencing maternal attack on conspecific intruders by lactating female "TO" strain mice. Aggr Behav 6: 103–117

Bond A, Lader M (1979) Benzodiazepines and aggression. In: Sandler M (ed) Psychopharmacology of aggression. Raven Press, New York, pp 173–182

Brain PF, Al–Maliki S (1979) A comparison of effects of simple experimental manipulations on fighting generated by breeding activity and predatory aggression in 'TO' strain mice. Behaviour 69: 183–200

Brain PF, Nowell NW (1970) Some observations on intermale aggression testing in albino mice. Commun Behav Biol 5: 7–17

Brain PF, Poole A (1974) Some studies on the use of "standard opponents" in intermale aggression testing in 'TO' albino mice. Behaviour 50: 100–110

Brain PF, Al–Maliki S (1979) Effects of lithium chloride injections on rank–related responses in naive and experienced 'TO' strain mice. Pharmacol Biochem Behav 10: 663–639

DaVanzo JP, Daugherty M, Ruckart R, Kang L (1966) Pharmacological and biochemical studies in isolation–induced fighting in mice. Psychopharmacologia 9: 210–219

Denenberg VH, Gaulin–Kremer E, Gandelman R, Zarrow MX (1973) The development of standard stimulus animals for mouse (Mus musculus) aggression testing by means of olfactory bulbectomy. Anim Behav 21: 590–598

DiMascio A (1973) The effects of benzodiazepines on aggression: Reduced or increased? Psychopharmacologia 30: 95–102

Edwards DA, Thompson ML, Burge KG (1972) Olfactory bulb removal vs peripherally induced anosmia: Differential effects on the aggressive behavior of male mice. Behav Biol 7: 823–828

Fox KA, Snyder RL (1969) Effect of sustained low doses of diazepam on aggression and mortality in grouped male mice. J Comp Physiol Psychol 69: 663–666

Gandelman R (1972) Mice: Postpartum aggression elicited by the presence of an intruder. Horm Behav 3: 23–28

Ginsberg BE, Allee WC (1942) Some effects of conditioning on social dominance and subordination in inbred strains of mice. Physiol Zool 15: 485–506

Grant EC, Mackintosh JH (1963) A comparison of the social postures of some common laboratory rodents. Behaviour 21: 246–259

Green JA (1978) Experimental determinants of postpartum aggression in mice. J Comp Physiol Psychol 92: 1179–1187

Hafez ES (1962) The behavior of domestic animals. Williams and Wilkins, Baltimore

Ieni JR, Thurmond JB (1985) Maternal aggression in mice: Effects of treatments with PCPA, 5–HTP and 5–HT receptor antagonists. Eur J Pharmacol 111: 211–220

Krsiak M (1975) Timid singly–housed mice: Their value in prediction of psychotropic activity of drugs. Br J Pharmacol 55: 141–150

Krsiak M (1979) Effects of drugs on behaviour of aggressive mice. Br J Pharmacol 65: 525–533

Lion JR, Azcarate CL, Koepke HH (1975) "Paradoxical rage reactions" during psychotropic medication. Dis Nerv Sys 36: 557–558

Miczek KA (1974) Intraspecies aggression in rats: Effects of d–amphetamine and chlordiazepoxide. Psychopharmacologia 39: 275–301

Miczek KA, Barry H (1976) Pharmacology of sex and aggression. In: Glick SD, Goldfarb J (eds) Behavioral Pharmacology. Mosby, St.Louis, pp 176–257

Miczek KA, O'Donnell J (1978) Intruder–evoked aggression in isolated and nonisolated mice: Effects of psychomotor stimulants and l–DOPA. Psychopharmacology 57: 47–55

Miczek KA, Krsiak M (1979) Drug effects on agonistic behavior. In: Thompson T, Dews PB (eds) Advances in behavioral pharmacology. Vol. 2, Academic Press, New York, pp 87–162

Miczek KA, Wooley J, Shlisserman S, Yoshimura H (1981) Analysis of amphetamine effects on agonistic and affiliative behavior in squirrel monkeys (Saimiri sciureus). Pharmacol Biochem Behav 14: 103–107

Miczek KA, Yoshimura H (1982) Disruption of primate social behavior by d–amphetamine and cocaine: Differential antagonism by antispychotics. Psychopharmacology 76: 163–171

Miczek KA, Thompson ML,Shuster L (1982) Opioid–like analgesia in defeated mice. Science 215: 1520–1522

Miczek KA (1983) Ethopharmacology: Primate models of neuropsychiatric disorders. Alan R. Liss Inc, New York

Miczek KA (1986) The psychopharmacology of aggression. In: Iversen LL, Iversen SD, Snyder SH (eds) Handbook of Psychopharmacology. Vol 19, Plenum Press, New York

Noirot E, Goyens J, Buhot M (1975) Aggressive behavior of pregnant mice towards males. Horm Behav 6: 9–17

Ogawa S, Makino J (1981) Maternal aggression in inbred strain of mice: Effects of reproductive states. Jap J Psychol 52: 78–84

Ogawa N, Yoshimura H (1983) Animal models for aggression research: An ethological approach. Clin Psychiatry (Seishin Igaku) 25: 235–241

Ogawa N, Yoshimura H, Takaoka N (1985) Ethopharmacology of postpartum aggression in female mice: Effect of psychoactive drugs. Paper presented at the 3rd Eur conf ISRA, Parma, Italy

Olivier B (1981) Selective anti–aggressive properties of DU27725: Ethological analysis of intermale and territorial aggression in the male rat. Pharmacol Biochem Behav 14 (S1): 61–77

Olivier B, Van Dalen D (1982) Social behavior in rats and mice: An ethologically based model for differentiating psychoactive drugs. Aggr Behav 8: 163–168

Olivier B, Van Aken H, Jaarsma I, Van Oorschot R, Zethof T, Bradford LD (1984) Behavioural effects of psychoactive drugs on agonistic behaviour of male territorial rats (resident–intruder model). In: Miczek KA, Kruk MR, Olivier B (eds) Ethopharmacological aggression research. Alan R Liss Inc, New York, pp 137–156

Olivier B, Mos J, Van Oorschot R (1985) Maternal aggression in rats: Effects of chlordiazepoxide and fluprazine. Psychopharmacology 86: 68–76

Poshivalov VP (1974) Pharmacological analysis of aggressive behaviour of mice induced by isolation. Zhurnal Vyssshei Nervnoi Deyatel'nosti imeni I.P. 24: 1079–1081

Ropartz P (1968) The relation between olfactory stimulation and aggressive behavior in mice. Anim Behav 16: 97–100

Scott JP (1966) Agonistic behavior of mice and rats: A review. Am Zool 6: 683–701

Scott JP, Fredericson E (1951) The causes of fighting in mice and rats. Physiol Zool 24: 273–309

St.John RS, Corning PA (1973) Maternal aggression in mice. Behav Biol 9: 635–639

Svare B (1981) Maternal aggression in mammals. In: Gubernick DJ, Klopfer PH (eds) Parental care in mammals. Plenum Press, New York, pp 179–210

Svare B (1983) Psychological determinants of maternal aggressive behavior. In: Simmel EC, Hahn ME, Walters JK (eds) Aggressive Behavior: Genetic and neural approaches, Lawrence Erlbaum Ass, New Jersey, pp 129–146

Svare B, Betteridge C, Katz D, Samuels O (1981) Some situational and experiential determinants of maternal aggression in mice. Physiol Behav 26: 253–258

Svare B, Gandelman R (1976a) Postpartum aggression in mice: The influence of suckling stimulation. Horm Behav 7: 407–416

Svare B, Gandelman R (1976b) Suckling stimulation induces aggression in virgin female mice. Nature 260: 606–608

Svare B, Mann M (1983) Hormonal influences on maternal aggression. In: Svare B (ed) Hormones and aggressive behavior. Plenum Press, New York, pp 91–104

Svare B, Mann M, Samuels O (1980) Mice: Suckling stimulation but not lactation important for maternal aggression. Behav Neural Biol 29: 453–462

Tedeschi RE, Tedeschi DH, Mucha A, Cook L, Mattis PA, Fellows EJ (1959) Effect of various centrally acting drugs on fighting behavior of mice. J Pharmacol Exp Therap 125: 28–34

Wilson EO (1975) Sociobiology. The new synthesis. Belknap Press, Cambridge

Yoshimura H (1981) Regional changes in brain cholinergic enzyme activities after bilateral olfactory bulbectomy in relation to mouse–killing behavior by rats. Pharmacol Biochem Behav 15: 517–520

Yoshimura H (1985) Aggressive behavior (Kougeki Koudoo). In: Ito R, Takahashi R, Honda N (eds) Animal models for developing new drugs (Doubutsu Moderu Riyou Shusei). Res Dev Plan, Tokyo, pp 10–19

Yoshimura H, Ogawa N (1982) Pharmaco–ethological analysis of agonistic behavior between resident and intruder mice: Effect of anticholinergic drugs. Jap J Pharmacol 32: 1111–1116

Yoshimura H, Ogawa N (1983) Pharmaco–ethological analysis of agonistic behavior between resident and intruder mice: Effect of ethylalcohol. Folia pharmacol jap 81: 135–141

Yoshimura H, Ogawa N (1984a) Pharmaco–ethological analysis of agonistic behavior between resident and intruder mice: Effect of psychotropic drugs. Folia pharmacol jap 84: 221–228

Yoshimura H, Ogawa N (1984b) The nature of the male partner during pregnancy as determinant of postpartum aggression in mice. Paper presented at the 6th biennial meeting of ISRA, Turku, Finland

Yoshimura H, Ogawa N (1985) Pharmaco–ethological analysis of agonistic behavior between resident and intruder mice: Effect of adenergic beta–blockers. Jap J Psychopharmacol 5: 223–229

Yoshimura H, Ueki S (1977) Biochemical correlates in mouse–killing behavior of the rat: Prolonged isolation and brain cholinergic function. Pharmacol Biochem Behav 6: 193–196

Yoshimura H, Ueki S (1981) Regional changes in norepinephrine content in relation to mouse–killing behavior by rats. Brain Res Bull 7: 151–156

Yoshimura H, Fujiwara M, Ueki S (1974b) Biochemical correlates in mouse–killing behavior of the rat: Brain acetylcholine and acetylcholinesterase after administration of delta-9-tetrahydrocannabinol. Brain Res 81: 567–570

Yoshimura H, Gomita Y, Ueki S (1974a) Changes in acetylcholine content in rat brain after bilateral olfactory bulbectomy in relation to mouse–killing behavior. Pharmacol Biochem Behav 2: 703–705

Yoshimura H, Ohdo S, Ogawa N (1985) Alteration in hypnotic effect of pentobarbital following agonistic confrontations in mice. Paper presented at the 3rd Eur conf of ISRA, Parma, Italy

THE UTILITY OF ETHOLOGICAL ASSESSMENTS OF MURINE AGONISTIC INTERACTIONS IN BEHAVIOURAL TERATOLOGY: THE FOETAL ALCOHOL SYNDROME

Paul F. Brain[1], Jamaan S. Ajarem[2] and Vesselin V. Petkov[3].

1 Biomedical and Physiological Research Group, Biological Sciences, University College of Swansea, Swansea, SA2 8PP, U.K.

2 Zoology Department, Faculty of Science, King Saud University, Riyadh, Saudi Arabia.

3 Institute of Physiology, Bulgarian Academy of Sciences, Acad G. Bonchev Str. bl 23, 1113 Sofia, Bulgaria.

INTRODUCTION OF TOPIC

Swaab and Mirmiran (1984) have reviewed the evidence indicating that early exposure to a wide variety of compounds alters the developing brain. As behaviour is essentially a product of the central nervous system, it has been suggested that behavioural measures provide more sensitive indices of teratological effects than changes in morphology or physiology (e.g. Spyker et al. 1972). Coyle et al. (1976) have reviewed the effects of administering a range of doses of potential teratogens to pregnant animals on the behaviour of their offspring. They suggest that four types of dose can be specified, namely:
1) low doses which have no measurable effects on behaviour;
2) teratogenic doses which modify behaviour;
3) doses lethal to the foetus, and
4) doses lethal to the mother.

Obviously, the last two types of dose are of little interest to behavioural science and remain in the realms of traditional teratology. It seems useful, however, to assess compounds at doses below these levels and below the concentrations that are known to alter anatomical indices to determine if type (2) actions can be identified. There are two reasons for this. One is simply an attempt to look at the potential dangers of materials using the most sensitive indices available and the other is to determine if such compounds can be implicated in behavioural disorders.

Many of the currently used behavioural teratological techniques are based on traditional learning paradigms (e.g. Grimm 1984) or on encounters in rodents assessed in the psychopharmacological manner (e.g. Yanai and Ginsburg 1977). Ethopharmacological analyses of complex social encounters although being more difficult to assess do have certain advantages. As they assess a wide spectrum of activities, they offer a maximal chance to determine in a small number of tests if a compound has behavioural activity. It also follows because the behavioural descriptions are inclusive, that they provide 'built in' controls for non–specific actions on the focal behaviour (for example, the "anti–aggressive" properties of a compound that simply sedates an animal is of limited theoretical and practical interest).

Benton et al. (1983) and Brain et al. (1986) have briefly reviewed their experiences in some preliminary studies involving the administration of a wide range of materials (including diazepam, corticosterone, cyproterone acetate, ethyl alcohol, glutamate, lithium chloride, medroxyprogresterone acetate, morphine sulphate, naloxone and thiourea) to pregnant female mice on subsequent behaviour of their male offspring in ethologically assessed 'standard opponent' encounters (Jones and Brain 1985).

Materials were injected before and/or after parturition, and involved (in some cases) a cross–fostering design. The approach made use of videotape recording techniques and employed microprocessors to assist with the analyses. This technique seems to have positive benefits in 'behavioural teratology'.

PRESENT STATUS OF ANIMAL STUDIES

Phenomena similar to the Foetal Alcohol Syndrome (FAS) have been demonstrated in animal tests after prenatal or early postnatal exposure via treatment of the mother. Exposed animals displayed a higher number of malformations of the CNS which appear comparable to some of the clinical symptoms sometimes described in humans (Volk 1984). Obviously, it is also easier to study precisely how alcohol produces lasting effects on behaviour in studies on animals (e.g. Elis and Krsiak 1975; Volk 1984) as such studies can be better controlled than clinical investigations and are not as constrained by ethical considerations. It generally appears that treatment of pregnant rodents with low doses of ethanol results in their offspring being hyper–aggressive in adult life whereas higher doses may depress this attribute (Yanai and Ginsburg 1977). The offspring of alcohol–treated mothers are said to show more attack, tail rattling and "aggressive unrest" than do controls, but other aspects of social behaviour are apparently unchanged (Elis and Krsiak 1975).

Many of the studies in which alcohol is given to pregnant rodents during the gestation period, report that this drug produces morphological changes in their offspring causing, for example, a decrease in mean pup weight and a delay in onset of fur development (Chernoff 1977; Diaz and Sampson 1980; Ewert and Cutler 1979; Randall and Taylor 1979; Swanberg and Crumpacker 1977; Volk 1984). Indeed, relatively low doses of alcohol administered chronically throughout pregnancy may change a mother's litter size and the average weight of her offspring (Abel 1978).

One should note, however, that some studies have failed to find significant effects on these measures using doses as high as 8 g/kg (Anadan et al. 1980). In spite of the occasional negative finding, care must be taken to ensure that any recorded behavioural changes are not simply consequences of reduced body size or retarded development in the offspring.

PRESENT STATUS OF HUMAN STUDIES

There are an increasing number of studies in which exposure to drugs produces lasting changes in the morphology and/or behavioural activities of children. For example, consumption of alcohol by pregnant women is said to be associated with a relatively high incidence of hyperactivity in their children (Goodwin et al. 1975). The main clinical features of the FAS include intrauterine death, retardation in growth and cardic malformations (Majewski 1980) but dysfunction of the CNS or mental retardation are also some common findings (Streissguth et al. 1980). Of specific relevance to this account is the finding that chronic drinking by pregnant mothers is clinically associated with physical and mental abnormalities (including hyperactivity and 'aggression') in their offspring (Bianchine and Taylor 1974; Ferrier et al. 1973; Jones and Smith 1973; Kaminski et al. 1975).

REMAINING QUESTIONS

Preliminary studies in this laboratory (Benton et al. 1986; Brain et al. 1986) have suggested that ethopharmacological techniques can be very usefully applied to behavioural teratological studies on a wide range of materials. Factors such as the dose, the stage of pregnancy/lactation over which the treatment is applied and whether or not maternal influences of the drug are ameliorated using a cross–fostering design can have major influences on the outcome of specific studies.

A series of studies are described in which the ethologically–inspired methodologies are assessed in studies on the offspring of pregnant mice treated with different doses of ethyl alcohol at varied stages of the reproductive process. The experiments described are far from constituting a complete assessment of the FAS in this species but were thought likely to reveal the merits and demerits of the assessment technique.

PRE–AND POST–PARTUM APPLICATION OF ETHYL ALCOHOL TO THE MOTHER

Gai and Grimm (1982) have shown that treatment with drugs in the period around parturition is most likely to be associated with long–lasting changes in the offspring behaviour of rats and mice.

Materials and Methods

Twenty primiparous pregnant Swiss mice bred and housed under highly controlled conditions (see Brain et al. 1985) were given daily SC injections of physiological (0.9%) saline or solutions of spectroscopic grade ethyl alcohol (BHD Ltd, Poole, Dorset) diluted in such saline. Injections were applied to the mother on the last four days of pregnancy (date of conception being estimated from the finding of a vaginal plug) and the first four days after the birth of her litter. The injection volume was always 0.1 ml per 15 g mouse but varying the concentrations of solution produced injections equivalent to 1, 2 or 4 g/kg of alcohol per day. Male offspring were tested after a short period of isolation with anosmic 'standard opponent' intruders. The 10 minute encounters have been described by Brain et al. (1985).

Tests were conducted under dim red lighting (circa 8 lux) in cleaned neutral type MI mouse cages (North Kent Plastics, England). Semi–permanent videotape records of encounters were made using a low light–sensitive camera (National Panasonic, Japan) which was positioned 2m vertically above the test cage on a tripod. A transparent perspex lid facilitated recording of the encounter. Dim red illumination was employed to make the situation as naturalistic as possible for this largely nocturnal species.

A Sony Trinitron videomonitor (Hitachi Electronics Limited, Japan) ensured correct alignment and focusing of the camera and a VTG–33 electronic timer display (For–A Company Limited, Japan) was superimposed on the screen to provide precise timings and to impose an identity code for the particular encounter. A model CR–6060E 'U'–matic videocassette recorder (Victor Company of Japan Limited) was used for recording the interactions.

Analysis of Videotapes

The selected "elements" of behaviour were based on the descriptions of Grant and Mackintosh (1963), but had been modified by Brain et al. (1984). Analysis of the videotapes involved assessing the total time (in seconds) allocated by subject to the broad behavioural categories of:

a)   Non–social activity (mainly exploration of the cage);
b)   Social investigation of the intruder;
c)   Defensive activity;
d)   Attack;
e)   Threat;
f)   'Displacement' activity (mainly digging and self–grooming).

The definitions of all postures and acts included in the various broad behavioural categories are given in Ajarem (1985).

Statistical Analysis and Presentation of Results
    Non–parametric Kruskal–Wallis tests were used to assess the variance in the behavioural measures over different treatment groups. Subsequently, paired comparisons were carried out using Mann–Whitney 'U' tests or Fisher's exact probability tests (Siegel 1956) to contrast behaviour in different treatment groups.

Results
    The median times allocated to each of the broad categories of behaviour in social encounters for each of the four treatment groups of males are given in Table 1. Kruskal–Wallis tests revealed significant treatment effects on social investigation and defensive behaviour in these offspring (H = 18.46, p<0.001 and H = 20.93, p<0.001, respectively).
    Indeed, paired comparisons with the Mann–Whitney 'U' tests showed significant increases in social investigation (U = 44.5, p<0.002 and U = 42.5, p<0.002 for 1, 2 and 4 g/kg, respectively) in male offspring of ethanol–treated mothers compared to counterparts from control–exposed mothers. Threat and attack were generally increased in subjects whose mothers were given ethanol, but the changes were complex and generally non significant.
    However, paired comparisons showed a significant increase in threat (U = 82, p<0.05) and attack (U = 79, p<0.05) in male offspring of 2 g/kg ethanol treated–mothers compared with controls. No other behavioural attribute was statistically changed by early ethanol treatment.

Discussion
    The data generally provide support for the basic conclusions of Elis and Krsiak (1975). One should note, however, that any changes in 'aggression' in males (not particularly revealed in the present study) may be a complex consequence of more basic alterations in the predispositions to show social investigatory activities (an increase of which could lead to 'aggression') and defensive behaviour (a reduction of which could also augment attack) in animals of this sex. Certainly, the utility of assessing the organism's entire behavioural repertoire is clearly revealed by the present study. Limiting the analysis to traditional measures of "aggression" would have suggested that early alcohol exposure has no influence on behaviour in male offspring.
    One should further note that alcohol studies should really include nutritional control groups (as out–lined by Smart 1981) to ensure that ethanol's effects are not simply related to its acting as a source of calories. It is worth recording, however, that there is little evidence to suggest that any lasting effects of early ethanol exposure are a consequence of its supplying calories (reviewed by Berry and Smoothy 1986).
    Tze and Lee (1975) noted that early ethanol exposure via treatment of the mother produced morphological changes, depressing mean pup weight and delaying development. With the exception of failing to find evidence of congenital malformation, mice in the present experiment also showed similar changes.

Table 1.   Effects of pre- and post-partum application of ethyl alcohol to the
mother on behaviour of their male offspring in a 'standard opponent'
test

Behav-iour Treatment	Median (with ranges) number of seconds allocated to:					
	Non-social investigation	Social investigation	Defense	Threat	Attack	Displacement
control saline	278 (81–407)	54 (0–113)	167 (45–519)	0 (0–192)	0 (0–189)	15 (0–56)
1 g/kg ethanol	273 (75–445)	89 (7–326)	72** (0–272)	51 (0–134)	79 (0–190)	23 (0–119)
2 g/kg ethanol	235 (107–304)	150*** (28–320)	22*** (0–290)	72* (0–136)	76* (0–246)	12 (2–33)
4 g/kg ethanol	248 (132–365)	165*** (24–309)	24*** (0–264)	48 (0–126)	60 (0–144)	15 (0–68)

*   differs from control category $p < 0.05$ Mann-Whitney 'U' test (two-tailed)
**   differs from control category $p < 0.02$ Mann-Whitney 'U' test (two-tailed)
*** differs from control category $p < 0.002$ Mann-Whitney 'U' test (two-tailed)

Administration of ethyl alcohol to the mothers during pregnancy and on the first
few days after giving birth, reduced body weight and increased relative brain
weight (it seems likely that this change is imply a consequence of the body weight
decline) in their male offspring. Similar findings were obtained by Phillips and
Stainbrook (1976) who administered wine to pregnant female rats. Such actions may
be a direct consequence of alcohol's effects on maternal behaviour rather than
reflecting calorie intake, which has also been suggested to cause these changes
(Ewart and Cutler 1979 and see above).

POST-PARTUM APPLICATION OF ETHYL ALCOHOL TO THE MOTHER

The spurt in whole brain growth occurs postnatally in rats and mice rather than
being the almost exclusively prenatal phenomenon seen in humans (Smart 1981).
Indeed, it has been demonstrated that rats exposed postnatally to alcohol show
retarded growth in a variety of brain regions. In addition, Rawat (1975) has
observed that pups suckling from ethanol-fed mothers have significantly lower
concentrations of cerebral DNA and RNA. Early postnatal exposure of rodents to
alcohol also increases infant mortality, retards growth and produces developmental
delays (Martin et al. 1977). In behavioural terms, such treatment increases
emotional reactivity (Abel 1975) and decreases aggressiveness (Yanai and Ginsburg
1977) in rodent studies.

In view of these above findings and noting that alcohol may attain high concentrations in milk, it seemed worthwhile using the methodology tested here to assess the effects of exposing the neonate to alcohol via treatment of its mother. This attempt essentially determined whether the changes recorded in the first experiment were due to actions in post–natal life.

Materials and methods

Sixteen pregnant Swiss mice were injected SC with physiological saline or solutions of ethyl alcohol in saline for the first four days after parturition. The injection volume and daily doses used were as in the previous experiment. Routine behavioural tests were applied to male offspring of these animals.

Results

Behavioural data for male offspring are given in table 2. Kruskal–Wallis tests revealed that early postnatal exposure to alcohol via the mother had little overall influence on non–social investigation, social investigation, attack, threat or displacement behaviour in male mice.

Table 2. Effects of post–partum application of ethyl alcohol to the mother on behaviour of their male offspring in a 'standard opponent' test

Behaviour Treatment	Median (with ranges) number of seconds allocated to:					
	Non–social investigation	Social investigation	Defense	Threat	Attack	Displacement
control saline	314 (200–413)	97 (0–276)	61 (0–226)	40 (0–81)	55 (0–115)	1 (0–24)
1 g/kg ethanol	301 (100–446)	11 (0–248)	30 (0–500)	25 (0–85)	32 (0–218)	12 (0–32)
2 g/kg ethanol	324 (101–378)	87 (4–475)	0 (0–209)	40 (0–87)	64 (0–169)	6 (0–57)
4 g/kg ethanol	353 (241–397)	146 (0–287)	60 (0–264)	0 (0–96)	0 (0–109)	0 (0–24)

NB: There are no meaningful differences between any of these results.

Discussion

The finding that early postnatal exposure to alcohol had no significant influences on subsequent social behaviour in male mice contrasts with the findings of Yanai and Ginsberg (1977) who found a significant reduction in aggression after early postnatal exposure to ethanol. In their study, mother mice ingested the material and it was suggested that the postnatal period was the most important time in which this alcohol could exert its lasting actions. In the Yanai and Ginsburg (op cit) study, however, there was a possibility of a confounding dose effect, since lactating mothers consumed greater quantities of alcohol than did pregnant individuals. Certainly, the present results suggest that early postnatal exposure via injection of the mothers with alcohol has a less dramatic effect on male offspring than similar more extended treatment applied pre and post-partum (as in experiment 1). It seems imperative, however, to use longer treatment periods (e.g. at least 6 days) which equates better with the third trimester of human pregnancy in terms of brain development (Grimm 1984).

Martin et al. (1977) have reported that direct early postnatal alcohol treatment retards growth and delays development in mice. In the present experiment some of the offspring of ethanol treated experimental animals appeared badly stunted, growing very slowly and being physically weak (mostly in the categories whose mothers were given low or high doses of ethanol). The subsequent relatively high mortality may account for the failure to find lasting behavioural effects (the weakest animals are likely to have died).

PRE-PARTUM APPLICATION OF ETHYL ALCOHOL TO THE MOTHER COMBINED WITH SUBSEQUENT CROSS-FOSTERING

The 'FAS' generally involves a failure to attain the appropriate body size after birth (Abel and Dintcheff 1978; Hanson et al. 1976; Leichter and Lee 1979), even when postnatal exposure to ethanol is avoided by fostering or other techniques. Lee and Leichter (1980) have also claimed that the postnatal retardation in growth and development is not due to postnatal under-feeding. These studies (as well as the cross-fostering investigation by Abel and Greizerstein (1979) on foetal growth) are consistent with the hypothesis that maternal alcohol consumption adversely affects the regulatory mechanisms for growth during embryonic or foetal development and that this effect persists after birth.

Alcohol's influences on offspring in the previous experiments could have been mediated by changes in maternal behaviour and/or lactational efficiency. Indeed, there is evidence that exposure to ethanol can depress the milk-ejection reflex in nursing mothers via the inhibition of oxytocin release (Fuchs 1969). Baer and Crumpacker (1977) have reported that mother mice forced to drink alcohol during the lactation period show increased cannibalism and have suggested that the effect of the drug on progeny survival is at least partially due to impaired maternal care. Such alterations in mother-pup interactions might account for long-term changes seen in the behaviour of surviving pups (Denenberg 1964). Although such effects would still be of interest in terms of assessing the aetiology of diseases related to early alcohol exposure, it seemed important to attempt to establish whether this compound had direct effects on the developing mice. This was done essentially by repeating the earlier part of the first experiment but incorporating a cross-fostering design.

Materials and Methods

Twenty pregnant Swiss mice were given daily SC injections of physiological saline or a solution of ethyl alcohol in saline on the last four days of pregnancy only (no further injections were given after birth). The injection volume was the same as in the first experiment but only the intermediate dose of alcohol used in the previous experiments (i.e. 2 g/kg) was employed in this study. The cross-fostering was carried out on the day of birth and involved some of litters [control (C-C) or drug-treated (E-E)] remaining with their original mothers whilst other animals were cross-fostered to mothers of the opposite treatment category (C-E or E-C i.e. control-exposed offspring that were cross-fostered to ethanol-treated mothers and ethanol exposed offspring that were cross-fostered to control-injected mothers, respectively). The subsequent procedures and behavioural tests were as described in the first experiment.

Results

Behavioural data for the male offspring in the 'standard opponent' test are given in table 3. Kruskal-Wallis tests revealed significant variance between the categories in defence (H = 8.6, p<0.02) and displacement (H = 9.32, p<0.02). Paired comparisons showed that the E-E males evidenced decreased defence (U = 66, p<0.02) and increased displacement (U = 70, p<0.02) compared to the C-C category.

Table 3. Effects of pre-partum application of ethyl alcohol to the mother followed by cross-fostering on behaviour of their male offspring

		Behaviour					
Treatment		Median (with ranges) number of seconds allocated to:					
Natural mother	Rearing Mother	Non-social investigation	Social investigation	Defence	Threat	Attack	Displacement
control	control	217 (79–388)	57 (3–361)	170 (0–398)	22 (0–128)	21 (0–272)	5 (0–24)
ethanol	control	283 (108–500)	68 (0–348)	78 (0–366)	0 (0–137)	0 (0–135)	8 (0–46)
ethanol	ethanol	250 (145–309)	127 (7–326)	22* (0–253)	41 (0–220)	57 (0–246)	14* (0–221)
control	ethanol	301 (105–436)	83 (2–442)	42 (0–374)	56 (0–181)	88 (0–205)	8 (0–33)

* differs from controls p <0.02 on the Mann-Whitney 'U' test (two-tailed)

Discussion

The indication that male C-E mice showed less social behaviour than similar alcohol-exposed males who were not cross-fostered (E-E) and that E-C males showed some increase in social and other behaviour compared with C-C counterparts suggests, that some but not all the influences of alcohol are mediated via changes in maternal behaviour and/or lactation. Martin et al. (1978), Abel (1979) and Bond and Diguisto (1978) have also suggested that the changes in the offspring's behaviour are not entirely due to altered maternal behaviour, but that early ethanol treatments influence regulatory mechanisms of behaviour during embryonic or foetal development. This effect persists after birth, even if post-partum effects are avoided by cross-fostering techniques (Abel and Dintcheff 1978; Abel and Greizerstein 1979; Lee and Leichter 1980).

The present cross–fostering experiment thus suggests that some of the lasting effects of ethanol application to the mother on the <u>behaviour</u> of their offspring in the first two experiments are mediated by indirect maternal influences whereas others are direct actions on the developing organism.

Perspectives

The quality of research that attempts to relate lasting behavioural effects of early exposure to drugs in humans and animals could be improved by:–
(a) assessing a wider variety of dose–ranges, treatment–durations and routes of administration and
(b) generating behavioural tests in which one can assess the full range of the organism's potential responses. The current increased usage of detailed 'ethological' analyses in clinical and animal studies represents a useful development in this direction.

Although the results described here are necessarily somewhat preliminary, a number of potentially useful situations are indicated, namely:–
(a) Pre–partum treatment of the mother followed by cross–fostering of the offspring to an untreated mother at parturition (this should detect any lasting influence of the drug directly on the CNS of the foetus);
(b) Post–partum treatment of the mother (this should enable one to assess actions mediated solely via changes in maternal behaviour and/or lactation in cases where the drug is <u>not</u> transmitted in the milk) and
(c) Post–parturition treatment of the neonate (this approach could determine actions of the drugs on the developing organism).

Obviously, there are still complicating factors possible in some of these situations e.g. drug treatment of the neonate may cause females to respond to them differently but a greater concentration on the behavioural complexity of each situation seems likely to pay dividends. Although it is difficult to get precise behavioural replications in such studies, the broad quality of action of particular compounds may be more faithfully indicated by the ethological approach.

Acknowledgements: Dr. J.S.Ajarem was supported throughout these studies by a grant from the King Saud University, Riyadh, Saudi Arabia. Professor V.V.Petkov was the recipient of an exchange scheme under the auspices of the Bulgarian Academy of Sciences and the Royal Society.

REFERENCES

Abel EL (1978) Effects of ethanol on pregnant rats and their offspring. Psychopharmacology 57: 5–11

Abel EL (1979) Effects of alcohol withdrawal and undernutrition on cannibalism of rat pups. Behav Neural Biol 25: 411–413

Abel EL, Dintcheff BA (1978) Effects of prenatal alcohol exposure on growth and development in rats. J Pharmacol Exp Ther 207: 916–921

Abel EL, Greizerstein HB (1979) Ethanol–induced prenatal growth deficiency:changes in fetal body composition. J Pharmacol Exp Ther 211: 668–671

Ajarem JS (1985) Lasting Behavioural Consequences of Perinatal Exposure of Laboratory mice to a Variety of Drugs. Ph.D. Dissertation, University of Wales

Anandan N, Felegi W, Stern JM (1980) In utero alcohol heightens juvenile reactivity. Pharmacol Biochem Behav 13: 531–535

Baer DS, Crumpacker DW (1977) Fertility and offspring survival in mice selected for different sensitivities to alcohol. Behav Genet 7: 95–103

Benton D, Ajarem JS, Brain PF (1985) The role of opiate receptors in behaviour. In: Brain PF, Ramirez JM (eds) Cross–Disciplinary Studies on Aggression, Psychobiology Series of University of Seville Press, Seville, pp 113–136

Berry MS, Smoothy R (1986) A critical evaluation of the claimed relationships between alcohol and aggression in infra–human animals. In: Brain PF (ed) Alcohol and Aggression, Croom–Helm, London, pp 84–137

Bianchine JW, Taylor BD (1974) Noonan syndrome and foetal alcohol syndrome. Lancet, p 933

Bond NW, Digiusto EL (1978) Avoidance conditioning and Hebb–Williams maze performance in rats treated prenatally with alcohol. Psychopharmacology 58: 69–71

Brain PF, Jones SE, Brain S, Benton D (1984) Sequence analysis of social behaviour illustrating the actions of two antagonists of endogenous opiates. In: Miczek KA, Kruk MR, Olivier B (eds) Ethopharmacological Aggression Research, Alan R Liss, Inc, New York, pp 43–58

Brain PF, Jones SE, Brain S, Benton D (1985) Ethological analysis of the effects of naloxone and the opiate antagonist ICI 154,129 on social interactions in male house mice. Behav Processes 10: 341–354

Brain PF, Ajarem JS, Petkov VV (1986) The application of ethopharmacological techniques to behavioural teratology: Preliminary investigations. Paper presented at 2nd International Symposium on the Pharmacology of Transmitter Interactions, Sofia, Bulgaria

Chernoff GF (1977) The fetal alcohol syndrome in mice: Animal model. Teratology 15: 223–230

Coyle I, Wayner MJ, Singer C (1976) Behavioral teratogenesis: A critical evaluation. Pharmacol Biochem Behav 4: 191–200

Denenberg VH (1964) Critical periods, stimulus input and emotional reactivity: A theory of infantile stimulation. Psychol Rev 71: 335–351

Diaz J, Samson HH (1980) Impaired brain growth in neonatal rats exposed to ethanol. Science 208: 751–753

Elis J, Krsiak M (1975) Effect of alcohol administration during pregnancy on social behaviour of offspring of mice. Act Nerv Sup 17: 281–282

Ewert FG, Cutler MG (1979) Effects of ethyl alcohol on development and social behaviour in the offspring of laboratory mice. Psychopharmacology 62: 247–251

Ferrier PE, Nicol I, Ferrier S (1973) Foetal alcohol syndrome. Lancet, p 1496

Fuchs AR (1969) Ethanol and the inhibition of oxytocin release in the lactating rat. Acta Endocrin 62: 346–455

Gai N, Grimm VE (1982) The effect of prenatal exposure to diazepam on aspects of postnatal development and behaviour in rats. Psychopharmacology 78: 225–229

Goodwin DW, Schulsinger F, Hermansen L, Guze SB, Winokur G (1975) Alcoholism and the hyperactive child syndrome. J Nerv Ment Dis 160: 349–353

Grant EC, Mackintosh JH (1963) A comparison of social postures of some common laboratory rodents. Behaviour 21: 246–259

Grimm VE (1984) A review of diazepam and other benzodiazepines in pregnancy. In: Yanai J (ed) Neurobehavioural Teratology, Elsevier North-Holland Biomedical Press, Amsterdam, pp 153–162

Hanson JW, Jones KL, Smith DW (1976) Fetal alcohol syndrome. Experience with 41 patients. J Am Med Assoc 235: 1458–1460

Jones KL, Smith DW (1973) Recognition of the foetal alcohol syndrome in early infancy. Lancet, p 999

Jones SE, Brain PF (1985) An illustration of simple sequence analysis with reference to the agonistic behaviour of four strains of laboratory mouse. Behav Processes 11: 365–388

Kaminski M, Rumeau-Rouquette C, Schwartz D (1975) Le consomation d'alcool chez les femmes enceintes et son effet sur issue de la grossesses. Anglo-French Symposium on Alcoholism, INSERM/MRC 54: 87–100

Lee M, Leichter J (1980) Effect of litter size on the physical growth and maturation of the offspring of rats given alcohol during gestation. Growth 44: 327–335

Leichter J, Lee M (1979) Effect of maternal ethanol administration on physical growth of the offspring in rats. Growth 43: 288–297

Majewski F (1980) Untersuchungen zur Alkoholembryopathie. Thieme, Stuttgart

Martin JC, Martin DC, Sigman G, Radow B (1977) Offspring survival, development and operant performance following maternal ethanol consumption. Dev Psychobiol 10: 435–446

Martin JC, Martin DC, Sigman G, Radow B (1978) Maternal ethanol consumption and hyperactivity in cross-fostered offspring. Physiol Psychol 6: 362–365

Phillips DS, Stainbrook GL (1976) Effects of early alcohol exposure upon adult learning ability and taste preferences. Physiol Psychol 4: 473–475

Randall CL, Taylor WJ (1979) Prenatal ethanol exposure in mice: teratogenic effects. Teratology 19: 305–312

Rawat AK (1975) Ribosomal protein synthesis and neonatal rat brain as influenced by maternal ethanol consumption. Res Commun Chem Pathol Pharmacol 12: 723–732

Siegel S (1956) Non–parametric Statistics for the Behavioral Sciences, McGraw–Hill, New York

Smart JL (1981) Undernutrition and aggression. In: Brain PF, Benton D (eds) Multidisciplinary Approaches to Aggression Research, Elsevier North Holland Biomedical Press, Amsterdam, pp 179–191

Spyker JM, Sparber SB, Goldberg AM (1972) Subtle consequences of methyl mercury exposure: behavioral deviations in the offspring of treated mothers. Science 177: 621–623

Streissguth AP, Landesman–Dwyer S, Martin JC, Smith DW (1980) Teratogenetic effects of alcohol in humans and laboratory animals. Science 209: 353–361

Swaab DF, Mirmiran M (1984) Possible mechanism underlying the teratogenic effects of medicines on the developing brain. In: Yanai J (ed) Neurobehavioural Teratology, Elsevier North–Holland Biomedical Press, Amsterdam, pp 55–71

Swanberg KM, Crumpacker DW (1977) Genetic differences in reproductive fitness and offspring viability in mice exposed to alcohol during gestation. Behav Biol 20: 122–127

Tze WJ, Lee M (1975) Adverse effects of maternal alcohol consumption on pregnancy and foetal growth in rats. Nature 257: 479–480

Volk B (1984) Neurohistological and neurobiological aspects of fetal alcohol syndrome in the rat. In: Yanai J (ed) Neurobehavioural Teratology, Elsevier North–Holland Biomedical Press, Amsterdam, pp 163–193

Yanai J, Ginsburg BE (1977) A developmental study of ethanol effect on behavioural and physical development in mice. Alcoholism 1: 325–333

ETHOPHARMACOLOGICAL AND NEUROPHARMACOLOGICAL ANALYSES OF AGONISTIC BEHAVIOUR

Vladimir P. Poshivalov. Division of Pharmacology, Pavlov Medical Institute, Leningrad 197089, U.S.S.R.

INTRODUCTION

Computerized ethological pharmacology applies specific ethological, mathematical, neuropharmacological and neurochemical principles to the analysis and control of agonistic behaviour. In recent years the most productive developments in ethopharmacology have been: 1) application of computerized mathematical methods for the investigation of intraspecific behaviour (Poshivalov 1986a, b); 2) adaptation and formalization of behavioural models for psychopharmacological research (Poshivalov 1986c); 3) combinations of neurochemical and ethological methods of investigation to the study of central nervous system functions (Kantak et al. 1980; Miczek et al. 1984; Olivier and Mos 1986; Poshivalov and Khodko 1984; Rodgers and Hendrie 1984). One of the remaining problems for ethopharmacology is the synthesis of molecular pharmacology and behaviour. The development of this direction of research could promote the better understanding of the interrelationships between behaviour, endogenous ligands and receptors (Nieminen and Poshivalov 1985; Sukchotina et al. in press).

The aim of the present review is to 1) develop and select appropriate ethological models and describe them mathematically for improved assessment of drug effects on behaviour of laboratory rodents; 2) apply methods of molecular pharmacology (radioligand assays) to the study of the mechanisms of agonistic behaviour at the receptor level; 3) investigate the psychotropic effects of endogenous ligands for benzodiazepine (BDZ), beta–carboline (BC) and opiate receptors in terms of their anxiogenic, anxiolytic, aggressogenic and anti–aggressive properties and 4) develop ethopharmacological standards and criteria for assessment and screening of new anti–aggressive compounds in tests of agonistic behaviour.

From the several experimental models of aggressive behaviour used in my laboratory to study drug effects (Poshivalov 1986a,b,c) two experimental paradigms were selected for analysis of agonistic behaviour. These included: 1) aggressive behaviour of male isolated mice encountering group housed intruders; and 2) timid–defensive behaviour of isolated male mice (subjected to attacks by an aggressive dominant mouse and to foot–shock) also encountering non–aggressive group housed intruders (Poshivalov 1986b,c).

The intraspecific behaviour of all animals was recorded and analysed by a previously described "Ethograph–Computer System" (see Poshivalov 1986a,b,c; Poshivalov et al. 1979). All acts and postures were scored according to an ethological atlas for laboratory rodents (Poshivalov 1978). Several mathematical methods have been developed and used to describe behavioural effects of psychotropic drugs (Poshivalov and Khodko 1984). The behavioural effects of drugs belonging to different drug classes (carbolines, benzodiazepines, opiate agonists and antagonists, antidepressants and serenics) were studied in different dose ranges. All drugs were given intraperitoneally.

In neurochemical experiments, the differences in the specific binding of [3H] flunitrazepam ([3H]FNZP) and [3H] β–carboline–3–carboxylate ethyl ester ([3H]BC–3–CEE; Amersham, UK) to membranes of different brain regions (brain stem, diencephalon, cerebral cortex) were analysed in aggressive and timid–defensive mice (Poshivalov 1986a; Sukchotina et al. in press). Computerized ethograms were generated for aggressive and timid–defensive mice in control and diazepam (5 mg/kg, IP, given during 14 days) treated animals and both groups were compared with the number and distribution of binding sites for [3H]FNZP and [3H]BC–3–CEE.

ETHOPHARMACOLOGY OF ANXIOGENIC DRUGS; PHARMACOLOGICAL FACILITATION OF TIMIDITY AND DEFENCE

Two properties of drugs in relation to "fear" can be differentiated: namely their intrinsic anxiogenic property and their ability to potentiate previously existing "fear" or "timidity". The currently used animal models of anxiety are particularly appropriate for testing "potentiation of anxiety" (Pellow and File 1984). It may be actually more fruitful, to analyse, using ethological methods, the ability of drugs to activate timidity and defence and the possibility that they can change aggression (offence) into defence.

Recently, several compounds have been reported to have "anxiogenic" effects. These include peptides (ACTH, α–MSH, MIF–1), benzodiazepine antagonists (RO 15–1788), β–carbolines (BC–3–CEE), GABA–antagonists (picrotoxin), kinurenines, pentylenetetrazol, yohimbine and in some cases caffeine and amphetamine. This list indicates that the sites causing anxiogenic effects may be quite varied and that specific receptors other than BDZ may be involved in the integrative processes of "anxiety" (e.g. β–carboline, GABA and 5–HT receptors). The discovery of the β–carbolines has provided a new perspective in the receptor pharmacology of "anxiety". BC–3–CEE is a more potent BDZ receptor ligand than diazepam (Braestrup et al. 1980) and there is evidence that beta–carbolines have intrinsic actions, generally in the opposite direction to those of diazepam (inverse agonism). These compounds provide us with an important new tool for studying BDZ and BC receptor mechanisms. One of the questions arising is whether some types of behavioural responses (aggression or timidity–defence) are correlated with specific distribution of BDZ and BC receptors in the brain.

In a series of experiments on mice it was shown (Table 1) that the number of specific binding sites (B_{max} for [3H]FNZP and [3H]BC–3–CEE) in aggressive and timid–defensive mice was significantly higher in the cerebral cortex than in other brain regions. The cerebral cortex of both aggressive and timid/defensive mice but the brain stem of only aggressive mice has a significantly higher number of specific binding sites for [3H]BC–3–CEE than for [3H]FNZP. These results may suggest that there are specific (independent from BDZ) BC receptors. Experiments (summarized in Table 1) show that subchronic injections of diazepam (5 mg/kg/day during 14 days) significantly decrease the affinity (K_D) for [3H]–BC–3–CEE in all the brain regions of mice independently of the subjects behavioural characteristics, whereas diazepam changes the number of BC–3–CEE binding sites only in timid/defensive mice.

The functional consequences of BC and BC–BDZ receptor activation in aggressive and timid–defensive mice was also investigated. The inverse BDZ agonists BC–3–CEE, BC–3–CME and BC–3–CPE, significantly decreased intraspecific sociability (Table 2) in aggressive mice and had a biphasic influence on aggressive behaviour. BC–3–CEE increased attacks at doses of 1–2.5 mg/kg and decreased aggression at doses of 5–10 mg/kg (not shown in table). It appears that psychotropic properties of β–carboline–3–carboxylate esters can be differentiated by computerized ethological techniques. For example, BC–3–CEE and BC–3–CPE suppressed sociability to a greater extent than BC–3–CME.

Table 1   [^{3}H]–Flunitrazepam (FNZP) and [^{3}H]–β–carboline–3–carboxylate ethyl ester (BC–3–CEE) specific binding in different brain regions of aggressive (AG) and timid–defensive (TD) mice, which are treated either with control (C) or Diazepam (D). K_D is expressed in nmol, B_{max} in fmol/mg protein.

Treatment	Type of animal	Brain region	[^{3}H]FNZP		[^{3}H]BC–3–CEE	
			K_D	B_{max}	K_D	B_{max}
C	AG	Brain stem	1.54 ± 0.13	443 ± 18+	1.08 ± 0.14	714 ± 37
	AG	Diencephalon	1.79 ± 0.12	892 ± 33	0.68 ± 0.11	777 ± 45
	AG	Cortex	1.57 ± 0.13	1198 ± 56+	1.35 ± 0.08	2197 ± 73
	TD	Brain stem	1.45 ± 0.11	675 ± 26+	1.17 ± 0.07	597 ± 17
	TD	Diencephalon	1.61 ± 0.09	922 ± 25	0.89 ± 0.08	873 ± 32
	TD	Cortex	1.52 ± 0.12	1224 ± 50+	1.21 ± 0.10	2203 ± 96
D	AG	Brain stem	1.74 ± 0.12	404 ± 12+	0.66 ± 0.11*	724 ± 40
	AG	Diencephalon	1.86 ± 0.15	422 ± 16+*	0.62 ± 0.08*	775 ± 38
	AG	Cortex	1.44 ± 0.12	1220 ± 52+	1.07 ± 0.06*	2077 ± 57
	TD	Brain stem	1.61 ± 0.07	874 ± 18+*	0.75 ± 0.04*	1480 ± 30*
	TD	Diencephalon	1.60 ± 0.08	802 ± 21+*	0.59 ± 0.06*	1561 ± 29*
	TD	Cortex	1.40 ± 0.09	978 ± 30+*	0.65 ± 0.03	1808 ± 28*

+ = Significant difference (p<0.01) between [^{3}H]–FNZP and [^{3}H]–BC–3–CEE specific binding. * = Significant difference (p<0.01) between control (C) and diazepam (D) treated mice.

Apparently, the level of drug–induced "anxiety" can modulate the level of attacks in aggressive mice which might also explain the biphasic effects of BC–3–CEE on aggression.

In timid–defensive mice only BC–3–CEE (2.5 mg/kg) significantly increased the frequency of upright postures in defence (Table 3), while BC–3–CME and BC–3–CPE had no distinct effect on this behaviour. It is possible, of course, that timidity in these animals was already maximal and could not be further increased.

Another potentially anxiogenic compound RO 15–1788 (BDZ antagonist), at a dose of 10 mg/kg, had no significant influence on the frequency of attacks in aggressive mice and seemed on these indices behaviourally inactive. At the same time RO 15–1788 modified the structure of behaviour (not shown). Inducing close associations between agonistic elements of behaviour as indicated by cluster analysis, RO 15–1788 increased the probability of defensive elements at a dose of 10 mg/kg in timid–defensive mice. Although RO 15–1788 potentiates timid–defensive behaviour, this compound cannot convert aggression into defence.

Tetrahydro–β–carbolines (THBC's) and dihydro–β–carbolines (DHBC's) may represent compounds with distinct aversive "interoceptive" properties which may explain some of their nonspecific actions. THBC's had no specific effect on intraspecies behaviour, indeed both aggression reducing and enhancing effects were found. A reduction of aggression by DHBC's, e.g. harmalol and 6–methoxy–harmalane, was observed in relative high doses and also seemed to lack

Table 2 Ethopharmacological profile of three β–carbolines; β–carboline–3–carboxylate ethyl ester (BC–3–CEE), β–carboline–3–carboxylate methyl ethyl ester (BC–3–CME) and β–carboline–3–carboxylate propyl ester (BC–3–CPE), in aggressive mice. The statistical probability of the appearance of each behavioural element is given. Δ=significant increase; ∇=significant decrease; compared to the corresponding vehicle condition (p<0.05 by Wilcoxon matched–pairs test, 2–tailed). As vehicle Tween–80 was used.

Behavioural categories and elements	vehicle	BC–3–CEE 2.5 mg/kg	vehicle	BC–3–CME 2.5 mg/kg	vehicle	BC–3–CPE 2.5 mg/kg
Aggression:						
Attack	0.048	0.144Δ	0.140	0.125	0.078	0.147Δ
Threat	0.048	0.036	0.074	0.070	0.049	0.072
Ambivalence:						
Tail rattle	0.016	0.010	0.026	0.019	0.024	0.020
Social investigation:						
Sniff–nose	0.084	0.050	0.029	0.014∇	0.053	0.029∇
Sniff–body	0.160	0.128∇	0.084	0.086	0.147	0.108∇
Sniff–genitals	0.101	0.060∇	0.080	0.068	0.076	0.077
Self groom	0.172	0.177	0.209	0.248Δ	0.200	0.274Δ

specificity. However, defensive attacks may be easily evoked by tactile influences from a partner and data indicate that the neurochemical mechanisms related to offensive attack are attenuated by DHBC's and some THBC's, whereas defensive attacks are not. This may indicate that the neurochemical system of β–carbolines in the CNS has no inhibitory control on defensive attacks.

The multiple opiate receptor systems (mu–, kappa– and delta–) may also have distinct functional significance for the integration of agonistic behaviour (Brain et al. 1984; Miczek et al. 1984; Poshivalov 1986a,c; see for an extensive description and figures Poshivalov 1986c)). Tifluadom, a selective agonist of kappa opiate receptors increased passive defence and decreased aggression and threats. In timid/defensive mice, tifluadom increased the frequency of sideways postures in defence and prolonged static sitting behaviour. Bremazocine decreased the active forms of defence (upright postures), while it increased passive defence – sideways postures and static sitting (Poshivalov 1986c). Opiate antagonists (naloxone, naltrexone) potentiate defensive behaviour and increase upright posture in defence in timid–defensive mice (Poshivalov 1986c). The different effects of kappa agonists and mu antagonists may reflect different integrative functions related to passive or active defence (cf. Poshivalov 1986c).

PHARMACOLOGICAL SUPPRESSION OF TIMIDITY AND DEFENCE

The identification of tranquillizing properties of drugs in ethological models is a current application of computerized ethopharmacology. Benzodiazepines (medazepam, diazepam, phenazepam), GABA analogues (phenibut, phenylpirrolidon) and GABA stimulating compounds (gamma–acetylenic–GABA, valproate, piracetam), are the most potent suppressors of defence and timidity. The

Table 3 Ethopharmacological profile of β–carbolines: (BC–3–CEE, BC–3–CME and BC–3–CPE) in timid–defensive mice. The statistical probability of the appearance of each behavioural element is given. Δ=significant increase; ▽=significant decrease; compared to the vehicle condition (p<0.05 by Wilcoxon matched–pairs test, 2–tailed). As vehicle Tween–80 was used.

Behavioural categories and elements	vehicle	BC–3–CEE 2.5 mg/kg	vehicle	BC–3–CME 2.5 mg/kg	vehicle	BC–3–CPE 2.5 mg/kg
Defense:						
Sideway posture	0.15	0.13	0.07	0.08	0.21	0.09
Upright posture	0.10	0.13Δ	0.04	0.06	0.08	0.09
Freezing	0.09	0.09	0.17	0.12	0.07	0.13
Social investigation:						
Sniff–nose	0.04	0.03▽	0.02	0.02	0.06	0.03▽
Sniff–body	0.21	0.16▽	0.09	0.08	0.15	0.11▽
Sniff–genitals	0.05	0.04	0.05	0.03	0.06	0.04
Individual behavior:						
Self grooming	0.21	0.21	0.34	0.40	0.22	0.30
Sitting	0.02	0.03	0.03	0.01	0.01	0.00

tranquillizing properties of substances can be detected by using laboratory rodents that develop timid or timid–defence behaviour as a result of isolation from conspecifics. For this purpose, isolated non–aggressive mice subjected to aggressive counterparts and exposed to foot–shock stimulation in a stochastic regime were used (Poshivalov 1986a).

Computerized ethograms of diazepam (3.5 mg/kg) obtained from timid isolated mice reveal the reduction of upright postures in defence, suppression of escape and increased social sniffing. Concomitant neurotoxicity tests suggest that the average effective ataxia dose for diazepam is 3.1 mg/kg. This suggests that the reduction of timid behaviour by diazepam is non–specific. Our experiments and those of others (Krsiak et al. 1981; Miczek and Krsiak 1979) show that mice treated with diazepam do not loose their ability to rear, assume upright postures or to move fast when escaping. Diazepam and medazepam increase the number of social sniffing acts directed towards the intruder. This suggests that the reduction of timidity by benzodiazepines action is not simple "ataxia".

PHARMACOLOGICAL FACILITATION OF AGGRESSION

It is possible to differentiate three kinds of drug facilitation of aggression. 1) the aggressogenic actions (drug–induced offence against a conspecific in previously non–aggressive animals), 2) restoration (or disinhibition) of aggression (e.g. of previously aggressive animals suppressed by an external stress factor) and 3) activation of offence in aggressive animals. In most experimental procedures only the activation of offence is observed (when using animals with low baselines of aggression) and more rarely the recovery of aggression (which may "simulate" an aggressogenic action). The identification of "pure aggressogenic" drug action demands a special experimental paradigm, therefore we will delimit ourselves here to the study of drugs that restore aggression.

ABILITY OF DRUGS TO RESTORE AGGRESSION

The ability of antidepressants to restore aggressive behaviour and to reduce timid–defensive behaviour was studied in dominant aggressive mice exhibiting behavioural depression caused by foot–shock stimulation and exposure to very aggressive dominant conspecifics for 14 days. The experiments show (Valdman and Poshivalov 1986) that trazodon, pyrazidol, clomipramine and zimeldine reduce the behavioural depression (Table 4). Pyrazidol, clomipramine and trazodone (10 mg/kg/day, intraperitoneally, for 7 days) restore to some extent intraspecific aggression and sociable behaviour in male mice.

Antidepressants differ in their ethological spectra of action as assessed by the computerized method. Indeed, this model might be useful to screen potential antianxiety properties of antidepressants to determine their ability to restore aggressive behaviour.

Table 4   Ethological spectrum of subchronical effects of antidepressants. Model of timid–defensive behaviour induced by prolonged nociceptice stimulation of aggressive mice.

Drugs	aggression recovery	defence	ambi- valent	sociability recovery	Individual dynamic	static
Trazodon 10 mg/kg	Δ	▽	▽	Δ	Δ	
Pyrazidol 10 mg/kg	Δ	▽	Δ	Δ		▽
Clomipramine 10 mg/kg	Δ	▽	Δ	Δ	▽	Δ
Zimeldine 10 mg/kg	▽	▽	▽	▽	▽	Δ

Δ–increase, ▽–decrease

PHARMACOLOGICAL SUPPRESSION OF AGGRESSIVE BEHAVIOUR

Experimental and clinical studies of various classes of psychotropic drugs show that a variety of compounds can exert "anti–aggressive" effects: haloperidol, diphenylhydantoine, propranolol, methylphenidate, carbamazepine, Sch–12679, buspirone, YG–19–256, fluprazine (DU 27716), diazepam etc. (Bradford et al. 1984; Miczek and Krsiak 1979; Olivier et al. 1984; Poshivalov 1986a,b,c). Although the problems of specificity and selectivity of such anti–aggressive actions have not yet been resolved some methodological and mathematical tools for the measurement of selective drug effects based on ethological models (Poshivalov and Khodko 1984) have been advocated. Recently some compounds of the phenylpiperazine chemical class have been shown to have relatively selective anti–aggressive properties, reducing offence without disrupting non–agonistic interactions (Bradford et al. 1984; Olivier et al. 1984). A comparative ethopharmacological study of the effects of the Duphar compounds DU 27716 (fluprazine) and DU 28412 was performed on 150 male mice in a variety of experimental models of agonistic behaviour including (1) aggression by isolated male mice; (2) foot–shock induced fighting in mice; (3)

128

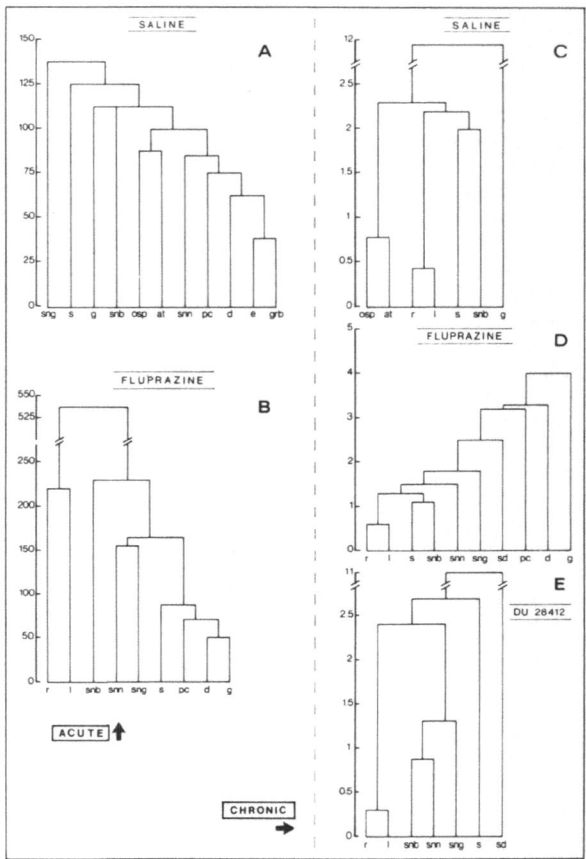

Fig. 1 Dendrograms by rows for (A) saline and (B) fluprazine (5 mg/kg) treated
aggressive mice after acute intraperitoneal administration. In the right
column dendrograms are shown after 14 days pretreatment (chronic) with
(C) saline, (D) fluprazine (5 mg/kg, IP) and (E) DU 28412 (5 mg/kg, IP).
Abbreviations: sng=sniffing genitals; s=sniffing; g=grooming; snb=sniffing
body; osp=offensive sideways posture; at=attack; snn=sniffing nose
pc=passive contact; d=digging; e=exploration; grb=grooming body;
r=rearing; l=locomotion; s=static sitting; sd=dynamic sitting.

attack by lactating female rats; (4) the tube-test in mice and (5) timid-defensive behaviour in defeated mice.

A single dose of fluprazine (2.5 mg/kg) decreased attacks and threats without any change in non-social elements of behaviour in aggressive isolated mice. Fluprazine suppressed offence (attacks) at a dose of 5 mg/kg and produced significant changes in the whole structure of behaviour. Cluster analysis shows (Fig. 1B) 1) changes in the spectrum of behavioural elements (lack of aggression elements, increased social elements); 2) changes in associations between elements; fluprazine produced closer associations between elements of sociability (SNN-SNB) and individual behaviour (R-L); 3) changes in the level of similarity: for example elements of sociability (active and passive elements) may become more differentiated from elements of individual behaviour. DU 28412 (5-10 mg/kg; single doses) did not display such a highly selective anti-aggressive effect in our experiments on aggressive isolated mice.

Chronic treatment of aggressive isolated mice with fluprazine (5 mg/kg/day over 14 days) also confirms the highly selective anti-aggressive effects of the compound. Cluster analysis (Fig. 1D) shows: 1) a broader spectrum of sociability elements; 2) close associations of these elements in distinct clusters (SNB, SNN, SNG) and 3) changes in the level of similarity between behavioural elements in control and under drug conditions. Fluprazine generally modifies the structure of intraspecific behaviour of isolated mice making it similar to that of non-aggressive grouped mice. Although the effects of DU 28412 in aggressive mice are not as selective, cluster analysis shows that after chronic treatment (Fig. 1E) the principal elements of the behavioural structure related to sociability are common for DU 28412 and fluprazine treatment.

Defensive foot-shock induced fighting in mice was suppressed by fluprazine at 5 mg/kg (isolation-induced aggression was inhibited by fluprazine at 2.5 mg/kg) and by DU 28412 at 10 mg/kg. Fluprazine significantly decreased attacks of lactating female rats towards male intruders at 5-7.5 mg/kg. Fluprazine decreased the contacts of the mother with the male intruder and increased the contacts of the mother with the pups. DU 28412 suppressed maternal aggression, sociability and contacts of the mother with the pups (Poshivalov 1986a). Non-differentiated hyperreactive biting in the "tube-test" was suppressed by fluprazine at 2.5-5 mg/kg and by DU 28412 at 10-15 mg/kg. Timid-defensive behaviour (in defeated mice) was increased by fluprazine at 5 mg/kg. Cluster analysis showed that fluprazine produced close associations between active and passive elements of defence (Poshivalov 1986a).

Comparing the effects of fluprazine and DU 28412 on different models of offence and defence show that fluprazine has more selective anti-aggressive properties than DU 28412. Fluprazine suppressed offence but inhibited defensive behaviour to a lesser extent. In some models these compounds can increase timid-defensive behaviour. Fluprazine is more potent in producing more social behaviour in aggressive mice.

CONCLUSIONS

Experiences in the Leningrad laboratory have shown that agonistic behaviour may be mathematically interpreted in terms of discrete and continuous statistical models. Thorough reevaluation of the models reveals that a discrete stationary model is most appropriate for assessment of changes in the structure of behaviour, for ethopharmacological screening of new compounds and for the development of drug standards.

A synthesis of ethopharmacology and molecular pharmacology seems especially useful for the analysis of relations between behavioural responses towards conspecifics and the distribution and affinity of particular kinds of specific recognition/binding sites in the brain (e.g. beta-carbolines and benzodiazepines).

Althcugh no direct correlations between specific changes at the receptor level and specific behaviours have been observed (Poshivalov 1986a; Sukchotina et al. in press), it seems likely that eventually we will be able to relate aggressive and timid–defensive behaviour to changes at the receptor level.

Methodologies need to be refined for the recognition of aggressogenic and anxiogenic drug properties. At the present time such techniques are less developed than counterparts for the assessment of anti–aggressive and anxiolytic properties. Restoration of aggression by antidepressants illustrates how drugs may have "pseudo–aggressogenic" properties. The increased timidity observed in the timid–defensive model of behaviour also illustrates that these drugs potentiate the previously observed "fear" and "timidity" and that there is no good evidence for a "pure" anxiogenic effect. The conversion of aggression or sociability into defence would perhaps be a more convincing expression of anxiogenic properties.

REFERENCES

Adams D (1979) Brain mechanisms for offense, defense and submission. Behav Brain Sci 2: 201–241

Adams D (1983) Hormone–brain interactions and their influence on agonistic behavior. In: Svare B (ed) Hormones and Aggressive Behavior. Plenum Press, New York, pp 223–245

Bradford LD, Olivier B, Van Dalen D, Schipper J (1984) Serenics – the pharmacology of fluprazine and DU 28412. In: Miczek KA, Kruk MR, Olivier B (eds) Ethopharmacological aggression research. Alan R Liss, New York, pp 191–208

Braestrup C, Nielsen M, Olsen CE (1980) Urinary and brain β–carboline–3–carboxylates as potent inhibitors of brain benzodiazepine receptors. Proc Natl Acad Sci 77: 2288–2292

Brain PF, Jones SE, Brain S, Benton D (1984) Sequence analysis of social behaviour illustrating the actions of two antagonists of endogenous opioids. In: Miczek KA, Kruk MR, Olivier B (eds) Ethopharmacological aggression research. Alan R Liss, New York, pp 43–58

Kantak KM, Hegstrand LR, Eichelman B (1980) Dietary tryptophan modulation and aggressive behavior in mice. Pharmacol Biochem Behav 12: 675–679

Krsiak M, Sulcova A, Tomasikova Z, Dlohozkova N, Kosar E, Masek K (1981) Drug effects on attack, defense and escape in mice. Pharmacol Biochem Behav 14: (S1) 47–52

Miczek K, DeBold JF, Thompson ML (1984) Pharmacological, hormonal and behavioral manipulations in the analysis of aggressive behavior. In: Miczek KA, Kruk MR, Olivier B (eds) Ethopharmacological aggression research. Alan R Liss, New York, pp 1–26

Miczek K, Krsiak M (1979) Drug effects on agonistic behavior. In: Thompson T, Dews PB (eds) Advances in behavioral pharmacology, Vol 2, Academic Press, New York, pp 87–162

Nieminen S, Poshivalov VP (1985) Ethopharmacology of agonists and antagonists of benzodiazepine receptors. In: Multidisciplinary approaches to conflict and appeasement in animals and man. 3rd Eur ISRA Conf, Parma, p 173

Olivier B, Mos J (1986) A female aggression paradigm for use in psychopharmacology: maternal agonistic behaviour in rats. In: Brain PF, Ramirez M (eds) Cross–disciplinary studies on aggression, University of Sevilla Press, pp 73–112

Olivier B, Van Aken H, Jaarsma I, Van Oorschot R, Zethof T, Bradford D (1984) Behavioural effects of psychoactive drugs on agonistic behaviour of male territorial rats (resident–intruder model). In: Miczek KA, Kruk MR, Olivier B (eds) Ethopharmacological aggression research. Alan R Liss, New York, pp 137–156

Pellow S, File S (1984) Multiple sites of action for anxiogenic drugs: behavioural, electrophysiological and biochemical correlations. Psychopharmacology 83: 304–315

Poshivalov VP (1978) Ethological atlas for pharmacological research in laboratory rodents. VINITI, Moscow, pp 3–42

Poshivalov VP (1981) Pharmaco–ethological analysis of social behaviour of isolated mice.Pharmacol Biochem Behav 14: (S1) 53–59

Poshivalov VP (1986a) Experimental psychopharmacology of aggressive behaviour. Nauka, Leningrad, pp 1–176

Poshivalov VP (1986b) Computerized ethological pharmacology: the new synthesis. In: Thompson T, Dews P, Barrett J (eds) Advances in behavioral pharmacology, Vol 6, Lawrence Erlbaum Ass, New York, pp 193–220

Poshivalov VP (1986c) Ethological pharmacology as a tool for animal aggression research. In: Brain PF, Ramirez M (eds) Cross–disciplinary studies on aggression, University of Sevilla Press, pp 17–49

Poshivalov VP, Khodko ST (1984) Mathematical description and experimental pharmaco–ethological analysis of animal intraspecific agonistic behaviour. In: Miczek KA, Kruk MR, Olivier B (eds) Ethopharmacological aggression research. Alan R Liss, New York, pp 59–80

Poshivalov VP, Khodko ST, Besov EV (1979) Ethograph–computer system for recording and analysis of zoosocial behaviour. J Higher Nerv Act 29: 420–423

Rodgers RJ, Hendrie CA (1984) On the role of endogenous opioid mechanisms in offense, defense and nociception. In: Miczek KA, Kruk MR, Olivier B (eds) Ethopharmacological aggression research. Alan R Liss, New York, pp 27–41

Sukchotina IA, Rojhanetz VV, Poshivalov VP (1986) Independent benzodiazepine and beta–carboline binding sites in the brain of aggressive and timid–defensive mice. Bull Exp Biol Med (in press)

Valdman AV, Poshivalov VP (1984) Pharmacological regulation of intraspecies behaviour. Meditsina, Leningrad, pp 1–208

Valdman AV, Poshivalov VP (1986) Pharmaco–ethological analysis of antidepressant drug effects. Pharmacol Biochem Behav 25: 515–520

PSYCHOPHARMACOLOGY OF SOCIAL PLAY

Jaak Panksepp, Larry Normansell, James F. Cox, Loring J. Crepeau and David S. Sacks. Department of Psychology, Bowling Green State University, Bowling Green, Ohio 43403, U.S.A.

INTRODUCTION TO THE STUDY OF PLAY

During the past decade, systematic methodologies have been developed for the experimental analysis of social play (Martin and Caro 1985; Meaney et al. 1985; Panksepp et al. 1984b; Thor and Holloway 1984a). For most of this century, play was considered a vacuous concept or a process of no great consequence for understanding brain organization of behaviour but now there is adequate evidence that play may be regarded as a fundamental neuro–behavioural category reflecting activity of a primal brain process. This category is seen as being elaborated by specific, genetically – ordained neural networks in subcortical areas of the brain. Although we presently know very little about the play circuits of the mammalian brain, we do know how to measure social play reliably, and how to bring it under reasonably precise experimental control. During the past decade, a preliminary psychobiological understanding of ludic processes has been slowly emerging.

PRESENT STATUS OF ANIMAL STUDIES

Although juvenile play can readily be detected in various species observed in their normal social environments (see Fagen 1981), those conditions, even when supplanted into the laboratory (e.g., the "focal–observation" procedure developed by Poole and Fish (1975, 1976) and Meaney and Stewart (1981)), are far from ideal for systematic experimental analysis.

As play is essentially a social activity, it seems desirable to have exacting experimental control over the social variables under which animals are reared and tested. The "paired–encounter" procedure (whereby play is enhanced by prior social deprivation and the interaction of matched pairs of animals placed into a neutral arena is observed for a limited amount of time) provides the maximal degree of experimental control (Panksepp 1979b, 1981b; Panksepp and Beatty 1980). This procedure also allows specific behaviours to be quantified by frequency and duration counts, thus eliminating ambiguous interpretive categories such as "play bouts". It is presently uncertain how many behaviours need to be scored to obtain an adequate overall picture of play, but it appears that they can be grouped as either appetitive or consummatory components.

Although the specific behaviours which need to be quantified differ among various mammalian species, most of the systematic experimental work has been done on the laboratory rat, so this account will be restricted to this species. When rats play, they appear to instigate or solicit play by pouncing on each other. This is often followed by chasing, sometimes in tighter and tighter nose–to–tail circles, until the animals end up in brief "wrestling bouts" involving considerable bodily contact. These bouts typically end with one animal "submissively" on its back and the other "victoriously" on top, yielding the measure we call the "pin". We have tried to quantify chasing, but have found it difficult to score reliably in a small test arena. It will probably require a more structured environment (perhaps one where play chambers are connected by tubes) to measure a chase sequence in an unambiguous fashion. We have also begun to place animals onto an activity platform for their play sessions so that an estimate of overall "rough–and–tumble" activity can be obtained. Although rats may solicit play interactions from each other in various ways (Thor and Holloway 1983b), the tendency of one animal to contact the dorsal surface of the other animal (with no immediate pin occurring) seems the most obvious and easily quantified index of play solicitation (i.e., the appetitive phase of ludic motivation), with the frequency and duration of pinning behaviour taken as the consummatory measure of play. It is believed that the distinction between appetitive and consummatory components should be fundamental to the analysis of play, although we have only recently adhered to this dictum by always measuring dorsal contacts in addition to pins during ongoing rough–and–tumble play. This has yielded several situations where pins are dramatically reduced with no reduction of dorsal contacts. For instance, anaesthetization of the anterior dorsal surface of rats yields that pattern of results (Sacks and Panksepp 1986; Siviy and Panksepp, in press). This suggests that the motivation for play is not compromised by diminished somatosensory input even though the information needed to consummate an ongoing play bout with a pin has been aborted.

We consider it essential to score the behaviour of each animal separately, since some type of dominance typically emerges in animals tested repeatedly with each other (Panksepp 1981b; Panksepp et al. 1985a). Also, for our statistics, we presently consider the score of each animal as a separate unit of measurement. It is considered more appropriate to have our measurement unit reflect the activity of each nervous system on the field of play, rather than be referenced to the social unit of the dyadic pair.

Although decortication experiments indicate that play circuitry is largely situated subcortically (Murphy et al. 1981; Normansell and Panksepp 1984), neither the brain–stem anatomy nor the neurochemistry of play has yet been revealed. Although a great number of pharmacological maneuvres modify play, the issue of pharmacological specificity has rarely been broached experimentally. The aim of this paper, in addition to providing an overall summary of the available pharmacological literature, is to highlight some important considerations for distinguishing relevant physiological controls from other influences which might also modify play.

PRESENT STATUS OF HUMAN STUDIES

There is little more than suggestive data concerning the biological controls of human play. Perhaps the most pertinent observation comes from the use of pharmacotherapy for childhood hyperkinesis. Clinicians have noted that stimulants, while reducing undesirable, "unfocussed" activity in such children, also tend to reduce the childish playfulness that is often deemed socially attractive and desirable by parents (Barkley 1977; Talmadge and Barkley 1983). This change may be due to heightened attentiveness, and resembles the marked reduction of social play that follows treatment of juvenile rats with stimulants (Beatty et al. 1982; Panksepp 1979b). It seems reasonable to assume, indeed, that the basic neurochemical controls of social play will be quite similar in all Mammalian species and may reflect our shared evolutionary heritage (MacLean 1985). Results of animal studies, therefore, might suggest neurochemical interventions for certain human clinical conditions where modification of play circuitry might be beneficial, such as early childhood autism (Black et al. 1975; Panksepp 1979a).

REMAINING QUESTIONS

Considering the brief history of systematic research on the psychobiology of play, it is certainly the case that most of the important questions in the area remain to be fully explored. From our perspective, the major concerns should be to identify the locations and trajectories of the neural circuits for play, and to identify the neurochemistries which modulate activity in those circuits. We also need to replace the vast storehouse of theoretical speculation concerning the functions of play with substantive experimental tests. Except for some preliminary work by Einon and colleagues (Einon et al. 1978), functional issues remain largely unexplored. In several unpublished studies, we have found no marked behavioural deficits in animals deprived of play during the juvenile period of life, suggesting that the functions of play may be more subtle than previously supposed.

PSYCHOPHARMACOLOGY OF PLAY

Overall, it should be emphasized that although the effects of a variety of pharmacological manipulations have been characterized, we are far from understanding the functional dynamics of the reported effects. There is no report in the literature that has fully analyzed the effects of a drug on a comprehensive list of objective behaviours that characterize play (e.g., chasing, soliciting, wrestling, pinning). Because of the paucity of other measures at the present time, most of this review is based on measurement of pinning behaviour during rough and tumble activity.

Opiates and Play

Our work on play emerged from research we initiated in the mid–1970's into the psychobiology of social emotions. The key postulate of that line of inquiry was that social emotions and social attachments are controlled in the brain by the same mechanisms which elaborate opiate dependence. That work has been reviewed several times (Panksepp 1981a; Panksepp et al. 1980; 1985b) and it will simply be noted that the theory generated the expectation that opiate receptor modulating drugs should have clear effects on play.

Both naloxone and morphine have distinct effects on play; opiate blockade decreases and morphine increases rough–and–tumble activity (Beatty and Costello 1982; Panksepp 1979; Panksepp et al. 1984b, 1985a; Siegel and Jensen 1986; Siegel et al. 1985). These drugs also yield clear effects on play dominance; an animal treated with naloxone pins less than its vehicle–treated partner (Panksepp et al. 1985a). In addition, indirect evaluation of central opioid activity via subtractive autoradiography suggests that brain opioids are released during play (Panksepp 1981a; Panksepp and Bishop 1981). More recently, however, we have found that morphine does not appear especially effective in increasing play during the late juvenile phase, perhaps indicating an ontogenetic change in drug sensitivity, a phenomenon which has also been reported by Spear et al. (1982).

The specific mechanisms by which opioids modulate play remain obscure. Effects could be due to simple activity changes as suggested by Beatty and Costello (1982), but we favor the hypothesis that modulation of opioid activity yields socio–affective changes in animals which can be either harmonious or incompatible with play (Panksepp et al. 1984b; Siegel and Jensen 1986). At the neural level, the effect of opiates could be mediated by changes in meso–limbic dopamine activity which controls jumpy/excitable behaviours (Joyce and Iversen 1979; Kelley et al. 1980). Parenthetically, lesions just dorsal to this area, in the parafasicular region of the meso–diencephalic junction, reduce pinning behaviour and also reduce the effects of opioid manipulations (Siviy and Panksepp 1985).

Cholinergic control of play

Extensive research has demonstrated that blockade of muscarinic cholinergic receptors can compromise a great number of behaviours, due perhaps to deficits in attention and other general processes needed for goal directed behaviours (see Panksepp 1986). Thus, it is not too surprising that anticholinergics such as scopolamine can very dramatically reduce play (Beatty 1983; Thor and Holloway 1983c). It is noteworthy, however, that doses as low as 0.25 mg/kg are quite effective, and that many doses which eliminate play can increase general locomotor activity. The scopolamine effect shows rapid tolerance at low doses (Thor and Holloway 1983c; 1984b), however, suggesting that the drug effect may be largely due to disruptive influences as opposed to specific synaptic controls within play circuits. This conclusion is reinforced by the failure of cholinergic agonists to increase play, as well as the inability of muscarinic agonists and antagonists to counteract each other (Wilson et al. 1986).

The analysis of nicotinic involvement in play has been less extensive, but it is clear that nicotine can reduce play. It is also suggestive that mecamylamine can increase play and that these two drugs can counteract each other (Panksepp et al. 1984a, b). Considerably more work needs to be done before the specific involvement of the cholinergic system in play can be claimed. It is noteworthy, however, that treatment of animals with anti-muscarinics provides a useful non-playful target animal which can be used as an appropriate stimulus for measurement of play solicitation (Thor and Holloway 1983b).

Indoleamines

From the analysis of other motivational systems, there is every reason to expect that facilitation of serotonergic activity will reduce play (Panksepp 1986). Indeed, receptor stimulation with quipazine yields clear dose-dependent reductions in pinning. These effects can be partially reversed by the antagonist methysergide at a dose which appears to have no effect on play (Normansell and Panksepp 1985b). Nevertheless, since in unpublished work we have been unable to facilitate play with either low tryptophan diets or PCPA treatment (fig. 1), it seems unlikely that serotonin systems have specific control over play circuitry. The systems are capable, however, of providing general modulatory control over all goal-directed behaviours, including play. Fenfluramine, a serotonin releasing agent, is very potent in reducing play. Whether this effect is due to serotonin release is not, however, certain, since the effect shows very rapid tolerance (i.e., the initial effect is diminished by almost 75% on the second test day) and is not reversed by pretreatment with PCPA (fig. 1). Work with more powerful and specific serotonin neurotoxins and/or depletors is clearly indicated.

Catecholamines

Although the role of catecholamines in the regulation of play is far from clear, a number of catecholamine-modulating agents do have effects on this behaviour. One such agent, clonidine, is an especially powerful inhibitor of play (Beatty et al. 1984; Normansell and Panksepp 1985a). This drug has a dose dependent action (with play virtually abolished at doses above 0.01 mg/kg) and is partially reversible with the α_2-antagonist yohimbine. Although clonidine's effect on social play appears to be specific for α_2-receptors, it is nearly certain that play reduction by clonidine cannot be attributed to downregulation of noradrenergic outflow via the α_2-autoreceptors on norepinephrine (NE)-containing neurons, since NE depletion has little or only modest effects on play.

Rats were treated neonatally with DSP4, a noradrenergic neurotoxin selective for locus coeruleus neurons (Jonsson et al. 1981). Parental DSP4 (50 mg/kg), which depleted cortical NE by about 70%, had no effect on pins, pin durations, solicitations, or overall rough-and-tumble activity. Furthermore, in other work (described in Panksepp et al. 1984b), rats were tested after treatment with

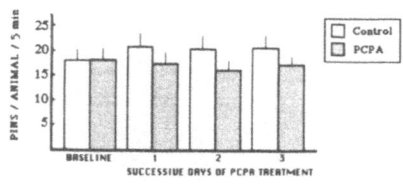

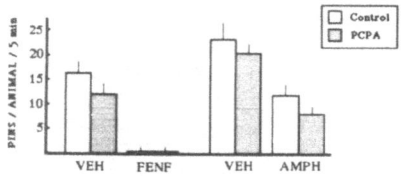

Fig. 1   Mean pins (±SEM) per animal for 5 min test sessions following either water or PCPA (100 mg/kg/day) treatment. Animals in each of those groups were subsequently tested after either fenfluramine (1 mg/kg) or amphetamine (1 mg/kg) administration. PCPA did not reliably reduce play. Both fenfluramine and amphetamine reliably reduced play (p's <.01), but not differentially in the two groups.

RO4–1284 and FLA–63, a pharmacological procedure which selectively depletes NE (Zolovick et al. 1982). This depleting regimen reduced pinning by about 30%, a modest effect compared to the clonidine–induced abolition of play. Other behavioural effects of clonidine also seem independent of presynaptic α_2–mediated reduction on NE outflow (Britton et al. 1984; Rossi et al. 1983), so a postsynaptic mechanism may be involved. There are further reasons to suspect that norepinephrine potentiation inhibits play circuitry. Brain noradrenergic systems appear to govern specific attentional processes (Foote et al. 1983; Segal 1985) which may serve to alert animals to novel or biologically significant stimuli, enabling external events to take precedence over internal motivations (Tucker and Williamson 1984). Certainly, animals that continued to play in the presence of predators and other novel or threatening stimuli would be unlikely candidates for longevity. Moreover, threatening stimuli can evoke augmented release of norepinephrine (Cassens et al. 1980; Iimori et al. 1982), and we have found that threatening stimuli are indeed potent inhibitors of play.

While it is likely that noradrenergic activity may inhibit play, the lack of directly acting α_1 and β agonists has prevented direct pharmacological tests of this hypothesis. Indirectly acting agonists, such as amphetamine and methylphenidate (Beatty et al. 1982), as well as ephedrine (Beatty et al. 1984), inhibit pinning behaviour, but this effect is not necessarily mediated noradrenergically.

Amphetamine–induced suppression appears to be specific for play. Measures of other social behaviours such as social investigation and grooming (Beatty et al. 1982), and time spent in proximity to a drugged conspecific (Sutton and Raskin 1986), remain unchanged or increased. The mechanism by which amphetamine reduces play has remained elusive despite considerable experimental attention. Since neither adrenalectomy nor chemical sympathectomy attenuate the effect of amphetamine (Beatty et al. 1983), the mechanism is presumed to be mediated centrally. A variety of centrally–acting noradrenergic antagonists, however, fail to reverse the play suppressive effects of amphetamine (Beatty et al. 1984; Cox et al. 1984).

Although none of the evidence supports noradrenergic involvement in amphetamine–induced reduction of play, it seems equally unlikely that dopamine systems are involved. For instance, the d– and l–isomers of amphetamine are roughly equipotent in suppressing play (Cox et al. 1984), while the d–isomer is generally more potent in stimulating dopamine–dependent behaviours (Rebec and Bashore 1984). Furthermore, amphetamine suppression of play exhibits modest tolerance (Cox et al. submitted) as opposed to the reverse tolerance that is often associated with dopamine–dependent effects (Robinson and Becker 1982).

Beatty et al. (1984) have also found inhibition of tyrosine hydroxylase to be ineffective in attenuating the amphetamine–induced suppression. Finally, haloperidol fails to block amphetamine–induced suppression and itself reduces play (Beatty et al. 1984). If anything, dopamine stimulation facilitates play since apomorphine, within a narrow dose range, slightly increases pinning (Beatty et al. 1984; Cox et al. 1984). In general, it remains unclear whether dopamine systems exert any specific control over play beyond their general roles in facilitating motivated behaviours.

Benzodiazepine/GABA

Low doses of the benzodiazepine chlordiazepoxide have no clear specific effects on play behaviour. Although initially the sedating effects of high doses modestly reduce play, animals show rapid tolerance to this sedation and play returns to control levels. Chlordiazepoxide (CDP), however, does appear to increase the likelihood of play among animals under stress. A recently conducted series of experiments utilizing a conditioned emotional response paradigm, recorded play behaviour for three consecutive 2–minute trials each day. Throughout the second trial a tone was presented, after which followed 2 sec of inescapable footshock (0.5mA). After 7 days of training, water–treated animals showed complete suppression of play during the tone presentation, while animals injected with CDP (5 mg/kg) exhibited only a 50% suppression, compared to unshocked controls (fig. 2).

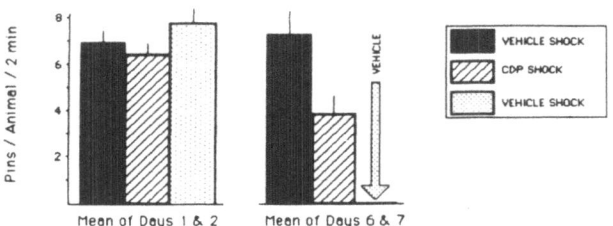

Fig. 2   Mean pins (±SEM) per animal for the 2 min sessions during tone presentation at the beginning and end of acquisition training. Animals were injected 20 min prior to testing with either water or chlordiazepoxide (5 mg/kg). Vehicle and CDP are reliably lower than control and differ from each other at p<.01 on last two days of testing.

In addition, drug treatment during acquisition, facilitated recovery of play during extinction training. CDP–treated animals were pinning at control levels

138

after only 3 days, whereas water–treated animals required 6 days of extinction before fully returning to control levels (fig. 3). These effects were noted regardless of the drug given during the extinction phase. In other words, animals treated with CDP during extinction, but water during acquisition were as slow to extinguish as animals treated with water during both phases. Since CDP did not elevate pain thresholds, it appears likely the maintenance of play in the stress situation occurred via an anxiety reducing mechanism.

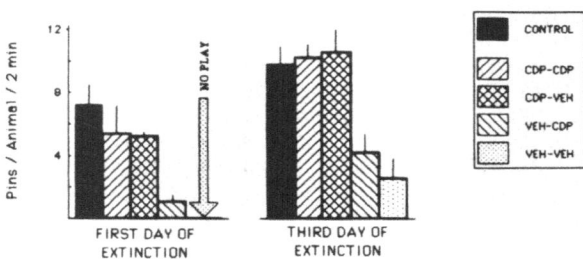

Fig. 3 Mean pins (±SEM) per animal for the 2 min sessions during tone presentation on day 1 and 3 of extinction. Control animals were injected with water and received no shock during training. Animals in the other groups were injected with either water or chlordiazepoxide (5 mg/kg) before their acquisition trials, and either water or chlordiazepoxide before extinction trials. CDP given during acquisition produced rapid and reliable (p<.01) inhibition of conditioned play suppression during extinction.

Since benzodiazepines enhance the inhibitory action of GABA (e.g., Costa et al. 1983), the effects on play of the GABA antagonist picrotoxin and a pharmacologically active GABA metabolite, gamma hydroxybutyrate (γ–OHBA) were investigated. Picrotoxin at low doses (0.125 – 0.25 mg/kg) had no effect on play, while a higher dose (0.5 mg/kg) reliably decreased pinning (from a mean of 15.8 to 9.1 per 5 min), but increased the number of dorsal contacts (from 17.8 to 20.3). γ–OHBA has a powerful effect on play, perhaps through its interactions with dopaminergic systems (Snead 1977). Both pins and dorsal contacts were reduced in a dose–dependent manner, while the average duration of pins was not affected (fig. 4).

The effect of pentobarbital on play has been recently evaluated. A dose of 5 mg/kg produced a reliable increase in pinning in groups of rats tested at either 22–25 (mean pins increased from 8.7 to 12.5) or around 55 (increase over control from 12.5 to 15.1 pins per 5 min) days of age. At a higher dose (20 mg/kg), play was almost completely eliminated (Donovan and Panksepp unpublished observations).

Low doses of ethanol also increase play. Animals were tested 20 min after intragastric intubation of 1, 2, or 4 g/kg ethanol in water. A dose of 1 g/kg reliably increased the mean number of pins (from 8.4 to 10.4), while the higher doses decreased pinning. Dorsal contacts were similarly affected.

Psychotherapeutic Agents, and Other Psychoactives

The play modulating effects of a number of drugs used clinically for the treatment of depression have been assessed. At a dose of 15 mg/kg, imipramine and des–methyl–imipramine all but eliminated play, while chloro–imipramine reduced

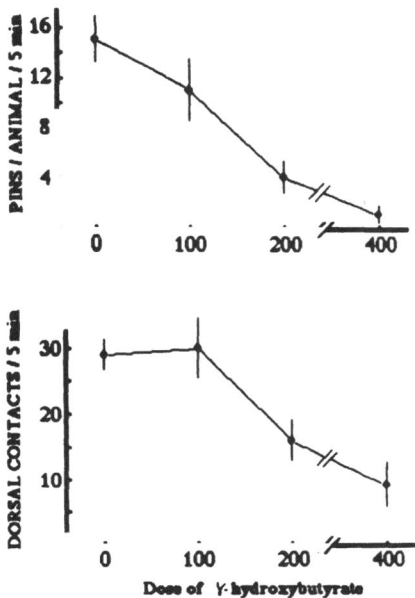

Fig. 4   Mean number of pins and dorsal contacts (±SEM) per animal following treatment with varying doses of γ–OHBA. Doses above 200 mg/kg reliably reduced both measures (p<.001)

pinning by about 75% (Knowles 1986). Like these tricyclics, the monoamine oxidase inhibitor pargyline (25 mg/kg) suppresses all measures of play, a reduction that was apparent over 5 days of testing, with no sign of tolerance developing.

The brain adenosine system has been implicated in the modulation in play. Acute administration of the antagonist caffeine has been shown to decrease both pinning and play solicitation (Holloway and Thor 1983, 1984, 1985; Thor and Holloway 1983a), while play increases have been reported following a chronic treatment regimen (Holloway and Thor 1984). Administration of the adenosine agonist 2–chloroadenosine also reduces pinning and play solicitation as well as the time spent in social investigation. This play suppressant effect of 2–chloroadenosine was partially reversed by administration of caffeine (Holloway and Thor 1985).

CONCLUDING REMARKS

Many neurochemical systems have been implicated in the control of play. Specific roles for the various influences will have to be dissected by additional research. At the present time, the most workable problem is the issue of what brain changes resulting from social deprivation invigorate rough–and–tumble activity, and what neurochemical consequences of play lead to gradual diminution of ludic activities. Our best guess is that reduced norepinephrine and serotonin activity, which are known to result from social deprivation (see Valzelli 1973), may be important ingredients in the invigoration of play resulting from isolation. Pharmacological manipulations which promote activity in these systems fairly uniformly yield animals which do not appear to be especially eager to play.

There are surely other mechanisms that can directly facilitate ludic motivation – such as increasing brain opioid, dopamine and acetylcholine activity. Presumably, these diverse influences mediate distinct functional controls on play, and will undoubtedly require substantial new experimental approaches to be clarified.

REFERENCES

Barkley RA (1977) The effects of methylphenidate on various types of activity level and attention in hyperkinetic children. J Abnorm Child Psychol 5: 351–353

Beatty WW (1983) Scopolamine depresses play fighting: A replication. Bull Psychonom Soc 21: 315–316

Beatty WW, Berry SL, Costello KB (1983) Suppression of play fighting by amphetamine does not depend on peripheral catecholaminergic influences. Bull Psychonom Soc 21: 407–410

Beatty WW, Costello KB (1982) Naloxone and play fighting in juvenile rats. Pharmacol Biochem Behav 17: 905–907

Beatty WW, Costello KB, Berry SL (1984) Suppression of play fighting by amphetamine: Effects of catecholamine antagonists, agonists and synthesis inhibitors. Pharmacol Biochem Behav 20: 747–755

Beatty WW, Dodge AM, Dodge LJ, White K, Panksepp J (1982) Psychomotor stimulants, social deprivation and play in juvenile rats. Pharmacol Biochem Behav 16: 417–422

Black M, Freeman BJ, Montgomery J (1975) Systematic observation of play behavior in autistic children. J Autism Childhood Schizophrenia 5: 363–371

Britton KT, Svensson T, Schwarz J, Bloom FE, Koob G (1984) Dorsal noradrenergic bundle lesions fail to alter opiate withdrawal or suppression of opiate withdrawal by clonidine. Life Sci 34: 133–139

Cassens G, Roffman M, Kuruc P, Orsulak P, Schildkraut J (1980) Alterations in brain norepinephrine metabolism induced by environmental stimuli previously paired with inescapable shock. Science 209: 1138–1139

Costa E, Corda MG, Epstein B, Forchetti C, Guidotti A (1983) GABA–benzodiazepine interactions. In: Costa E (ed) The benzodiazepines: from molecular biology to clinical practice, Raven Press, New York, pp 117–136

Cox J, Schoen L, Normansell L, Rossi J, Siviy S, Panksepp J (1984) Dopaminergic substrates of play. Soc Neurosci Abstr 10: 1177

Cox JF, Strope TDL, Panksepp J (1986) d–Amphetamine effects on social play: tolerance and withdrawal. Manuscript submitted for publication.

Einon DF, Morgan MJ, Kibbler CC (1978) Brief periods of socialization and later behavior in the rat. Dev Psychobiol 11: 213–225

Fagen RM (1981) Animal Play Behavior. Oxford University Press, New York

Foote SL, Bloom FE, Aston–Jones G (1983) Nucleus locus coeruleus: New evidence of anatomical and physiological specificity. Physiol Rev 63: 844–914

Holloway WR, Thor DH (1983) Caffeine: Effects on the behaviors of juvenile rats. Neurobehav Toxicol Teratol 5: 127–134

Holloway WR, Thor DH (1984) Acute and chronic caffeine exposure effects on play fighting in the juvenile rat. Neurobehav Toxicol Teratol 6: 85–91

Holloway WR, Thor DH (1985) Interactive effects of caffeine, 2–chloroadenosine and haloperidol on activity and social behaviors of juvenile rats. Pharmacol Biochem Behav 22: 421–426

Iimori K, Tanaka M, Kohno Y, Ida Y, Nakagawa R, Hoaki Y, Tsuda A, Nagasaki N (1982) Psychological stress enhances noradrenaline turnover in specific brain regions in rats. Pharmacol Biochem Behav 16: 637–640

Jonsson G, Hollman H, Ponzio F, Ross S (1981) DSP4 (N–(2–chloroethyl)–N–ethyl-2–bromobenzylamine)—useful denervation tool for central and peripheral noradrenergic neurons. Eur J Pharmacol 72: 173–188

Joyce EM, Iversen SD (1979) The effect of morphine applied locally to mesencephalic dopamine cell bodies on spontaneous motor activity in the rat. Neurosci Lett 14: 266–272

Kelley AE, Stinus L, Iversen SD (1980) Interactions between d–ala–met–enkephalin and A10 dopaminergic neurons and spontaneous behavior in the rat. Behav Brain Res 1: 3–24

Knowles PA (1986) Effects of activated sleep deprivation (ASD) on play behavior of juvenile rats. Unpublished doctoral dissertation, Bowling Green State University, Bowling Green, Ohio

MacLean PD (1985) Brain evolution relating to family, play, and the separation call. Arch Gen Psychiat 42: 405–417

Martin P, Caro TM (1985) On the functions of play and its role in behavioral development. In: Rosenblatt JS, Beer C, Busnel MC, Slater PJB (eds) Advances in the study of behavior, vol 1, Academic Press, New York, pp 59–103

Meaney MJ, Stewart J (1981) A descriptive study of social development in the rat (Rattus norvegicus). Anim Behav 29: 34–45

Meaney MJ, Stewart J, Beatty WW (1985) Sex differences in social play: The socialization of sex roles. In: Rosenblatt JS, Beer C, Busnel MC, Slater PJB (eds) Advances in the study of behavior, vol 1, Academic Press, New York, pp 1–58

Murphy MR, MacLean PD, Hamilton SC (1981) Species–typical behavior of hamsters deprived from birth of the neocortex. Science 213: 459–461

Normansell L, Panksepp J (1984) Play in decorticate rats. Soc Neurosci Abstr 10: 612

Normansell L, Panksepp J (1985a) Effects of clonidine and yohimbine on the social play of juvenile rats. Pharmacol Biochem Behav 22: 881–883

Normansell L, Panksepp J (1985b) Effects of quipazine and methysergide on play in juvenile rats. Pharmacol Biochem Behav 22: 885–887

Panksepp J (1979a) A neurochemical theory of autism. Trends Neurosci 2: 174–177

Panksepp J (1979b) The regulation of play: Neurochemical controls. Soc Neurosci Abstr 5: 172

Panksepp J (1981a) Brain opioids – a neurochemical substrate for narcotic and social dependence. In: Cooper SJ (ed) Theory in psychopharmacology, vol 1, Academic Press, London, pp 149–175

Panksepp J (1981b) The ontogeny of play in rats. Devel Psychobiol 14: 327–332

Panksepp J (1986) The neurochemistry of behavior. Ann R Psychol 37: 77–107

Panksepp J, Beatty WW (1980) Social deprivation and play in rats. Behav Neural Biol 30: 197–206

Panksepp J, Bishop P (1981) An autoradiographic map of [3H]–diprenorphine binding in rat brain: Effects of social interaction. Brain Res Bull 7: 405–410

Panksepp J, Herman BH, Vilberg T, Bishop P, DeEskinazi F (1980) Endogenous opioids and social behavior. Neurosci Biobehav Rev 4: 473–487

Panksepp J, Jalowiec J, DeEskinazi F, Bishop P (1985a) Opiates and play dominance in juvenile rats. Behav Neurosci 99: 441–453

Panksepp J, Sahley T, Normansell L (1984a) Cholinergic control of social play. Soc Neurosci Abstr 10: 1177

Panksepp J, Siviy S, Normansell L (1984b) The psychobiology of play: Theoretical and methodological perspectives. Neurosci Biobehav Rev 8: 465–492

Panksepp J, Siviy S, Normansell L (1985b) Brain opioids and social emotions. In: Reite M, Fields T (eds) The psychobiology of attachment and separation, Academic Press, New York, pp 3–49

Poole TB, Fish J (1975) An investigation of playful behaviour in Rattus norvegicus and Mus musculus (Mammalia) J Zool 175: 61–71

Poole TB, Fish J (1976) An investigation of individual, age and sexual differences in the play of Rattus norvegicus (Mammalia: Rodentia). J Zool 179: 249–260

Rebec GV, Bashore TR (1984) Critical issues in assessing the behavioral effects of ampehetamine. Neurosci Biobehav Rev 8: 153–159

Robinson TE, Becker JB (1982) Behavioral sensitization is accompanied by enhancement in amphetamine–stimulated dopamine release from striatal tissue in vitro. Eur J Pharmacol 85: 253–254

Rossi III J, Sahley TL, Panksepp J (1983) The role of brain norepinephrine in clonidine suppression of isolation–induced distress in the domestic chick. Psychopharmacology 79: 338–339

Sacks DS, Panksepp J (1986) Neonatal capsaicin increases play deficit produced by xylocaine. Soc Neurosci Abstr 12: (in press)

Segal M (1985) Mechanisms of action of noradrenaline in the brain. Physiol Psych 13: 172–178

Siegel MA, Jensen RA (1986) The effects of naloxone and cage size on social play and activity in isolated young rats. Behav Neural Biol 45: 155–168

Siegel MA, Jensen RA, Panksepp J (1985) The prolonged effects of naloxone on play behavior and feeding in the rat. Behav Neural Biol 44: 509–514

Siviy SM, Panksepp J (1985) Dorsomedial diencephalic involvement in the juvenile play of rats. Behav Neurosci 44: 509–514

Siviy SM, Panksepp J (1987) Somatosensory modulation of juvenile play in rats. Devel Psychobiol 20: in press

Snead OC (1977) Minireview. Gamma hydroxybutyrate. Life Sci 20: 1935–1944

Spear LP, Horowitz GP, Lipovisy J (1982) Altered responsivity to morphine during the periadolescent period in rats. Behav Brain Res 4: 279–288

Sutton ME, Raskin LA (1986) A behavioral analysis of the effects of amphetamine on play and locomotor activity in the post–weaning rat. Pharmacol Biochem Behav 24: 455–461

Talmadge J, Barkley RA (1983) The interactions of hyperactive and normal boys with their fathers and mothers. J Abnorm Child Psychol 11: 565–580

Thor DH, Holloway WR (1983a) Play soliciting in juvenile male rats: Effects of caffeine, amphetamine, and methylphenidate. Pharmacol Biochem Behav 19: 725–727

Thor DH, Holloway WR (1983b) Play–solicitation behavior in juvenile male and female rats. Anim Learn Behav 11: 173–178

Thor DH, Holloway WR (1983c) Scopolamine blocks play fighting in juvenile rats. Physiol Behav 30: 545–549

Thor DH, Holloway WR (1984a) Social play in juvenile rats: A decade of methodological and experimental research. Neurosci Biobehav Rev 8: 455–464

Thor DH, Holloway WR (1984b) Social play in juvenile rats during scopolamine withdrawal. Physiol Behav 32: 217–220

Tucker DL, Williamson PA (1984) Asymmetric neural control systems in human self–regulation. Psych Rev 91: 185–215

Valzelli L (1973) Psychopharmacology: An introduction to experimental and clinical principles, Spectrum Publications, Flushing, New York

Wilson LI, Bierley RA, Beatty WW (1986) Cholinergic agonists suppress play fighting in juvenile rats. Pharmacol Biochem Behav 24: 1157–1159

Zolovick AJ, Rossi J, Davies RF, Panksepp J (1982) An improved pharmacological procedure for depletion of noradrenaline: Pharmacology and assessment of noradrenaline–associated behaviors. Eur J Pharmacol 77: 265–271

THE RELATIONSHIP BETWEEN ETHANOL AND AGGRESSION: STUDIES USING ETHOLOGICAL MODELS

Robert J. Blanchard and D Caroline Blanchard. Department of Psychology, University of Hawaii, U.S.A.

INTRODUCTION

Sociological and criminological research has consistently demonstrated a strong positive relationship between alcohol consumption, violence, and violent crime, both in the street and in the family (Pernanen 1976; Shupe 1954; Straus et al. 1980; Wolfgang and Strohm 1956). This relationship is not consistent over all cultural groups, but appears to be largely complexly related to factors of social stress (Levinson 1981) and individual affective and cognitive changes produced by alcohol (Williams 1966).

Understanding of these interlocking phenomena requires unusually precise and ecologically valid experimental models. These models must permit the full development of social relationships, including those which engender stress, in a setting which allows such social and stress factors to produce their normal and usual consequences. The models must permit a description of the effects of alcohol ingestion on aggressive and fearful/anxious behaviours for subjects of different sex, age, and status within the group. Finally, they must provide an evaluation of the extended interaction of stress, alcohol consumption, and aggressive and emotional behaviours of subjects in functioning social groups.

A colony model (Blanchard and Blanchard 1977; 1984) of social behaviour and aggression provided these requisite features. Extensive use of this model in studies on mice (cf. Brain 1981) and rats (Blanchard et al. 1977) has shown that it provides for measurement of social status of individuals living in groups (Blanchard et al. 1985) and is sensitive to changes in environmental features (Lore and Flannelly 1981). One extremely important feature of this well–established and extensively tested animal model is that it provides an aggression analysis which may be applicable to human as well as animal behaviour (Blanchard and Blanchard 1984). Comparative analysis of aggression in different Mammalian species including cats (Leyhausen 1979), bears (Blanchard et al. 1978), ungulates (Geist 1971) and primates (Adams and Shoel 1981) as well as rodents (Adams 1980), suggests a striking, detailed, cross–species continuity of the relationships between aggression and fear, and the separability of offensive vs. defensive forms of attack in these species. In humans, fear–based and anger–based aggression, with different eliciting stimuli, emotional patterns, and inhibiting variables appear to represent analogues to these mammalian models (Blanchard and Blanchard 1981, 1984). The importance of this distinction is now widely accepted by both behavioural and pharmacological researchers (Miczek et al. 1984).

An offence–defence analysis of aggression has already proved useful in several studies of the effects of ethanol on aggression in mice (Borgesova et al. 1971; Brain 1981; Krsiak 1976; Miczek and O'Donnell 1980; Smoothy 1985) and rats (Krsiak and Borgesova 1973; Miczek and Barry 1977; Miczek et al. 1984). The present paper reports a series of studies of the effects of ethanol on a variety of agonistic behaviours of rats. This series used the offence–defence analysis and procedures designed to elicit and measure specific components of the offence and defence patterns. These studies provide a number of significant findings detailing specifics of the relationship between alcohol ingestion and aggressive behaviour in rats.

EFFECTS OF ETHANOL IN FLIGHT AND DEFENSIVE ATTACK

Ethanol effects on fear and defence may influence the perceived relationship between fear and aggression in two very different ways. First, defensive behaviours have often been confused with aggression, despite the fact that offensive and defensive behaviours are elicited, modulated, and controlled by very different stimulus antecedents and consequences, as well as different historic/organismic factors (Blanchard and Blanchard 1984). If ethanol alters specific defensive behaviours which are mistaken for aggression, the ethanol–defence link may then be viewed as a relationship between ethanol and aggression. A second possible mechanism by which ethanol effects on fear/defence could impact an ethanol–aggression relationship is by fear inhibiting (as it normally does) offensive behaviours (Blanchard and Blanchard 1984). A general reduction in fear or in response–inhibition components of the defence pattern following ethanol ingestion might therefore enhance offence by removing this form of inhibition.

There is, however, a confounding factor which must be considered in parametric studies of alcohol and defence. Alcohol has analgesic effects (Bovier et al. 1984; Pohorecky 1977), and increases shock thresholds (Jeavons and Taylor 1985). Procedures for eliciting the full range of defensive behaviours in laboratory rats rely heavily on shock or other pain, since these animals have been selectively bred for a reduction of defensiveness to nonpainful stimuli (Blanchard et al. 1984). The use of subjects which show a full range of defensive behaviours to nonpainful stimuli is therefore necessary in order to provide an unambiguous view of the relationship between alcohol ingestion and changes in the topography of defence.

Wild rats show a defence pattern which is similar to that of domesticated rats (Blanchard et al. 1986; Takahashi and Blanchard 1982), with the exceptions that active defensive behaviours such as flight and defensive attack are proportionately more prominent than for domesticated rats, and, each component of the defence pattern can be elicited by appropriate nonpainful threat stimuli. In order to make these comparisons, and additionally for the purpose of measuring changes in defence following brain manipulations, we have developed (Blanchard et al. 1981a; Blanchard et al. 1981b) a set of tasks designed to elicit a wide range of defensive activities in both wild and laboratory rats.

These tests measure flight, freezing, boxing, vocalization and jump attacks to a human experimenter, as well as flinching and jumping to taps on the back, and boxing, vocalization and jump attacks to vibrissae stimulation and to an anaesthetized conspecific. The test battery therefore taps an array of passive and active components of the rat's defensive repertoire to a variety of threatening stimuli.

We used 18 male and 39 female wild–trapped Rattus rattus, allocated to four groups in these tasks. Thirty minutes prior to the onset of testing, subjects received an equal–volume IP injection of 0 (saline: n=12), 0.3 (n=12), 0.6 (n=11) 1.2 (n=11) or 1.8 (n=11) g/kg ethanol. Tests were run in the following order: Avoidance and escape to an approaching human experimenter in an oval runway 6 m long on each long side; freezing and defensive attack to a human experimenter as a function of distance between subject and experimenter (defensive distance) measured in an alley runway 5 m long and 1 m wide; flinch and jump reactions to hand claps and, to being tapped lightly on the back; defensive attack to vibrissal stimulation, and toward a terminally anaesthetized conspecific which was brought up to face the subject and reactions (struggling, vocalization, biting) to being picked up by the experimenter.

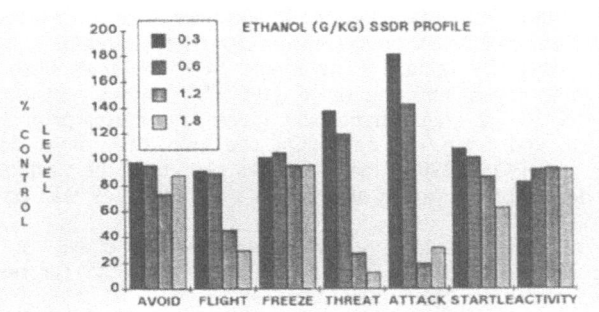

Fig. 1   Defensive behaviours of wild rats under 0.3, 0.6, 1.2 and 1.8 g/kg ethanol, as percentage of control (saline dose) levels of these behaviors.

Figure 1 presents a profile of ethanol–dose related changes in species–typical defensive behaviours for wild rats in these tasks. Note that the scores for each dose level are given in term of the saline level (=100%). As these measures suggest, activity, the percentage of avoidances, and freezing duration, did not change substantially with increasing ethanol doses. Flight speed and startle to hand claps and to taps on the back were stable at the two lower doses but did decline somewhat after the 1.2 and 1.8 g/kg doses. In contrast, defensive threat, vocalization and attack, measured to vibrissae stimulation, and to an anaesthetized conspecific (scores given in fig. 1 are a combination of these tests), increased substantially with the low ethanol doses, and then declined with the higher doses.

For the vibrissae stimulation tests, reliable differences existed among the dose level groups on jump attack: $F(4,52)=3.57$, $p<.05$; biting: $F(4,52)=7.80$, $p<.01$; and vocalization: $F(4,52)=5.64$, $p<.01$. The enhancement of vocalization for the low dose groups approached, but failed to reach, an acceptable level of statistical significance ($t(22)=1.81$, $.10>p>.05$, for saline vs. 0.3 g/kg group comparisons). When an anaesthetized conspecific was used as the eliciting stimulus, jump attack group differences were significant ($F(4,52)=4.55$, $p<.01$). Subsequent t–tests indicated that the increase for the 0.3 g/kg group was statistically reliable ($t(22)=2.12$, $p<.05$), while the decrease for the 1.2 g/kg group in comparison to controls approached but failed to reach an acceptable level of statistical significance ($t(21)=2.04$, $.10<p<.05$).

These findings clearly do not suggest a general decrement in fear or defensiveness for rats given low to moderate ethanol doses: What they do suggest is that such doses may selectively enhance defensive threat and attack, perhaps as part of a more general shift involving decrements in freezing and passive or inhibitory aspects of defence. While an ethanol–related freezing decrement was not seen in wild rats, these animals show relatively little freezing, compared to laboratory animals and, as will be seen later, another study in this series does suggest freezing decrements in laboratory rats with 0.3 and 0.6 g/kg ethanol doses. If such a shift does occur, it might also be expected to have considerable effect on offence.

MATERNAL AGGRESSION

Maternal aggression in particular, and female aggression generally, are coming to be regarded as a behavioural phenomenon somewhat different in terms of physiological control (Svare and Gandelman 1973) as well as in their patterning

(Blanchard et al. 1984) from male attack. It was therefore of considerable interest to examine the effect of ethanol on aggression using such models.

The subjects were 36 female Long–Evans rats, in three dose–level groups, tested in their home cages, two and three days after giving birth to pups. An IP injection of 0, 0.5, or 1.2 g/kg ethanol was given 15 minutes prior to these tests, and the pups removed from the cage. On the two test days, either a large (400–450 g) or a small (250–290 g) male rat was placed in the females's cage for a 30–min test. Order of presentation of the two stimulus sizes was balanced within the different groups.

Frequencies of offensive behaviours (lateral attack, chase, on–top–of) were analyzed during successive 5–min blocks of the 30–min test, for tests made with small or large male rats (fig. 2 and 3).

Statistical analysis indicated that both drug ($\underline{F}$ (2,33) = 5.04, $\underline{p}$<.02), and intruder size ($\underline{F}$ (1,66) = 13.17, $\underline{p}$<.002) effects were reliable. The interaction of these effects was also reliable ($\underline{F}$ (2,66) = 6.88, $\underline{p}$<.005). Additional analysis indicated that drug levels produced a reliable effect for tests involving small intruders ($\underline{F}$ (2,33) = 6.91 $\underline{p}$<.005), but not for large intruder tests ($\underline{F}$ (2,33) = 1.89, $\underline{p}$<.05). For the small intruder tests only, aggression for the 0.5 g/kg group was reliably higher than for the 1.2 g/kg group, but were not significantly lower than those of the saline control group ($\underline{t}$ (33) = 1.22, $\underline{p}$.>05).

Thus, while all groups showed somewhat higher offence toward the smaller males (see also Erskine et al. 1978), this difference was exaggerated for the 0.5 g/kg group. This dramatic and reliable increase suggests that ethanol enhancement of offence may be most apparent in situations in which the motivation to attack is just balanced by opposing or inhibitory factors, rather than when inhibition is much stronger. However, the very high level of variability in this phenomenon additionally suggests that some animals may be more susceptible to the effects of ethanol here than are others.

The effect of time (successive 10–min. blocks in the 30 min. test situation) was also reliable ($\underline{F}$ (2,66) = 3.42, $\underline{p}$<.05), as was the time by drug level interaction ($\underline{F}$ (4,66) = 2.67, $\underline{p}$<.05). For the first 10–min. block only, aggression scores for the 0.5 g/kg group were reliably higher than those for the saline group ($\underline{t}$ (33) = 3.03, $\underline{p}$<.005), while the saline group was higher, but not reliably so, than the 1.2 g/kg group ($\underline{t}$ (33) = 1.09, $\underline{p}$>.05).

These data suggest that offence was very differently distributed over time for females of the saline and 0.5 g/kg ethanol groups: Females receiving 0.5 g/kg ethanol showed significantly higher offence on the introduction of the male intruder (especially the small one), than did the saline controls: However, offence for this group declined over time relative to its initial high level. In contrast, the saline injected animals showed their highest levels of offence near the end of the 30–min session. Animals receiving 1.2 g/kg ethanol displayed very few offensive behaviours at any time. These findings suggest that the observation that rats given low ethanol doses show an increase in aggression is especially likely when the control–ethanol difference is maximal just after confrontation with the offence–eliciting stimulus, thus showing up most strongly in short–duration tests.

A TEST OF THE DISINHIBITION HYPOTHESIS IN MALE RATS: ETHANOL EFFECTS ON OFFENCE AND FREEZING AFTER EXPOSURE TO A CAT

An important view of the mechanism of ethanol effects on aggression is that ethanol increases overt aggressive behaviour by disinhibiting fear (Higgins and Marlatt 1973).

If fear can be reduced by ethanol, thus "releasing" aggression, then an ethanol enhancement of aggression should be very clear in situations in which fear levels

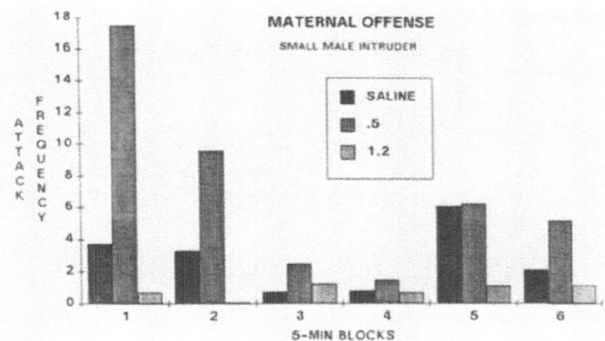

Fig. 2   Attack frequencies during successive 5–minute blocks of a 30–minute test session, for post–parturitional female rats confronted by a small male intruder into their nesting cages, as a function of ethanol dose level.

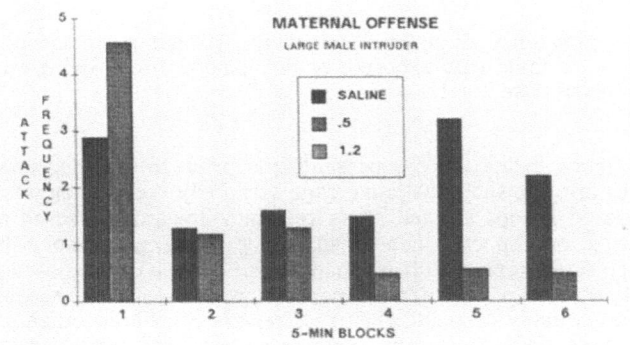

Fig. 3   Attack frequencies during successive 5–minute bloks of a 30–minute test session, for post–parturitional female rats confronted by a large male intruder into their nesting cages, as a function of ethanol dose level.

are experimentally manipulated and this manipulation reduces offence. Brief exposure of a dominant rat to a cat just prior to confrontation with a strange male intruder has been demonstrated to reduce offence to the intruder (Blanchard et al. 1984). Cat presentation was therefore used in combination with saline control or differing ethanol levels, to examine the interaction of these factors with reference to the subject's aggression toward a conspecific.

In this procedure, 2 randomly selected groups of male rats were housed with females and given 4 pretest sessions with intruder males. On the test days they were injected IP with saline, 0.3, 0.6 or 1.2 g/kg ethanol before being placed in their home cage (a 37 x 21 x 18 cm plastic pan cage) into a cat compartment and exposed for 2 min to either a live, 4.5 kg cat or to a stuffed toy cat of approximately the same size and colour. Thirty seconds after the stimulus cat was removed, a 10–15 % smaller male conspecific was placed in the subject's cage for a

150

30 min test session. Offence to the male intruder and freezing were both scored from video tape records of these sessions.

Figure 4 presents offence frequencies and bite latencies for the 2 groups under varying ethanol levels.

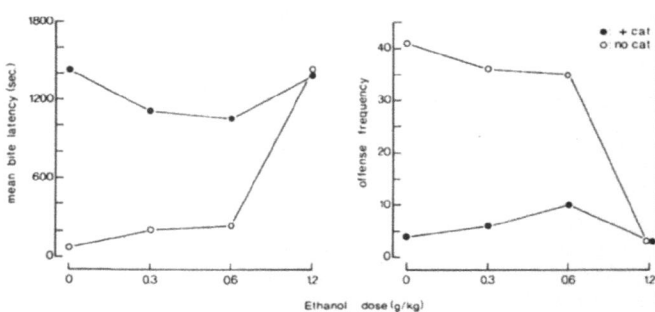

Fig. 4   Mean frequency of offence (right panel), and mean latency to bite (left panel) for male rats exposed, or not exposed, to a living cat, as a function of ethanol dose level.

As this figure indicates, cat presentation prior to introduction of the strange male intruder into the subject's home cage strikingly reduced aggressive behaviour. The cat exposed groups showed some tendency toward increased offence (lateral attack, chasing, on top of) when given 0.3 or 0.6 g/kg ethanol. The bite latency data also suggested some specific enhancement for the cat–exposed groups at these doses. However, none of the offence increases were statistically significant and the behaviour was reliably reduced when rats received 1.2 g/kg ethanol. In contrast to the lack of a reliable change in offence at low and intermediate dose levels, increasing doses of ethanol potently reduced freezing in the cat–exposed rats (fig. 5).

The differences in magnitude of the effects on freezing and offence, plus the poor correlations obtained for the two for individual animals at most ethanol levels, suggest that a general reduction in fear cannot account for any ethanol potentiation of offensive behaviours.

A TEST OF THE DISINHIBITION HYPOTHESIS IN MALE RATS: ETHANOL EFFECTS ON OFFENCE TO A STRANGER IN A NOVEL ARENA

The cat may have been too potent a fear–eliciting stimulus to use in demonstrating ethanol enhancement of offence. A second study was therefore added in which the fear–eliciting stimulus (a neutral arena) was much milder. We felt this would also be useful to provide a comparison with previous work, since the low–dose ethanol enhancement of offensive aggression has been more often demonstrated in a new or neutral arena (Krsiak 1976; Miczek and O'Donnell 1980). Thirty–three adult male Long–Evans rats, housed for a week with females, were randomly divided into two groups receiving an injection of either saline, or 0.5 g/kg ethanol before confronting a somewhat smaller intruder in a neutral arena.

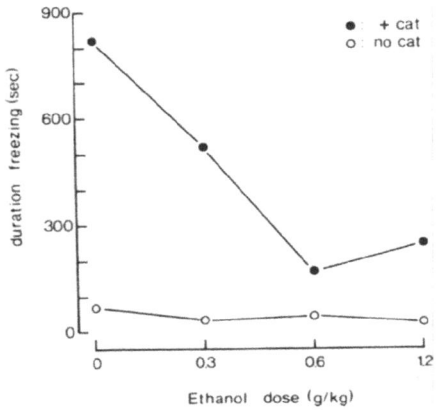

Fig. 5. Mean duration of freezing (sec.) for rats exposed, or not exposed, to a cat, as a function of ethanol dose level.

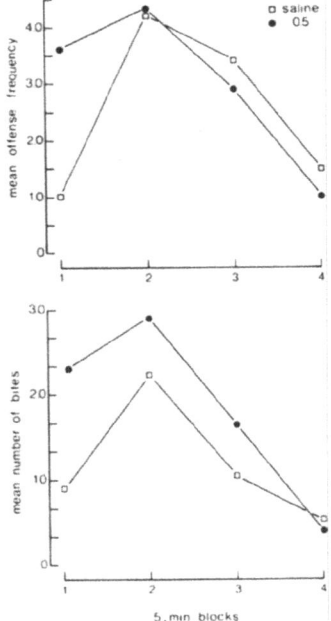

Fig. 6 Mean offence frequency (upper graph) and number of bites (lower graph) for male rats confronted by a slightly smaller male stranger in a neutral arena, over 5–min blocks in the 20–min test period, as a function of ethanol dose level.

Figure 6 presents the distribution of offensive behaviours and bites across blocks of time in the test session. As with the maternal aggression study, these data suggest a time–based difference between the saline and ethanol groups, with the control animals showing little offence in the first 5 minutes in the neutral arena, in contrast to the high offence scores of the ethanol–drugged rats. However, as in virtually all the ethanol–offence work done in this laboratory, the neutral arena study was characterized by unusual intersubject variability. Neither the overall tests, nor tests involving only the initial 5–min block showed statistically reliable increases in offence at the 0.5 g/kg ethanol dose.

ETHANOL EFFECTS ON ANIMALS DIFFERING IN INTENSITY OF OFFENSIVE ATTACK

The variability in the previous study suggests that the tendency of some subjects, to show an ethanol potentiation of attack, might be associated with pre–existing subject characteristics. Krsiak (1976) and Miczek and Barry (1977) have also suggested that the enhancement of offence at low doses of ethanol may be dependent on the subject's normal level of such behaviour. This led to two attempts to demonstrate such a rate–dependency of the ethanol effect.

An initial pilot study involved selection of subjects on the basis of bites made on an opponent in a preliminary test. This produced no substantial evidence that bites, in these relatively naive subjects, were predictive of the effects of ethanol. In fact, such bites were also rather poorly correlated with initial attack behaviours (lateral attack, chasing, and on–top–of) which we (Blanchard and Blanchard 1977; 1984) have analyzed as strategic movements enabling the attacker to position itself for a successful bite. In highly experienced animals, there is a clear relationship among the strategic and the terminal components of this pattern, but this coordination appears to depend to a large degree on fighting experience.

The study was repeated, using the initial, and more common, components of the attack pattern (frequency of lateral attack, chase, and on–top–of) as the group selection criterion. Thirty seven adult male Long–Evans rats served as subject. These were given two preliminary bouts to provide some agonistic experience, and ther were observed in four 10–min tests with naive male intruders placed into the subject's home cage. Intraperitoneal injections (0, 0.3, 0.6, or 1.2 g/kg ethanol) were given 30 minutes prior to each session, with order of drug dose level counterbalanced. The subject's data was divided into three groups on the basis of the combined frequencies of lateral attack, chase, and on–top–of behaviours during the 10 min saline test; nonaggressive (0 frequency), low to intermediate (frequency of 1 through 15 defensive acts), or, high aggressive (more than 15). Since the non aggression (0) group could only change in the direction of becoming more aggressive, its scores were examined separately: Essentially, this group showed no aggression with any ethanol dose level. Change scores (dose level score minus saline score) were analyzed for the two remaining groups (fig. 7).

The high aggressive group tended to show less aggression under all ethanol doses, but the low to intermediate aggression group showed a biphasic effect, with increased aggression under 0.3 and 0.6 g/kg ethanol and a decrease with the highest (1.2 g/kg) dose. Change scores (subject score at a given level minus subject score under saline) at the different levels were significantly different for the two groups $(F (1,27) = 23.50, p<.001)$. The dose level effect was also significant $(F (2,54) = 11.74, p<.001)$. The interaction of group and dose level was not significant $(F (2,54) = .17, p<.05)$. Individual t's for correlated measures indicated that, while the low to intermediate group's increases at the 0.3 and 0.6 levels were statistically significant $(\underline{t} (16) = 2.40$ and 2.22, respectively, $p<.05$ for each), the high aggressive group's lower scores at these same levels approached but failed to reach statistical

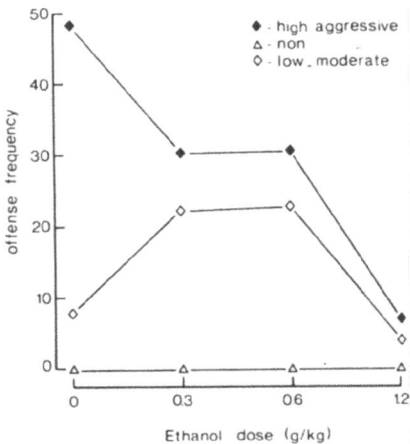

Fig. 7 Mean offence frequency for non aggressive, low to moderately aggressive, and highly aggressive (under saline) rats, as a function of ethanol dose level.

significance ($\underline{t}$ (11) = 2.10 for the 0 – 0.3 group comparison, and 1.83 for the 0 – 0.6 group comparison, .05 <p<.10 for each). The scores for both groups under 1.2 g/kg ethanol were significantly lower than their saline scores (Wilcoxon T = 12 and 17, p<.001 and p<.05, respectively for the high aggressive and low to intermediate aggressive groups).

Analysis of variance on the offence duration change scores indicated that the group difference was significant ($\underline{F}$ (1, 27) = 11.99, p<.001), as was the dose level effect ($\underline{F}$ (2, 54) = 14.08), but not the interaction of group and dose level ($\underline{F}$ (2, 54) = .25, p<.05), Subsequent $\underline{t}$'s indicated that the decreases in duration of offence for the high aggressive group at the 0.3 and 0.6 dose levels were not significant ($\underline{t}$ (11) = .71 and .88, respectively, p>.05), but that the decline seen with the 1.2 g/kg dose was significant (T (12) = 0, p<.001). For the low–intermediate aggressive group, the increases seen at the 0.3 and the 0.6 dose levels were statistically significant ($\underline{t}$ (16) = 2.22 and 2.54, respectively, p<.05 for each). The decline seen for this group at the 1.2 dose level was also significant (Wilcoxon T (17) = 27, p<.01). A similar, though nonsignificant, pattern was obtained for biting.

This study, then, suggests a robust and consistent enhancement of offence by low to moderate doses of ethanol, but, only for those animals initially displaying low to moderate offence tendencies. The patterns shown by nonaggressive, or, highly aggressive animals may be different, or even opposite to, this effect. These differences provide an obvious explanation for the extreme variability of the ethanol enhancement effect, and the consequent difficulty of obtaining a statistically reliable enhancement of offence by ethanol in studies using unselected animals.

In fact, the differences among various components of the attack pattern (i.e. bites vs. frequency of strategic attack movements), in terms of relative dependency on individual experience, suggest that care must be used in selecting the behavioural criterion for different levels of offence, at least when using subjects who are not highly experienced fighters.

EFFECTS OF ETHANOL ON AGGRESSIVE BEHAVIOUR DURING GROUP FORMATION

Further studies in this series examined ethanol effects in the extended colony situation. Colony studies add a new level of complexity, and potential relevance to human behaviour to the study of ethanol effects on aggression. This situation permits the analysis of attack by treated males on untreated males and on females, as well as attack by untreated animals on males receiving different ethanol doses, on untreated males, or on females. It also permits analysis of the interaction of aggression with other social behaviours toward both familiar and unfamiliar animals.

In an initial study we measured ethanol effects on offence by male rats in social groups. In order to avoid the additional complexity of a preexisting social order, this first study involved offence during the formation of colony groups. While this situation does not involve previous mutual social experience of the animals, it affords relatively high levels of intermale fighting. The subjects were 102 male and 102 female adult Long–Evans rats, placed 6 (3♂, 3♀) to a colony enclosure. A single, randomly selected, male in each colony was given either the control (saline) or on ethanol dose, IP, 20 min prior to placement of all animals in the colonies. Of the 34 colonies used, 10 received saline and 8 each received 0.3, 0,6, or 1.2 g/kg ethanol, given to only a single male of the group. Aggressive and sexual behaviours were scored from videotapes made during the first 6 hours after group formation.

The overall frequency of offence declined systematically and significantly over increasing ethanol doses in treated colony males showing no tendency toward enhancement at lower doses. However, in terms of saline – individual dose comparisons, the decline was statistically significant only for the 1.2 g/kg group. Another, and very intriguing, change was in the target of these attacks (Table 1).

Table 1   Colony Formation Study Percentages of Offence and Bites toward Females By the Treated Males of Each Dose–Level Group

Group (g/kg IP)	Percentage offence	Percentage Bites
0 g/kg ethanol	23.1	22.7
0.3 g/kg ethanol	16.2	24.0
0.6 g/kg ethanol	51.1	35.7
1.2 g/kg ethanol	59.8	66.7

Each treated male could either attack the other 2 (untreated) colony males, or, the 3 females. Saline treated males and the 0.3 group attacked females less than one–quarter as often as they attacked other males. In the 0.6 and the 1.2 g/kg groups, however, the proportion of attack directed at females rose to over 51%, and 59%, respectively. This increase in relative frequency of female attack does not appear to be associated with increased sexual activity, since the most sexually active group, those receiving 0.3 g/kg ethanol, showed the lowest proportion of female attacks.

Thus, while this study did not provide any indication of an overall enhancement of offensive attack at any of the ethanol doses used, it did suggest a relative increase in a form of offence which is normally inhibited in male rats, namely attack on females. Previous reports (Miczek et al. 1984) indicate that ethanol dosed animals are the targets of enhanced attack from nontreated animals. In this study, such an effect was not apparent. It should be noted, however, that the present ethanol levels were lower than those previously shown (1.7 g/kg) to reliably produce this effect in rats.

EFFECTS OF ETHANOL ON ATTACK TOWARD DIFFERENT TARGETS BY ANIMALS IN ESTABLISHED COLONIES

Another study investigated ethanol effects on social interactions among animals of different age and sex classification in established colonies as well as attack toward different age/sex classes of unfamiliar intruders. This long–term, data–intensive study is still in the process of analyses, but several specific analyses can be briefly described here. The colonies of this study were maintained in a visible burrow system in which a series of Plexiglass burrows are provided in addition to an open space. Previous work indicates that this system promotes the rapid development of a strong dominance hierarchy among male rats of mixed–sex groups (Blanchard et al. 1985). A series of intruders including adult males, adult females and juvenile males were placed individually into such burrow system habitats while the habitat animals had all been given either 0, 0.5 or 1.2 g/kg ethanol. Offensive attacks and bites were scored from videotape records in terms of the targets of attack; colony animals or intruders, divided in each case into sex/maturity classifications.

The analyses completed for this study suggest that the only significant enhancement of attack seen toward any specific target, was toward colony females. Frequency of offence to colony females by colony members was reliably higher under 0.5 as opposed to 0 g/kg ethanol, and bite frequencies followed the same pattern although these were not significant. Colony males also showed some enhanced tendency to attack other colony males under 0.5 g/kg ethanol, and male attack toward male and female colony animals combined, was significantly higher under 0.5 g/kg ethanol as opposed to saline.

While the intruder attack data provided several interesting new findings, such as much higher attack toward juvenile intruders by colony females as opposed to attacks by males, there was no evidence of an ethanol–related increase in attack toward any non–colony target, male, juvenile, or female. This pattern is particularly interesting since it confirms the finding of the previous study showing females may be an especial target of offence by intoxicated males. These data also suggest that an ethanol enhancement of offence may be generally more visible among familiar animals than toward strangers, again supporting a view that some form of relatively mild inhibition is involved in this phenomenon.

FORCED ETHANOL CONSUMPTION IN ESTABLISHED COLONIES

A final study using social groups, involved use of the liquid Lieber–DeCarli (1982) diet to force consumption of ethanol as part of the regular diet. The 5 colonies used each consisted of 3 male and 3 female adult rats in a standard 9m x 9m enclosure. The animals were left undisturbed except for routine maintenance, for 75 days prior to the use of the liquid diet. A baseline test of within–colony agonistic interactions was videotaped immediately prior to removal of normal food and water and substitution of the 5% ethanol Lieber–DeCarli diet. The three–hour videorecording procedure was repeated 24 and 72 hours later, while the ethanol–containing diet was the only one available, and, during withdrawal on day 5, immediately after the liquid diet was replaced by standard laboratory chow and water. Figure 8 presents the mean frequency of lateral attack, chase and on–top–of behaviours during baseline, ethanol, and withdrawal tests.

There was a substantial increase in offence from baseline to the mean of the two days of forced ethanol ingestion, and a drop during withdrawal. The increase from baseline to drug test was statistically highly reliable. Mortalities among colony rats were also noted. These mortality data add to the view that abnormally high levels of offence are associated with ethanol ingestion. Two subordinate males died during the four ethanol days and two others died in the next three weeks. As a point of comparison, two colony males had died in the 75 days between colony formation and the ethanol tests (note also that the Lieber–DeCarli diet, used for longer periods with a number of pregnant females, has produced no female mortality in this laboratory).

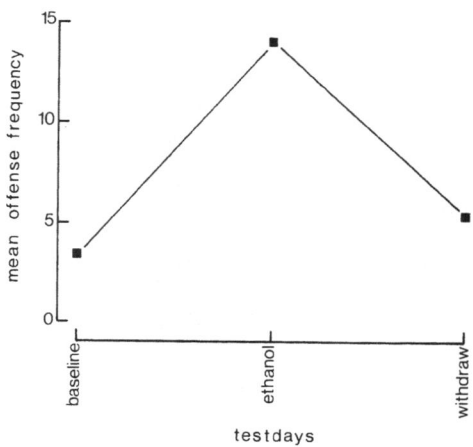

Fig. 8   Mean offence frequency in mixed–sex colonies during a baseline period, in tests while the only diet contained 5% ethanol, and during withdrawal from the ethanol diet.

Since deaths of subordinate males appear to specifically reflect social stress associated with high levels of offence by the dominant male (Blanchard et al. 1985), this apparent increase in subordinate male mortality is consonant with the finding of higher levels of offence within the colonies during ethanol ingestion.

The data from this study are interesting in a number of respects. First, they indicate that such factors as situational novelty, and encounters with a strange conspecific, such as occur in many aggression tests, are not required for ethanol enhancement of attack. This may also be seen in the findings of the previous study, suggesting a greater ethanol enhancement of attack for colony, as opposed to intruder, animals. Moreover, since the colonies were not disturbed for filming and no new conditions were imposed at the start of the tests, the finding that ethanol treated animals show an initial burst of offence on confrontation with an offence–eliciting situation, when controls do not (see also earlier studies), cannot be a major factor in the ethanol enhancement of attack seen in this study. Moreover, this study involve neither selected animals nor a large number of subjects, yet the differences were highly reliable.

Several possibilities are suggested: a) experienced animals may show less variability in offence; b) ethanol enhancement of attack may be more visible among long–term cohabitants because such animals do apparently develop inhibitions involving attacking each other and c) oral and/or chronic ingestion may have different, more powerful, or more specific effects than IP injection. For example, there is evidence that chronic, or repeated, ethanol intake increases biting in a tube–restraint situation (Tramill et al. 1981; Tramill et al. 1983; Weitz 1974), whereas acute ethanol in the same task, does not (Al–Hazmi and Brain 1984; Smoothy and Berry 1984; Tramill et al. 1980; Tramill et al. 1983). This study thus suggests many possibilities for further research.

DISCUSSION

These findings suggest that when important features of the normal social and physical environment are modelled for rodent groups, the particular subject, situational, and target variables which influence the alcohol–aggression link are more specific than is generally believed, and may provide striking parallels to the factors influencing alcohol and aggression in humans. First, these results suggest that the effects of ethanol on defence and the defensive emotions (fear, anxiety) are not global, but differ for specific components of this pattern: alcohol increases defensive threat and attack but either fails to change (wild rats) or reduces freezing/movement arrest aspects of defence. This last effect may perhaps be involved in a linkage between alcohol and impulsive behaviours. Both chronic early alcohol ingestion (Gold and Sherry 1984) and acute intake (Schuckit 1973) may lead to impulse–control problems associated with human aggression.

A second implication of these studies is that alcohol potentiation of aggression occurs selectively in only a subgroup of male rats; those showing low to intermediate levels of pre–drug aggression. Similarly, some human males are consistently nonaggressive when drinking, while others are consistently hostile, argumentative, or rowdy (Robbins 1979). Bailey (1982) found that human males showing low levels of aggression before consuming alcohol showed an increase after drinking, while high aggressive subjects tended to be unchanged after consuming alcohol.

Third, these studies suggest that alcohol ingestion may selectively potentiate aggression targeted toward familiar females. Although there is no direct evidence available of a specific targeting of aggression toward women by intoxicated men, there is considerable evidence that spouse abuse is strongly linked to alcohol ingestion (Gelles 1972; Straus and Hotaling 1980; Straus et al. 1980) and that many wife–abusing alcoholics do not offend while sober (Corenblum 1983). In fact the human parallel may be even more specific: On a U.S. country–wide basis, about 35% of female homicide victims are the wives, mothers, sisters, daughters or girlfriends of their killers. Only about 11% of male homicide victims are similarly related to their killers (Sourcebook of Criminal Justice Statistics 1983). Homicide is a crime very often committed under the influence of ethanol: Estimates in the United States consistently suggest that about 60% of the perpetrators of homicide have been drinking at the time of the crime (Shupe 1954; Voss and Hepburn 1968; Wolfgang and Strohm 1956). Taken together, these two sets of statistics suggest that ethanol may selectively increase aggression directed toward familiar females.

The view that alcohol reduces some emotional/inhibitory mechanism such as anxiety or tension has long been viewed as important in the etiology of drinking behaviour (Conger 1956; Horton 1943; Kingham 1958) as well as in "releasing" aggression (Pernanen 1976). Taken together the studies described here provide results which are readily interpretable in terms of a disinhibition mechanism mediating the effects of alcohol on aggression. Selective ethanol effects on defence, on the timing of aggression in response to aggression–eliciting stimuli, on attack toward specific targets for which some mild attack inhibition is plausible or in subjects normally displaying reduced attack levels, all contribute to this view.

Nevertheless, these studies, and others currently underway in other laboratories, provide only a beginning in terms of specification of this mechanism.

Moreover, since the details of the link between alcohol and aggression outlined in these studies appear to be strikingly similar to aspects of human alcohol–aggression phencmena, it seems especially important that further work on this mechanism be done. These studies have demonstrated that the use of specific, extensively analyzed animal models can provide contexts which allow experimental manipulation of important variables in the analysis of the mediating mechanisms for these relationships. We hope that further use of these models will provide a greatly expanded basis for understanding of the complex relationships between alcohol and aggression in mammals.

Acknowledgement: The authors would like to express their thanks to Dr. Kevin Flannelly, who participated in the design and running of many of these studies. This research was supported by NIH grant AA06220.

REFERENCES

Adams DB (1980) Motivational systems of agonistic behavior in muroid rodents. A comparative review and neural model. Aggr Behav 6: 295–346

Adams DB, Shoel WM (1981) Motor patterns and motivational systems of social behavior in male rats and stumptail macaques—Are they homologous? Aggr Behav 7: 267–280

Al–Hazmi M, Brain PF (1984) Effects of age, habituation and alcohol administration on tube restraint–induced attack by Swiss mice. Aggr Behav 10: 145

Bailey L (1982) Differential effects of alcohol consumption on aggression in a reaction time competition task for high and low aggressive subjects. Unpublished Master's Thesis, Kent State University, Kent, Ohio

Blanchard RJ (1984) Pain and aggression reconsidered. In: Flannelly KJ, Blanchard RJ, Blanchard DC (eds) Biological Perspectives on Aggression, Alan R Liss Inc, New York, pp 1–26

Blanchard RJ, Blanchard DC (1977) Aggressive behavior in the rat. Behav Biol 21: 197–224

Blanchard RJ, Blanchard DC (1981) The organization and modelling of animal aggression. In: Brain PF, Benton D (eds) The Biology of Aggression, Sijthoff/Noordhoff, Alphen aan den Rijn, pp 529–561

Blanchard DC, Blanchard RJ (1984) Affect and aggression: An animal model applied to human behavior. In: Blanchard RJ, Blanchard DC (eds) Advances in the Study of Aggression, Vol I, Academic Press, New York, pp 1–62

Blanchard RJ, Blanchard DC, Flannelly KJ (1985) Social stress, mortality and aggression in colonies and burrowing habitats. Behav Processes 11: 209–213

Blanchard DC, Blanchard RJ, Lee EMC, Williams G (1981a) Taming in the wild Norway rat following lesions in the basal ganglia. Physiol Behav 27: 995–1000

Blanchard RJ, Blanchard DC, Takahashi T, Kelley MJ (1977) Attack and defensive behaviour in the albino rat. Anim Behav 25: 622–634

Blanchard DC, Blanchard RJ, Takahashi T, Suzuki N (1978) Aggressive behaviors of the Japanese Brown Bear. Aggr Behav 4: 31–41

Blanchard RJ, Flannelly KJ, Blanchard DC (1986) Defensive behaviors of laboratory and wild Rattus Norvegicus. J Comp Psychol 100: 101–107

Blanchard DC, Fukunaga–Stinson C, Takahashi LK, Flannelly KJ, Blanchard RJ (1984a) Dominance and aggression in social groups of male and female rats. Behav Processes 9: 31–48

Blanchard RJ, Kleinschmidt CK, Flannelly KJ, Blanchard DC (1984b) Fear and aggression in the rat. Aggr Behav 10: 309–315

Blanchard DC, Williams G, Lee EMC, Blanchard RJ (1981b) Taming of wild Rattus Norvegicus by lesions of the mesencephalic gray. Physiol Psychol 9: 157–163

Borgesova M, Kadlecova O, Krsiak M (1971) Behavior of untreated mice to alcohol– or chlordiazepoxide–treated partners. Act Nerv Sup 13: 206–207

Bovier PH, Broekkamp CL, Lloyd KG (1984) Ethyl alcohol on escape from electrical periaqueductal gray stimulation in rats. Pharmacol Biochem Behav 21: 353–356

Brain PF (1981) Differentiating types of attack and defence in rodents. In: Brain PF, Benton D (eds) Multidisciplinary Approaches to Aggression Research. Elsevier/North–Holland Biomedical Press, Amsterdam, pp 53–77

Conger J (1956) Alcoholism: theory, problem and challenge – II. Reinforcement theory and the dynamics of alcoholism. Quart J Stud Alc 17: 296–305

Corenblum B (1983) Reactions to alcohol–related marital violence. J Stud Alc 44: 665–674

Erskine MS, Denenberg VH, Goldman BC (1978) Aggression in the lactating rat: effects of intruder age and test arena. Behav Biol 23: 52–66

Geist V (1971) Mountain Sheep: A Study in Behavior and Evolution. Chicago: University of Chicago Press

Gelles R (1972) The Violent Home. Beverly Hills: Sage

Gold S, Sherry L (1984) Hyperactivity, learning disabilities and alcohol. J Learn Disab 17: 3–6

Higgins R, Marlatt G (1973) The effects of anxiety arousal upon the consumption of alcohol by alcoholics and social drinkers. J Cons Clin Psychol 41: 426–433

Horton D (1943) The functions of alcohol in primitive societies. A crosscultural study. Quart J Stud Alc. 4: 292–303

Hutchinson RR (1983) The pain–aggression relationship and its expression in natural settings. Aggr Behav 9: 229–243

Jeavons CM, Taylor SP (1985) The control of alcohol–related aggression. Redirecting the inebriate's attention to socially appropriate conduct. Aggr Behav 11: 93–102

Kingham R (1958) Alcoholism and the reinforcement theory of learning. Quart J Stud Alc 19: 320–330

Krsiak M (1976) Effect of ethanol on aggression and timidity in mice. Psychopharmacology 52: 75–80

Krsiak M, Borgesova M (1973) Effect of alcohol on behavior of pairs of rats. Psychopharmacologia 32: 201–208

Levinson D (1981) Alcohol use and aggression in American subcultures. In: Room R, Collins G (eds) Alcohol and disinhibition: Nature and Meaning of the Link. Research Monograph–12 National Institute on Alcohol Abuse and Alcoholism, Washington, D.C., U.S. Government Printing Office

Leyhausen P (1979) Cat behavior: The predatory and social behavior of domestic and wild cats. New York, Garland STPM Press

Lieber CS, DeCarli LM (1982) The feeding of liquid diets: Two decades of applications and 1982 update. Clin Exp Res 6: 523–531

Lore RK, Flannelly KJ (1981) Comparative studies of wild and domestic rats: Some difficulties in isolating the effects of genotype and environment. Aggr Behav 7: 253–257

Miczek KA, Barry H III (1977) Effects of alcohol on attack and defensive–submissive reactions in rats. Psychopharmacology 69: 39–44

Miczek KA, Kruk MR, Olivier B (eds) (1984) Ethopharmacological Aggression Research, New York: Alan R Liss

Miczek KA, O'Donnell JM (1980) Alcohol and chlordiazepoxide increase suppressed aggression in mice. Psychopharmacology 69: 39–44

Miczek KA, Winslow JT, DeBold JF (1984) Heightened aggressive behavior by animals interacting with alcohol–treated conspecifics; Studies with mice, rats and squirrel monkeys. Pharmacol Biochem Behav 20: 349–353

Pernanen K (1976) Alcohol and crimes of violence. In: Kissin B, Begleiter SH (eds) The Biology of Alcoholism: Social Aspects of Alcoholism, Vol 4, New York, Plenum Press

Podhorecky LA (1977) Biphasic action of ethanol. Biobehav Rev 1: 221–240

Robbins R (1979) Alcohol and the identity struggle. Some effects of economic change on interpersonal relation. In: Marshall M (ed) Beliefs, Behaviors and Alcoholic Beverages: A Cross–Cultural Survey. Ann Arbor, University of Michigan Press

Robinson R (1984) Norway rat. In: Mason IL (ed) Evolution of Domesticated Animals, London, Longman

Schuckit MA (1973) Alcoholism and sociopathy–diagnostic confusion. Quart J Stud Alc 35: 157–164

Shupe LM (1954) Alcohol and crimes: A study of the urine alcohol concentration found in 882 persons arrested during or immediately after the commission of a felony. J Crim Law Criminol 44: 661–665

Smoothy R (1985) Videotype analysis of the effects of alcohol on aggressive and other behaviour in laboratory mice (Mus musculus L.) Ph.D.Thesis, University of Wales

Smoothy R, Berry MS (1984) Effects of ethanol on murine aggression assessed by biting of an inanimate target. Psychopharmacology 83: 268–271

Sourcebook of Criminal Justice Statistics 1983 (1984) U.S. Department of Justice, Bureau of Justice Statistics, Washington D.C.

Straus MA, Gelles RJ, Steinmetz SK (1980) Behind Closed Doors: Violence in the American Family. New York, Anchor/Doubleday

Straus MA, Hotaling GT (1980) The Social Causes of Husband–Wife Violence. Minneapolis, University of Minnesota Press

Svare BB, Gandelman R (1973) Postpartum aggression in mice: Experimental and environmental factors. Horm Behav 4: 323–334

Takahashi LK, Blanchard RJ (1982) Attack and defence in laboratory and wild Norway and black rats. Behav Processes 7: 49–63

Tramill JL, Gustavson K, Weaver MS, Moore SA, Davis SF (1983) Shock–elicited aggression as a function of acute and chronic ethanol challenges. J Gen Psychol 109: 53–58

Tramill JL, Turner PE, Sisemore DA, Davis SF (1980) Hungry, drunk and not real mad: the effects of alcohol injections on aggressive responding. Bull Psychon Soc 15: 339–341

Tramill JL, Wesley AL, Davis SF (1981) The effects of chronic ethanol challenges on aggressive responding in rats maintained on a semideprivation diet. Bull Psychon Soc 17: 51–52

Voss HL, Hepburn JL (1968) Patterns in criminal homicide in Chicago. J Crim Law Criminol Police Sci

Weitz MK (1974) Effects of ethanol on shock–elicited fighting behavior in rats. Quart J Stud Alc 35: 953–958

Williams AF (1966) Social drinking, anxiety and depression. J Personal Soc Psychol 3: 689–693

Wolfgang ME, Strohm EB (1956) The relationship between alcohol and criminal homicide. Quart J Stud Alc 17: 411–425

SEROTONERGIC MODULATION OF AGONISTIC BEHAVIOUR

Berend Olivier, Jan Mos, Jan van der Heyden, Jacques Schipper, Martin Tulp, Bas Berkelmans and Paul Bevan.

Dept. of Pharmacology, Duphar B.V., P.O. Box 2, 1380 AA Weesp, The Netherlands.

INTRODUCTION

Over the last fifteen years several hypotheses have emerged concerning the neurochemical control of aggressive behaviour. A variety of single neurotransmitters were suggested to control aggression e.g. the "aggressive monoamines" (Eichelman and Thoa 1973), acetylcholine (Smith et al. 1970) and serotonin (Valzelli and Garattini 1968). Later, the theories of single neurotransmitter control were extended to multi-transmitter modulation of aggressive behaviour (Avis 1974; Daruna 1978; Pradhan 1975; Reis 1974).

Although multiple-transmitter modulation is certainly more likely to be the case for complex sets of behaviours like aggression, many aspects of the important modulatory role of serotonin (5-HT) in several types of agonistic behaviour in animals remain to be elucidated (Miczek 1987; Miczek and Barry 1976; Valzelli 1984) Early work on 5-HT and aggression indicated that a general 5-HT activation decreased aggression whereas an overall inactivation of 5-HT (achieved by various means) led to enhanced aggression (e.g. Valzelli 1981). However, recent evidence indicates a much more complicated pattern.

The anatomical distribution and localization of cell groups of 5-HT in the CNS and their differential projections (cf. Pazos and Palacios 1985; Pazos et al. 1985; Steinbusch 1981) suggest that it is no longer justifiable to speak simply about "the role of 5-HT" in any aggression paradigm. Moreover, the recent differentiations of 5-HT receptor binding sites (Peroutka 1986; Richardson and Engel 1986) and their distinct functional roles further delineate the possibility of a functional discrimination in the 5-HT-system in the CNS with regard to different kinds of agonistic behaviour.

The present contribution first demonstrates the problems in the interpreting 5-HT turnover and aggressive behaviour in two rat strains. Following this, some pharmacological tools currently available to modify 5-HT activity in the brain will be briefly described and their application in animal models exemplified. The use of a variety of animal models to study aggression in different species facilitate a more precise description of the mode of action of 5-HT in aggression.

5-HT TURNOVER AND SPONTANEOUS AGONISTIC BEHAVIOUR IN RATS

Support for the relationship between neural 5-HT activity and aggressive behaviour in animals is not unequivocal. This is largely due to the lack of direct correlations between the spontaneous activity of (subsets of) 5-HT neurones in the CNS and measures of ongoing aggressive behaviour. One part of the available evidence is the correlation between levels of 5-HT and 5-HIAA (or 5-HT-turnover) in certain specific brain areas with some forms of aggressive behaviour (c.f. Daruna 1978). More evidence stems from different pharmacological manipulations of the 5-HT system and the effects on some parameters of aggression (for a review see Miczek 1987). Thus it was observed that depleting 5-HT in the brain, facilitated or elicited various kinds of aggressive behaviour in the rat, e.g. mouse–killing (Applegate 1980; Valzelli et al. 1981; Vergnes and Kempf 1982, 1983) and shock–induced fighting (defence) (Kantak et al. 1981; Knutson et al. 1979; Sheard and Davis 1976). Vergnes et al. (1986) showed in a resident–intruder paradigm, that depletion of 5–HT by PCPA administration enhanced offence when the resident was treated, but had no effect on defensive behaviour of the intruder when the latter was treated. This suggests that 5–HT plays a role in modulating offence, but not defence or submissive behaviour.

Attempts were made in our laboratory to assess the relationship between spontaneous 5–HT turnover in the CNS and different forms of aggressive behaviour. The purpose of the experiments was to determine whether rat strains with different propensities to both kill mice and engage in more intensive agonistic interactions with conspecifics, varied in terms of their 5–HT activity. As a measure of 5–HT activity, the ratio between 5–HIAA/5–HT was measured, giving a good indication of the real 5–HT turnover (Shannon et al. 1986). Two rat strains were used, the TMD–S3 and the Wistar. Animals were housed singly for 4 weeks and tested once weekly for mouse killing and aggression towards a naive male intruder. Both strains were compared with regard to 5–HT turnover in certain brain regions and aggressive behaviour (mouse–killing behaviour and resident–intruder aggression). Moreover, within–strain correlations were computed for 5–HT turnover and aggression.

Fig. 1 shows the lack of correlation in individual rats between the level of 5–HT–turnover (in the striatum) and two measures of aggressive behaviour, namely the total time spent on fighting in a resident–intruder paradigm and the killing latency in the muricide data. S3 rats show a significantly higher incidence of mouse killing than Wistar rats (63% vs 13%) but also a higher 5–HT–turnover. This contradicts the general inhibitory role of 5–HT in (predatory) aggression as proposed by Valzelli (1981) and actually suggests a facilitatory role of the neurotransmitter. However, the within–strain analysis shows that there is no such correlation within the S3 strain between muricidal behaviour and 5–HT–turnover. Thus neither the between strains comparison nor the within–strain analysis reveals a definite relationship between this measure of 5–HT turnover and muricidal responses.

In the Resident–Intruder paradigm, no clearcut difference was noted in the time spent on aggression by S3 and Wistar rats against intruders. There was also no correlation between individual 5–HT–turnover and aggression levels in either strain. Although S3 rats did not spend more time on aggression, they were more vigorous fighters as they caused more wounds on the opponents. These data do not indicate a direct relationship between this measure of 5–HT activity and aggression, although in other areas in the CNS such correlations may exist. Other measures of 5–HT turnover, e.g. 5–HIAA accumulation after probenecid treatment may reveal correlations not revealed with the present techniques. Although currently technically impossible, it would be more relevant to measure the actual 5–HT turnover during aggressive interactions.

Another, and perhaps more fruitful way to investigate the role of 5–HT in aggression may be the manipulation of central 5–HT activity with different pharmacological tools. This approach will be elaborated in the remainder of this chapter. Essentially, 5–HT drugs with different mechanisms of action were used and their effects on aggression established using several behavioural paradigms.

NEUROCHEMICAL PROFILE OF 5–HT DRUGS

The following section first provides a short description of the neurochemical profiles of some of the serotonergic drugs used. Three functional parameters are given, namely receptor binding data for $5-HT_{1A}$ and $5-HT_{1B}$ and $5-HT_2$,

164

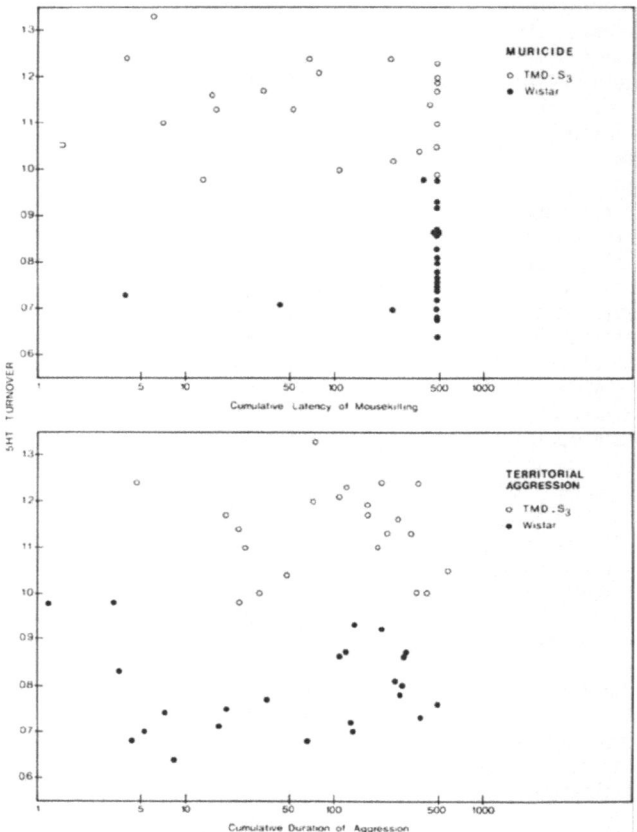

Fig. 1 In the upper panel, the 5–HT turnover in the striatum is plotted versus the cumulative killing latency in 4 consecutive tests for the two strains of rats. Wistar rats show virtually no spontaneous mouse–killing behaviour, therefore no correlation with 5–HT turnover could be computed. S3 rats spontaneously kill mice, but this index is not correlated with 5–HT turnover. In the lower panel, the correlation between 5HIAA/5–HT ratio in the striatum and the % of total time spent on aggression during 4 consecutive intruder tests is plotted for the rat strains. No significant correlation was found between 5–HT turnover and aggression. 5–HT turnover in S3 rats is higher than in Wistars, but the total time spent on aggression did not differ significantly.

5–HT agonistic activity in vitro on the presynaptic autoreceptor and modulation of 5–HT turnover in vivo.

Receptor binding

Several compounds have a high affinity for $5-HT_{1A}$ binding sites relative to $5-HT_{1B}$ or $5-HT_2$ (5–Me–O–DMT, 8–OH–DPAT, buspirone and ipsapirone (TVX–Q 7821)), others have about equal affinity for $5-HT_{1A}$ and $_{1B}$ sites (RU24969, DU 28853), whereas TFMPP and quipazine exert some preference for $5-HT_{1B}$ sites. In general, affinity of these compounds for $5-HT_2$ receptors is relatively low. Of course, several compounds also have affinity for other receptors, but for the sake of simplicity such data is omitted.

Autoreceptor modulation of 5–HT release

5–HT release from nerve terminals is subjected to a negative feedback via presynaptic 5–HT receptors. It has been proposed by various authors (Engel et al. 1986; Middlemiss 1984) that these 5–HT autoreceptors are of the $5-HT_{1B}$ type as defined by radioligand binding studies.

Table 1 Affinity for 5–HT binding sites, in vitro 5–HT release and ED_{50} for the 5–HTP accumulation of some 5–HT agonists and antagonists. n.t.=not tested

	Affinity[1] K_i (nM)			5–HT release in cortex[2]		5–HTP formation[3]
	$5-HT_{1A}$	$5-HT_{1B}$	$5-HT_2$	pD_2	α	ED_{50} mg/kg
5–HT	4.2	4.5	1,300	7.7	1.0	–
RU24969	8.8	8.6	1,500	8.0	0.8	1.0
5–Me–O–DMT	6.3	51	2,300	5.5	0.9	n.t.
TFMPP	200	29	690	7.7	0.7	1.9
DU 28853	37	59	1,700	7.6	0.4	0.3
Fluprazine	510	1,500	1,800	<5.5	0	n.t
8–OH–DPAT	2.5	1,600	1,100	<5.0	0	0.2
Quipazine	1,400	160	1,100	<5.0	0	n.t.
Buspirone	12	5,900	540	<6.0	0	n.t.
Ipsapirone	6.7	3,400	3,200	<6.0	0	n.t.

1 The affinity of several compounds was determined by receptor binding studies according to procedures described by Gozlan et al. (1983) for 5–HT–1A, Sills et al. (1984) for 5–HT–1B and Creese and Snyder (1978) for 5–HT–2 receptors.
2 The effects of putative 5–HT agonists on K^+ (20 mM)–evoked ^3H–5–HT release from rat cortex slices. Neuronal 5–HT uptake was blocked during superfusion by fluvoxamine (10 µmol/l). Agonists were added to the superfusion medium 15 min before stimulation. pD_2 values (nM) and intrinsic activity (α) are calculated from the concentration–response curves.
3 Effects of 5–HT agonists on 5–HTP formation in striatum of rats. Agonists were given 60 min and NSD 1015 (100 mg/kg IP) 30 min before decapitation. 5–HTP levels were determined in striatal tissue by use of HPLC with electrochemical detection. Doses necessary to induce half–maximal inhibition of 5–HTP levels were determined from the dose–response curves. All compounds were given orally except 5–Me–O–DMT and 8–OH–DPAT which were given intraperitoneally.

Several 5–HT agonists inhibit the 5–HT release in vitro as measured by the potassium or electrically–stimulated release of pre–stored (^{3}H) 5–HT from brain slices (Göthert 1980). As indicated by the pD_2 values in table 1, 5–HT, RU24969, 5–Me–O–DMT, TFMPP and DU 28853 act as agonists at the presynapticautoreceptors. Compounds with a high affinity for the 5–HT$_{1A}$ binding site, such as 8–OH–DPAT, buspirone and ipsapirone, do not affect 5–HT release, indicating a lack of agonistic activity on the 5–HT autoreceptor. The same holds true for quipazine. The phenylpiperazines TFMPP and DU 28853 show activity comparable to 5–HT, but their intrinsic activities classify them as partial agonists.

5–HT turnover in vivo

After inhibition of the aromatic–L–amino–acid decarboxylase by NSD 1015, 5–hydroxytryptophan (5–HTP) accumulates in brain regions containing 5–HT terminals. The rate of accumulation of 5–HTP has been used as an index of 5–HT neuronal activity (Carlsson et al. 1972). 5–HT agonists inhibit 5–HTP accumulation and it has been suggested that this is due to a negative feedback mechanism (Arvidsson et al. 1986).

The 5–HT agonists that inhibited 5–HT release in vitro also reduced 5–HTP accumulation in vivo. Interestingly, compounds such as 8–OH–DPAT, buspirone and ipsapirone also reduce 5–HTP accumulation. The in vitro data suggest that it is unlikely that these 5–HT$_{1A}$ compounds reduce 5–HT neuronal activity via presynaptic receptors mediating 5–HT release. It is possible that another presynaptic feedback mechanism exists via 5–HT$_{1A}$ receptors, which modulates synthesis rather than the release mechanism. On the other hand, one can not exclude the possibility that a postsynaptic 5–HT$_{1A}$ receptor is involved, resulting in a "long–loop" feedback mechanism as known for e.g. dopaminergic neurones (Arvidsson et al. 1986). Further research will be necessary to unravel the mechanisms involved in the modulation of 5–HT neuronal activity via 5–HT receptor subtypes.

SEROTONERGIC DRUGS AND AGONISTIC BEHAVIOUR

Isolation–induced aggression in mice

When a male mouse is isolated for some time and subsequently confronted with a male group–housed intruder mouse, intense agonistic interactions may occur, including threat, chasing and biting (see Miczek 1987 for a review of the important variables determining this kind of aggression). Although this paradigm has been suggested to model a psychopathology (Garattini and Valzelli 1981; Valzelli 1973), more recent studies, assessing ethological, pharmacological and endocrinological features strongly suggest that such isolated males also occur naturally; they are highly similar to adult male mice defending their territory (Blanchard et al. 1979; Brain 1975; Crawley et al. 1975; Miczek and O'Donnell 1978).

The behaviour of aggressive isolated males has been considered as offensive (cf. Miczek 1987; Olivier et al. 1986) and effects of drugs can be easily studied in this paradigm, (e.g. Janssen et al. 1960; Malick 1979; Miczek 1987; Olivier and Van Dalen 1982). The effects of 5–HT drugs were studied in two set–ups. First, in a relatively simple screening test, the ED_{50} for inhibition of aggression was determined (according to the methods of Tedeschi et al. 1959).

Although the results given in table 2, show that a substantial number of drugs inhibit aggression, and others do not, it is unclear from these values which behavioural inhibitory mechanisms are involved. Therefore, more detailed ethological studies were performed, which, by taking into account a broader repertoire of behaviours, may indicate more specifically how aggression was reduced.

The methodology used has been described in detail in Olivier and Van Dalen (1982) and Olivier et al. (1986). Briefly, the behaviour of the isolate was recorded using 16 categories to describe the ongoing behaviour (during 5 min confrontations between a 4–8 week isolated male mouse and a group–housed male intruder). This approach permits a direct, simultaneous measurement of drug effects on aggressive and non–aggressive behaviours, revealing the specificity of drug action (cf. Miczek 1987; Miczek and Krsiak 1979). Figure 2 summarizes how six 5–HT drugs affect such behaviour.

Table 2 ED_{50}–values for isolation–induced aggression in male mice. Drugs were given orally 60 min before testing suspended in 1% tragacanth.

Drug	ED_{50} (mg/kg PO)
RU24969	0.7
5–Me–O–DMT	4.2
TFMPP	0.2
DU 28853	0.4
Fluprazine	1.2
8–OH–DPAT	0.3 (IP)
Quipazine	>38
Buspirone	>20
Ipsapirone	>20
Fluvoxamine	70
Fenfluramine	10
Methysergide	>10
Ritanserine	>10

DU 28853 exhibited a nice dose–dependent decrease in aggression with a concomitant increase in social interest, whereas non social activities (mainly exploration) were also somewhat enhanced. It is especially noteworthy that no sedation was evident even at the dose of 20 mg/kg, which is 40 times the ED_{50} of isolation–induced aggression. Although defence seems to be dramatically enhanced, it has to be remembered that under control conditions, defence hardly exists, whereas at doses where aggression no longer exists (1 mg/kg and higher) animals defend themselves against obtrusive intruders. A similar reasoning may hold for avoidance.

A rather similar pattern emerges for fluprazine and TFMPP, (for full descriptions see Olivier et al. 1986). Fluvoxamine, a specific 5–HT reuptake blocker (Claassen et al. 1979) also decreases aggression, but in a less specific way because it is accompanied by decreases in social interest and avoidance. 8–OH–DPAT, a specific 5–HT$_{1A}$–agonist (Middlemiss and Fozard 1983) reduced aggression but also decreased non–social activities somewhat, and reduced defence (at 1 mg/kg) and avoidance. RU 24969, a mixed 5–HT$_{1A/1B}$–agonist (Tricklebank 1985) reduces aggression at the highest dose tested, concomitantly increasing avoidance.

Resident–intruder aggression in rats

Rats may be considered as socially living animals (Barnett 1975; Timmermans 1978) and male rats defend territories (Barnett 1975) against the appearance of unfamiliar intruders. This territorial behaviour can be easily demonstrated under semi-natural laboratory conditions using domesticated and inbred strains of rats (Blanchard and Blanchard 1984; Lehman and Adams 1977; Olivier 1977). Introduction of a strange male in such territories evokes a complete agonistic repertoire, similar to that occurring under natural circumstances (Barnett 1975; Timmermans 1978). Therefore, the resident–intruder paradigm has major advantages in this species over more artificial paradigms involving e.g. isolated housing, restraint or footshock because it relies on the basic natural social structure. The occurrence of a rich, more natural behavioural repertoire is used in the ethopharmacological approach. It is possible by recording a wide variety of behavioural elements to detect very specific changes and gain information about the mechanism of action of drugs (Miczek and Krsiak 1981; Miczek and Winslow 1986; Olivier et al. 1984a, 1986a,b).

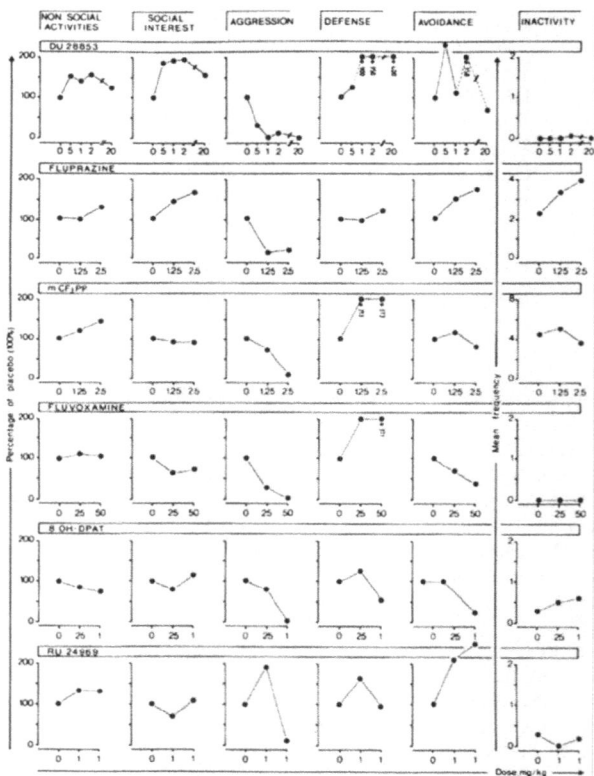

Fig. 2 Effects of DU 28853, fluprazine, metaCF$_3$phenyl piperazine (mCF$_3$PP or TFMPP), fluvoxamine, 8–OH–DPAT and RU 24969 on six behavioural categories in intermale aggression in mice.

Details of the methods are provided in Olivier (1981) and Olivier et al. (1984a). Briefly, male rats (residents) living together with a female in a large territory cage, were confronted during 10 min with a strange male intruder. Behaviour was recorded using ethograms as described before (Olivier 1981). Drugs were given 30 min (IP) or 60 min (PO) before testing. Animals were tested once a week and treatments (vehicle and 3 doses of a drug) were randomized according to a Latin square design.

The results for DU 28853, TFMPP, 5–Me–O–DMT, fluvoxamine and buspirone are shown in fig. 3. The behavioural effects are shown on four main behavioural categories: aggression, social interest, exploration and inactivity. Each category comprises several behavioural elements (cf. Olivier et al. 1984a, 1986) and adequately reflects the general effects of drugs on (social) behaviour. DU 28853 dose–dependently reduces aggression without a concomitant decrease in social interest. Exploration is not changed or even somewhat enhanced. Although inactivity is not significantly enhanced (at the highest dose) there is a tendency for this aspect to increase, presumably indicating that it replaces the time normally spent on aggression. A similar pattern occurs after TFMPP and (not shown) fluprazine. 5–Me–O–DMT reduces aggression in a nonspecific way, indicated by a simultaneous decrease in social interest and an increase in inactivity. Fluvoxamine has no dramatic effects at the doses used but the general pattern indicates a non–specific inhibitory effect on aggression. Buspirone very nonspecifically reduces aggression as the compound is extremely sedative (cf. Olivier et al. 1984a) in such tests.

Maternal aggression in rats

Lactating females of almost every Mammalian species defend their pups and nests against threatening objects e.g. strange intruders. Lactating female rats attack male conspecific intruders with very short latencies and with a high intensity form of attack, predominantly targeted at the head and upper back (Olivier and Mos 1986a; Olivier et al. 1985). Fig. 4 shows the course of aggression over the postpartum lactation period, measured by the number of bite attacks and attack latency.

The lactation period 3–12 days after birth appears a relatively stable time to perform aggression tests using the females as their own controls (Olivier et al. 1985, 1986; Olivier and Mos 1986a; Mos and Olivier 1986). Furthermore, a test duration of five minutes suffices, because after that period aggression stabilizes at a low level (fig. 4C). The behavioural structure of maternal aggressive behaviour is shown in fig. 5. It appeared that this structure did not change over time during day 3 to 12. This structure illustrates the purely offensive motivation of the female. All her behaviour is oriented towards the intruder and her behaviour does not strongly depend on the qualities (male, female or castrated) of the strange intruder or its behaviour (Mos et al. 1987). This suggests, contrary to earlier suggestions (Svare and Mann 1983) that maternal aggression is one of the purest forms of offensive behaviour (cf. Van der Poel et al. 1984), although its primary function may be conceived as defence of the young.

Several drugs were tested using this paradigm, during test periods of 5 minutes. Each female was repeatedly tested on alternate days between post–partum days 3 to 12. Testing occurred every other day with vehicle or drug (several doses) alternated with "wash–out" days. For the sake of comparison the mean number of bite attacks/minute is presented, a measure which is corrected for the latency to the first attack (fig. 6).

This representation of the data shows that most serotonin–modulating drugs tested thusfar inhibit aggression, although there are vast differences between drugs. 8–OH–DPAT, a specific $5-HT_{1A}$ agonist (table 1) has a somewhat peculiar

170

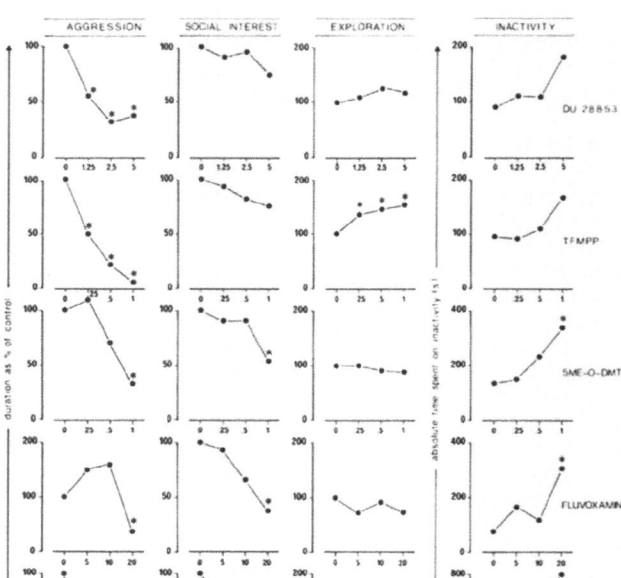

RESIDENT INTRUDER AGGRESSION

Fig. 3   The effects of DU 28853, TFMPP, 5–Me–O–DMT, fluvoxamine and buspirone on four behavioural categories in resident–intruder aggression in rats.

profile in that aggression suddenly decreases without a clear dose–response distribution. Behaviourally this inhibition is also manifest in a sudden drop in interest for the intruder and a high increase in inactivity (see fig. 7).

Other 5–HT agonists (of the 1B, mixed 1A/1B and mixed 1/2–types) all have regular dose–dependent, inhibitory effects on aggression (fig. 6). However, more refined behavioural analyses reveal substantial differences between the drugs involved (fig. 7). Quipazine, a nonspecific 5–HT–agonist (table 1), clearly has a very nonspecific behavioural profile: decreases in aggression and social interest and a strong increase in inactivity (cf. Olivier and Mos 1986). The 5–HT$_{1B}$–agonist, TFMPP and both mixed 5–HT$_{1A/1B}$ agonists RU24969 and DU 28853 (and also the putative weak 5–HT$_{1A/1B}$ agonist fluprazine (Bradford et al. 1984)), have a similar profile in this female paradigm namely reducing aggression whilst increasing social interest, exploration, pup care and inactivity.

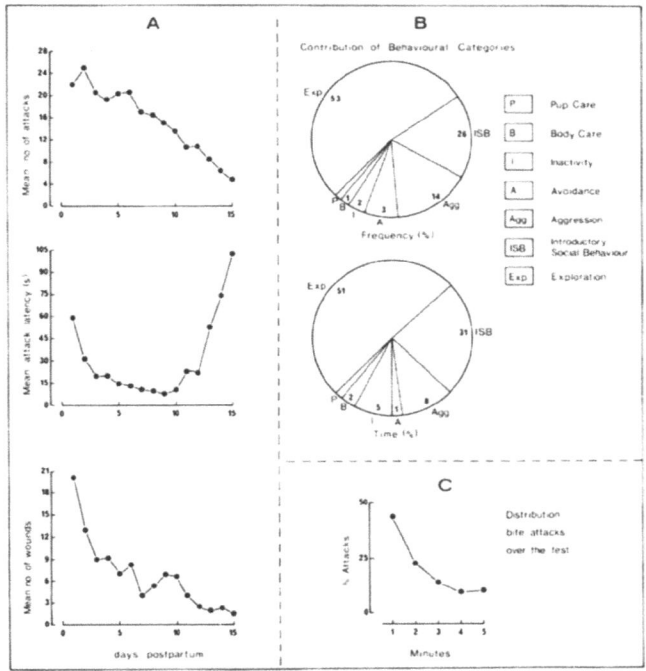

Fig. 4   The left panel (A) shows the mean number of attacks, the mean attack latency and the mean number of wounds inflicted upon intruders over the first 15 postpartum days in the maternal aggression test in rats. The right panel (B) shows the distribution of the different behavioural categories over the total observation period. Both the frequency and the duration distribution are shown. Fig. 4C shows the distribution (in %) of the bite attacks within one test period of 5 minutes.

MURICIDAL (MOUSE KILLING) BEHAVIOUR IN RATS

When rats are confronted with a mouse, some individuals will attack and kill it, although most strains show a very low (<25%) incidence of such spontaneous behaviour (Walsh 1982). Mouse killing is classed as predatory aggression (Karli 1956; Rossi 1975) on the basis of its behavioural topography (seizing in the neck followed by cervical dislocation), its very short latency and the fact that it is mostly followed by eating the prey (Rossi 1975). Although this interspecies model certainly reflects other aspects than intraspecific agonistic behaviour, it is of value in determining psychoactive properties of drugs. Moreover, by assessing the concomitant behavioural effects (scoring of sedation) a subjective measure can be obtained about the specificity of a drug's inhibitory effect.

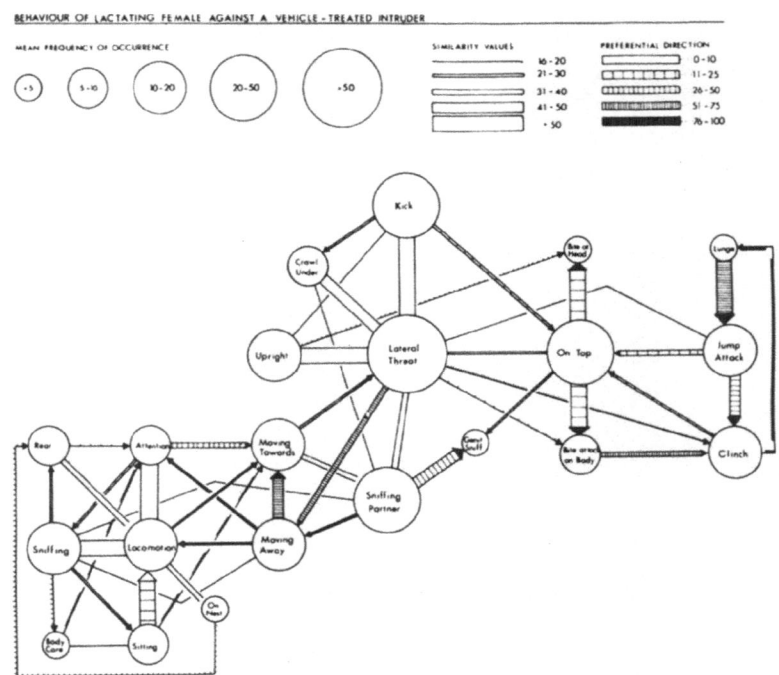

BEHAVIOUR OF LACTATING FEMALE AGAINST A VEHICLE - TREATED INTRUDER

Fig. 5 Behavioural structure of maternal aggression. The frequency of occurrence of the behavioural elements is depicted by the radius of the circles. Similarities and preferential directions are calculated using the methods of Olivier (1981).

In this study, rats of the TMD–S3 strain were used, a strain which has a high spontaneous mouse killing frequency (Olivier et al. 1984b; Walsh 1982). These experienced rats kill a mouse immediately upon confrontation; all killing latencies are reliably smaller than 1 min. By measuring during a 30 min test, the killing latency after giving the drug 30 min. prior to the test (IP), one can estimate the lowest effective dose (LED) which significantly inhibits muricide.

Table 3 shows the LED's of several serotonergic compounds. Serenics (fluprazine, TFMPP, DU 28853) inhibit muricidal behaviour in a specific way, both in males and females. 8–OH–DPAT does not inhibit muricide. RU 24969 effectively inhibits muricide, although some stimulatory action of the compound may interfere with the killing behaviour. Quipazine, fenfluramine and fluvoxamine inhibited muricide but not in a behaviourally specific way. Methysergide and ritanserine (5–HT antagonists) had no inhibiting effects on mouse–killing whereas 5–Me–O–DMT (a 5–HT$_1$ agonist) exerted a quite specific anti–muricidal effect. Finally buspirone nonspecifically inhibited mouse–killing, only at higher doses whereas ipsapirone, a structurally closely related compound, had no such effect.

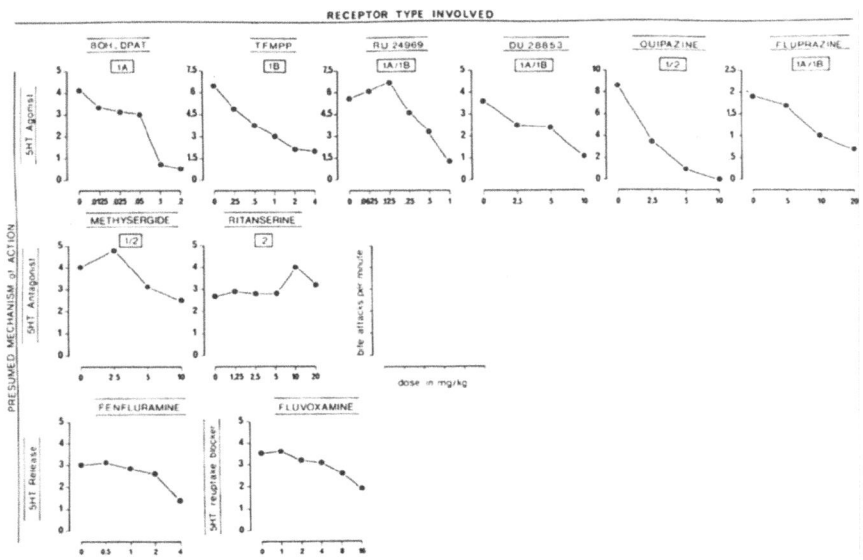

Fig. 6 The mean number of bite attacks/minute (+ SEM) is shown for a number of serotonergic drugs in the maternal aggression test in rats. This measure is corrected for the latency to first attack.

Table 3 Effects of serotonergic drugs on muricide in rats. Data are expressed as LED required to inhibit mouse killing.

Drug	LED (mg/kg IP)		Specificity of inhibiting effect *
	in males	in females	
Fluprazine	8	40	+
TFMPP	2	0.5	+
8–OH–DPAT	>5	>5	−
RU24969	1	2	+/−
DU 28853	5	20	+
Quipazine	4	4	−
Fenfluramine	2	4	−
Fluvoxamine	20	10	+/−
Methysergide	>20	>30	0
Ritanserine	>20	>20	0
5–Me–O–DMT	1	5	+
Buspirone	10	15	−
Ipsapirone	>10	>10	0

* + means behaviourally specific effects, – means nonspecific inhibition by e.g. sedation or motoric disturbances; 0 no inhibition.

174

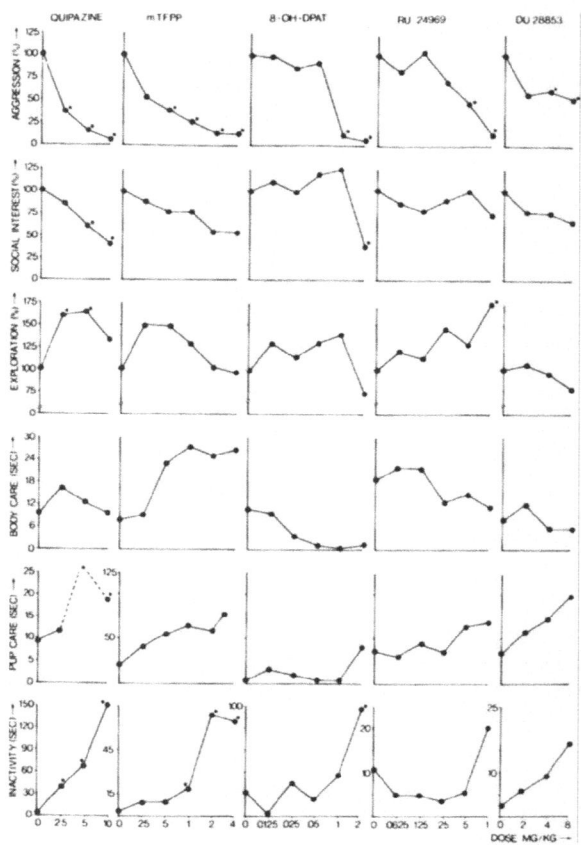

Fig. 7 Effects of five serotonergic drugs on the mean duration of six behavioural categories in maternal aggression in rats. Data are expressed either as % of vehicle (0 mg/kg) or as mean duration (sec).

In general, drugs which increase 5-HT-activity (5-HT$_1$/5-HT$_2$ agonists, 5-HT-release, 5-HT-reuptake block) decrease mouse-killing behaviour. The data indicate that the 5-HT$_{1A}$-receptor or binding site is not necessarily involved in this inhibition.

PLAY–FIGHTING IN RATS

When juvenile rats are observed in a neutral arena, after some initial minutes of predominantly exploration, they start intense play and play–fighting, largely consisting of wrestling and pinning (rough and tumble) (see Panksepp et al. 1984). There is ample evidence that play–fighting represents an early form of later social interaction and agonistic behaviour (Hinde 1974).

We tested play–fighting in 28–30 days old male juvenile rats which were housed isolated from weaning at day 21. Two weight–matched, unfamiliar animals, were placed for 10 minutes in a large cage (40x20x30 cm) and the number of pinnings and their activity (number of crossings of a midline) was determined. Animals were injected intraperitoneally 30 min before testing with either vehicle (saline) or one of 4 doses of DU 28853 (0.5, 1.0, 2.0 and 4.0 mg/kg).

Figure 8 shows that DU 28853 strongly inhibits play–fighting, with an ED_{50} of approx. 0.5 mg/kg. Concomitant measurement of activity shows that this decrease in play–fighting coincides with an increased activity which wanes at higher doses, but is still at the vehicle level at 4 mg/kg indicating the specificity of the effect on play–fighting.

AGONISTIC BEHAVIOUR IN PIGS

When young pigs unfamiliar to each other are mixed, e.g. at weaning, intense fighting may occur (Dantzer and Mormede 1980, McGlone et al. 1981) which may eventually lead to severe wounds or even death. Often, agonistic interactions continue until a clear hierarchy has been established, which may last for as long as 48 hours (McGlone 1985). Fraser (1974) recognized two distinct patterns of aggressive behaviour during agonistic interactions, one involving biting, and the other, less intense, involving butting and pushing.

In livestock farms the neuroleptic azaperone is often used to reduce the undesired aggression observed after mixing unfamiliar pigs (Symoens and Van den Brande 1969). Azaperone is a strong sedative drug (Porter and Slusser 1985) and acts only by postponing the hierarchy fights because after the drug effects wane, these fights still occur.

Treatment of pigs with a specific anti–aggressive drug, which would not sedate the animals, may have advantages in that the treated animals could interact socially and become familiar with each other. This could mean that a hierarchy is settled without the intense fighting which normally occurs during mixing. DU 28853, which exerts specific anti–aggressive (serenic) effects in several aggression models in rats and mice (Olivier et al. 1986b) was therefore tested in a porcine aggression paradigm, which includes mixing of unfamiliar piglets. Azaperone was also tested for comparison.

Three litters of female and castrated male piglets, about 10–14 weeks old and 15 to 40 kg bodyweight at the beginning of the experiment, were used. Animals within a litter (11 per litter) were familiar to each other, but unfamiliar to animals of other litters. The three groups were housed in one of the piggeries of Duphar's farm in Muiden with each group being in a pen of 3x4 m. Food and water were freely available and social encounters took place after an adaptation period of 1 week. Observations started between 8:30 and 10:00 hrs and lasted 4 hours.

Social encounters were arranged in a neutral pen (3x3 m). From each of the three groups 3 animals were introduced into the pen resulting in a mixed group of 9 animals. 15 minutes before encounters, animals were injected intramuscularly in the neck with saline, DU 28853 (2.5 and 10 mg/kg) or azaperone (1.5 mg/kg) at a volume of 1 ml/20 kg body weight. All 9 animals of a mixed group received the same treatment. The experiments were conducted in such a way that animals of the three litters used for mixing were always unfamiliar to each other. Each animal had an identification number on its back to facilitate behavioural observations. The observation and analysis took place by video–recording and direct observations by three observers who recorded: a) the piglet taking the initiative to fight and the piglets who suffered attacks; and b) the duration of the interaction. For a description of the social ethogram of pigs we refer to McGlone (1985). Briefly, agonistic interactions consist of bites (at the ears, neck, shoulder and face), pushes (head under push, head up push), head–jumps, body turns and flight. When one or more of these elements occurred it was scored as aggression.

176

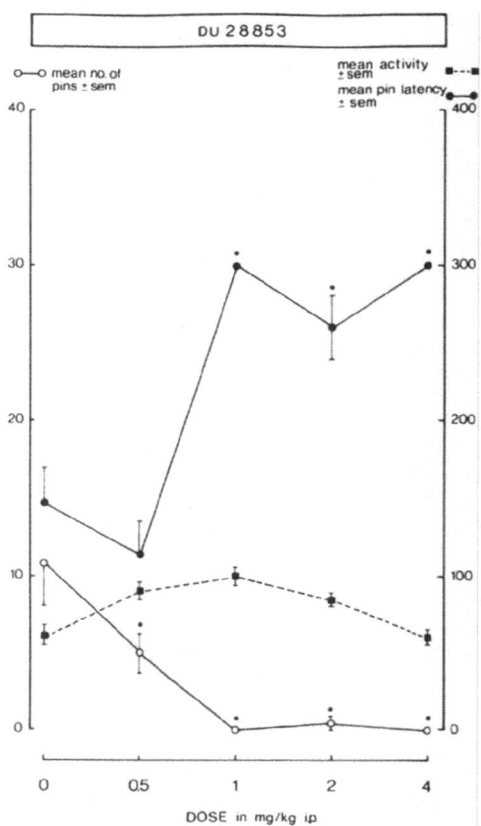

Fig. 8   The effects of DU 28853 on the frequency of pinning, the latency of the
first pin and the mean activity of the pair of juvenile rats in play–fighting.

The results of treatment with saline, DU 28853 (10 mg/kg, IM) and azaperone
(1.5 mg/kg, IM) are summarized in figures 9 and 10 as three aspects of agonistic
behaviour, namely the number of agonistic interactions/15 minutes, the percentage
of total time spent on agonistic behaviour per 15 minutes and the bout duration of
agonistic interactions (the mean time spent on each interaction).

In both of the figures (9 and 10) saline–treatment is shown in order to facilitate
the comparison with the respective drug treatments.

Under saline conditions, agonistic interactions start immediately after mixing
of the pigs (latency <1 minute) and during the first 15 minutes observation time,
almost 60% of total time is spent on aggression (fig. 9C). During this time the
highest number of agonistic interactions occurs (fig. 9B) involving almost all
individual piglets. Typically most interactions are relatively short–lasting (fig. 9A).

After this period, aggression waxes and wanes until about 2.5 hours after
starting when apparently some hierarchy becomes settled as indicated by a lower
number of interactions and short bout durations. Moreover, during this last phase

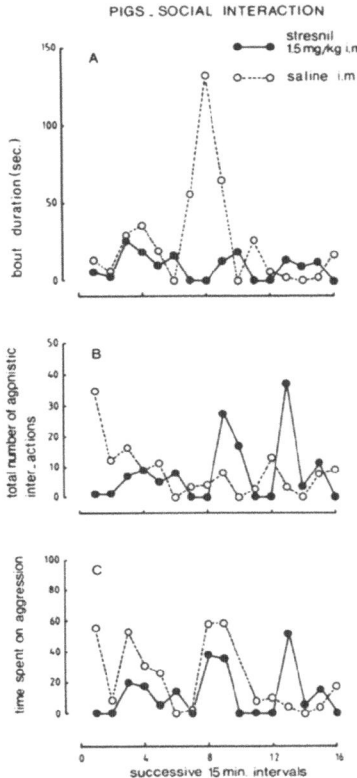

Fig. 9   The effects of azaperone (Stresnil: 1.5 mg/kg IM) on social interactions in pigs are shown in terms of the bout duration (panel A), the total number of agonistic interactions (panel B) and the time spent on aggression (panel C). All items are expressed per 15 min observation period.

only a limited number of pigs initiate interactions, in contrast to the initial phase. After 4 hours, most pigs have sustained many wounds and scratches, mainly on the ears, head and behind the ears, as a result of the agonistic interactions.

Azaperone inhibits aggression only for a relatively short period (30 minutes). For the first half hour after azaperone, animals lie flat and sedated, dispersed through the observation cage. After this period aggression reappears, although it remains at a lower level than saline treated animals for some 2 hours more.

The low dose of DU 28853 (2.5 mg/kg, IM) only had a very short–lasting effect, inhibiting aggression for around 30 minutes (figure not shown). After this period, heavy and long–lasting fighting occurred, leading to many wounds and scratches. During the initital inhibition–phase, animals were not sedated and showed all kinds of exploratory behaviour.

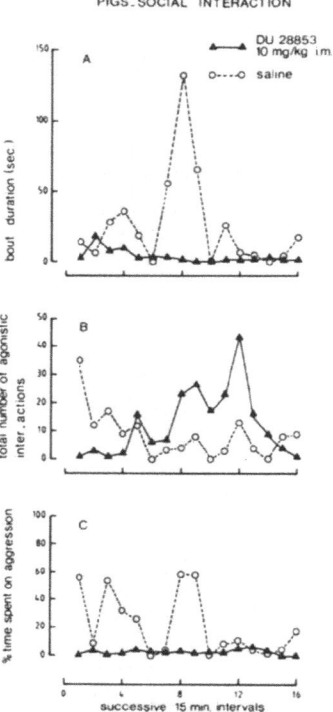

PIGS. SOCIAL INTERACTION

Fig. 10 The effects of DU 28853 (10 mg/kg, IM) on social interactions in pigs are shown on the bout duration (panel A), the total number of agonistic interactions (panel B) and the time spent on aggression (panel C). All items are expressed per 15 min observation period.

The higher dose of DU 28853 (10 mg/kg, IM) reduced aggression almost completely during the 4 hour confrontation (fig. 10). Only a very small amount of time (less than 6%) was spent on aggression in any period, whereas the nature of the interactions was dramatically shifted from the severe and damaging fighting seen in saline treated animals towards subtle and nondamaging interactions with DU 28853 treatment. Typically, DU 28853 animals were not sedated and showed no other debilitating signs interfering with the performance of agonistic behaviour in a nonspecific manner. Instead, animals showed high levels of social (introductory agonistic) interactions especially in the second part of the observation period. This suggests that no serious fighting would follow and that the hierarchy had been settled.

DISCUSSION

The present contribution outlines the behavioural effects of several serotonergic drugs with different mechanisms of action on diverse aggression

Table 4 Effects of serotonergic drugs on several aggression paradigms in mice (m) and rats (r).

Drug	isolation induced aggression (m)	intermale aggression (m)	footshock induced defence (m)	resident-intruder aggression (r)	maternal aggression (r)	EBS (r)	muricide (r)	Putative 5-HT mechanism of action
Fluprazine	①	①	①	①	①	①	①	weak 1A,1B and 2 agonist
TFMPP	①	①	nt	①	①	①	①	1B–agonist (partial)
DU 28853	①	①	–	①	①	①	①	1A,1B–agonist (partial)
RU24969	①	①	nt	nt	↓	nt	①	1A,1B–agonist
5–Me–O–DMT	①	nt	nt	↓	nt	nt	①	1A–agonist
Quipazine	–	nt	nt	↓	↓	①	↓	weak 1,2–agonist
8–OH–DPAT	①	↓	nt	nt	↓	–	–	1A–agonist
Buspirone	–	nt	nt	↓	↓	nt	↓	1A–agonist
Ipsapirone	–	nt	nt	nt	↓	nt	–	1A–agonist
Fluvoxamine	①	↓	nt	↓	①	↓	↓	reuptake block
Methysergide	–	nt	nt	nt	–	–	–	1,2–antagonist
Ritanserine	–	nt	nt	nt	–	nt	–	2–antagonist
Fenfluramine	①	nt	nt	↓	↓	nt	↓	release

①:specific behavioural decrease; ↓:nonspecific behavioural decrease; – no effect; nt: not tested. EBS=Electrical brain stimulation–induced aggression.

paradigms using largely mice and rats. This approach has several pitfalls, e.g. it is assumed that the drugs used exert their actions on receptors in the brain, although 5–HT receptors are also located abundantly in the periphery (see Richardson and Engel 1986). Moreover, after peripheral administration of drugs pharmacokinetic processes may lead to unexpected pharmacodynamic results. It is therefore quite conceivable that data obtained with in vitro receptor binding do not directly predict actual in vivo (behavioural) effects. This should be kept in mind when interpreting behavioural effects although as long as clearly–distinct behavioural effects are noted, one is apt to neglect the above–mentioned consideration.

The data obtained in this paper are related to the hypothesis that the behavioural changes induced by the drugs are in some way related to their in vitro pharmacological profile (see table 1 for a short summary of some 5–HT aspects of the drugs used). Table 4 summarizes, in a very general way, how the different 5–HT drugs tested exert their behavioural effects in several aggression paradigms in mice and rats. The last column indicates the most likely serotonergic mechanism of action based on a) our own research (table 1 and unpublished results) and b) on the available literature on binding studies (e.g. Arvidsson et al. 1986) and several behavioural studies including drug discrimination (e.g. Arvidsson et al. 1986; Glennon 1986; Glennon et al. 1982, 1984). Although there are still no drugs with certain specific 5–HT properties, e.g. 5–HT$_2$–agonists and 5–HT$_1$(A and B) antagonists, the available data summarized in table 4 suggests that the 5–HT$_{1B}$ binding site plays an important and specific modulatory role in aggressive behaviour. The 5–HT$_{1A}$ site does not seem to play an important role because 8–OH–DPAT, 5–Me–O–DMT, buspirone and ipsapirone have either no anti–aggressive activity or display a nonspecific effect. Moreover, the

$5-HT_2$-site also seems not to be directly involved in the modulation of aggressive behaviour, although the lack of specific $5-HT_2$-agonists means that this statement cannot be verified.

Of course, almost every compound will eventually cause an inhibition of aggressive behaviour, but it is assumed that such an effect must be specific on aggression and should not involve effects on other behavioural systems, like exploration, motor activity, defensive capabilities and social interest. The use of ethological methodologies shows that it is possible to simultaneously measure these several aspects of agonistic behaviour occurring between competing conspecifics, whether they are mice or rats (see Olivier et al. 1984b, 1986b). On the basis of this simultaneous measurement of all ongoing behaviour, we have developed a new class of psychoactive drugs termed "serenics" (Olivier et al. 1986; Olivier and Mos 1986; Bradford et al. 1984). Serenics, represented here by fluprazine and DU 28853, appear to have their main effects on the 5-HT-neuronal systems, especially via the $5-HT_{1B}$ binding sites, although a contribution from the $5-HT_{1A}$ site cannot be completely ruled out. Therefore, the general hypothesis that 5-HT activity is inversely correlated with aggression, seems untenable. A more refined (5-HT) hypothesis should specify a role for $5-HT_{1B}$ receptors. Before this can be done, a considerable amount of work has to be performed and new pharmacological tools (especially specific $5-HT_{1A}$ and $5-HT_{1B}$ antagonists and $5-HT_2$ agonists) have to be made available. The presence of $5-HT_{1B}$ receptors in several areas known to be involved in the modulation of agonistic behaviour (Albert et al. 1982) lends support to a direct regulation of such behaviour by a $5-HT_{1B}$-related mechanism. The different 5-HT receptor types have been shown to be differentially localized in the CNS using quantitative autoradiography (Pazos and Palacios 1985; Pazos et al. 1985). $5-HT_{1B}$ sites were present in high density in the globus pallidus, dorsal subiculum, substantia nigra and the olivary pretectal nucleus, whereas the highest density of $5-HT_{1A}$ sites was found in the dentate gyrus of the hippocampus and the lateroseptal nucleus. The neocortex and the hypothalamus also showed high concentration of both types of receptors. In contrast, the density of $5-HT_1$ receptors in the brainstem and spinal cord was low.

It should not be forgotten that 5-HT is very widely involved in many other kinds of behavioural and motoric systems (see reviews by Green 1985; Osborne 1982; Soubrie 1986). Recent evidence shows that 8-OH-DPAT enhances food intake (Dourish et al. 1985) and sexual behaviour (Ahlenius et al. 1981). In contrast, specific $5-HT_{1B}$ agonists like TFMPP decrease food intake and sexual behaviour (Olivier 1987). This holds also for the mixed $5-HT_{1A/1B}$ agonists DU 28853 and RU24969 (Olivier, unpublished results). In view of these contrasting effects of activation of the $5-HT_{1A}$ and $5-HT_{1B}$ receptors on feeding and sexual behaviour (and on other systems too) one could even question whether specific $5-HT_{1A}$ agonists might enhance aggressive behaviour or at least antagonize the anti-aggressive effects of $5-HT_{1B}$ agonists. 8-OH-DPAT did not increase aggression in any paradigm tested so far, but it may be possible that the aggression levels were too high to show such effects (cf. Mos and Olivier, this volume, on benzodiazepine effects). In preliminary experiments on maternal aggression, 8-OH-DPAT failed to antagonize the anti-aggressive effects of DU 28853.

It is clear that further studies await the development of more specific, 5-HT-pharmacological tools and localized brain studies which will facilitate the unravelling of the complex modulation by 5-HT receptor subtypes of agonistic and other behaviours.

Acknowledgement: We thank Mr.E.Baloch, Mr. A.Vedder and Mrs.A.Veen for assistance in the pig studies, Mr. A.Rademaker and Mr. R. van Oorschot for technical support and Mrs. M.Mulder for typing the report.

REFERENCES

Ahlenius S, Larsson K, Svensson L, Hjorth A, Carlsson A, Lindberg P, Wikstrom H, Sandchen D, Arvidsson LE, Hacksell U, Nilsson JLG (1981) Effects of a new type of 5-HT receptor agonist on male rat sexual behaviour. Pharmacol Biochem Behav 15: 785-792

Albert DJ, Walsh ML (1982) The inhibitory modulation of agonistic behavior in the rat brain: a review. Neurosci Biobehav Rev 6: 125-143

Applegate CD (1980) 5,7-Dihydroxytryptamine-induced mouse killing and behavioral reversal with ventricular administration of serotonin in rats. Behav Neural Biol 30: 178-190

Arvidsson LE, Hacksell U, Glennon RA (1986) Recent advances in central 5-hydroxytryptamine receptor agonists and antagonists. Progress in Drug Research. Vol 30, Birkhaüser Verlag, Basel, Boston, Stuttgart, pp 365-471

Avis HH (1974) The neuropharmacology of aggression: A critical review. Psychol Bull 81: 47-63

Barnett SA (1975) The Rat. A Study in Behavior. The University of Chicago Press: Chicago and London.

Blanchard DC, Blanchard RJ (1984) Affect and aggression: an animal model applied to human behavior. In: Blanchard RJ, Blanchard DC (eds) Advances in the Study of Aggression, Vol I. Academic Press, Orlando, pp 1-62

Blanchard RJ, O'Donnell V, Blanchard DC (1979) Attack and defensive behaviors in the albino mouse (Mus musculus). Aggr Behav 5: 341-352

Bradford LD, Olivier B, Van Dalen D, Schipper J (1984) Serenics: the pharmacology of fluprazine and DU 28412. In: Miczek KA, Kruk MR, Olivier B (eds) Ethopharmacological Aggression Research. Alan R Liss, New York, pp 191-207

Brain PF (1975) What does individual housing mean to a mouse? Life Sci 16: 187-200

Carlsson A, Davis JN, Kehr W, Lindquist M, Atack CV (1972) Simultaneous measurement of tyrosine and tryptophan hydroxylase activities in brain in vivo using an inhibitor of the aromatic amino acid decarboxylase. Naunyn-Schmiedeberg's Arch Pharmacol 275: 153-168

Claassen VC, Davies JE, Hertting G, Placheta P (1979) Fluvoxamine, a specific 5-hydroxytryptamine uptake inhibitor. Br J Pharmacol 60: 505-516

Crawley JN, Schleidt WM, Contrera JF (1975) Does social environment decrease propensity to fight in male mice? Behav Biol 15: 73-83

Creese I, Snyder SH (1978) (^{3}H) Spiroperidol labels serotonin receptors in rat cerebral cortex and hippocampus. Eur J Pharmacol 49: 201-202

Dantzer R, Mormede P (1979) Effects of lithium on aggressive behaviour in domestic pigs J Vet Pharmacol Therap 2: 299-303

Daruna JH (1978) Patterns of brain monoamine activity and aggressive behavior. Neurosci Biobehav Rev 2: 101–113

Dourish CT, Hutson PH, Curzon G (1985) Characteristics of feeding induced by the serotonin agonist 8–hydroxy–2–(Di–n–propylamino) tetralin (8–OH–DPAT). Brain Res Bull 15: 377–384

Eichelman BS, Thoa NB (1973) The aggressive monoamines. Biol Psychiatry 6: 143–164

Engel G, Göthert M, Hoyer D, Schlicker E, Hillenbrand K (1986) Identity of inhibitory presynaptic 5–hydroxytryptamine (5–HT) autoreceptors in the rat brain cortex with 5–HT$_{1B}$ binding sites. Naunyn–Schmiedeberg's Arch Pharmacol 332: 1–7

Fraser D (1974) The behaviour of growing pigs during experimental social encounters. J Agric Sci Cambridge 147–163

Garattini S, Valzelli L (1981) Is the isolated animal a possible model for phobia and anxiety? Progr Neuro–Psychopharmacol 5: 159–165

Glennon RA (1986) Site selective serotonin agonists as discriminative stimuli. Psychopharmacology 89 (S1): 135

Glennon RA, McKenney JD, Young R (1984) Discrimination stimulus properties of the serotonin agonist 1–(3–trifluoromethylphenyl) piperazine (TFMPP). Life Sci 35: 1475–1480

Glennon RA, Rosecrans JA, Young R (1982) The use of the drug discrimination paradigm for studying hallucinogenic agents. In: Colpaert FC, Slangen JL (eds) Drug discrimination: Applications in CNS pharmacology, Elsevier Biomedical Press, Amsterdam, pp 69–98

Göthert M (1980) Serotonin receptor mediated modulation of Ca^{2+} dependent 5–hydroxytryptamine release from neurones of the rat brain cortex. Naunyn–Schmiedeberg's Arch Pharmacol 314: 223–230

Gozlan H, El Mestikawy S, Pichat L, Glowinsky J, Hamon M (1983) Identification of presynaptic serotonin autoreceptors using a new ligand: ^{3}H–PAT. Nature 305: 140–142

Green AR (1985) Neuropharmacology of Serotonin. Oxford University Press, Oxford, pp 1–436

Hinde RA (1974) Biological Bases of Human Social Behaviour. McGraw–Hill, New York

Janssen PAJ, Jagenau AHM, Schellekens KHJ (1960) Chemistry and pharmacology of compounds related to 4–(4–hydroxy–4–phenyl–piperidino) – butyrophenone. Part IV. Influence of haloperidol (R 1625) and of chlorpromazine on the behaviour of rats in an unfamiliair "open field" situation. Psychopharmacologia 1: 389–392

Kantak KM, Hegstrand LR, Eichelman B (1981) Facilitation of shock–induced fighting following intraventricular 5,7–dihydroxytryptamine and 6–hydroxy dopa. Psychopharmacology 74: 157–160

Karli P (1956) The Norway rat's response to the white mouse: an experimental analysis. Behaviour 10: 81–103

Karli P (1981) Conceptual and methodological problems associated with the study of brain mechanism underlying aggressive behaviour. In: Brain PF, Benton D (eds) The Biology of Aggression. Sijthoff and Noordhoff, Alphen a.d. Rijn, The Netherlands, pp 323–361

Knutson JF, Kane NL, Schlosberg AJ, Fordyce DJ, Simansky KJ (1979) Influence of PCPA, shock level, and home cage conditions on shock–induced aggression. Physiol Behav 23: 897–907

Lehman MN, Adams DB (1977) A statistical and motivational analysis of the social behaviors of the male laboratory rat. Behaviour 61: 238–275

Malick JB (1979) The pharmacology of isolation–induced aggressive behaviour in mice. In: Essman WB, Valzelli L (eds) Current developments in Psychopharmacology. New York, Spectrum Publications Inc., Vol. 5: 1–27

McClone JJ, Kelley KW, Gaskins CT (1981) Lithium and porcine aggression. J Anim Sci 51: 447–455

McClone JJ (1985) A quantitative ethogram of aggressive and submissive behaviors in recently regrouped pigs. J Anim Sci 61: 559–565

Miczek KA (1987) The psychopharmacology of aggression. In: Iversen LL, Iversen SD, Snyder SH (eds) Handbook of Psychopharmacology. Vol. 19: Behavioural Pharmacology. New York: Plenum Press (in press)

Miczek KA, Barry H (III) (1976) Pharmacology of sex and aggression. In: Glick SD, Goldfarb J (eds) Behavioral Pharmacology. C.V. Mosby, St Louis, pp 176–257

Miczek KA, O'Donnell JM (1978) Intruder–evoked aggression in isolated and nonisolated mice: effects of psychomotor stimulants and l–dopa. Psychopharmacology 57: 47–55

Miczek KA, Krsiak M (1979) Drug effects on agonistic behaviour. In: Thompson T, Dews PB (Eds) Advances in Behavioral Pharmacology. New York, Academic Press Vol. 2: 87–162

Miczek KA, Winslow JT (1986) Psychopharmacological research on aggressive behaviour. In: Greenshaw AJ, Dourish CT (eds) Experimental Psychopharmacology. Humana Press, Clifton, New Jersey, pp 27–113

Middlemiss DN (1984) 8–hydroxy–2–(di–n–propylamino) tetralin is devoid of activity at the 5–hydroxytryptamine autoreceptor in rat brain. Naunyn–Schmiedeberg's Arch Pharmacol 327: 18–22

Middlemiss DN, Fozard JR (1983) 8–hydroxy–2–(di–n–propylamino) tetralin discriminates between subtypes of the $5-HT_1$ recognition sites. Eur J Pharmacol 90: 151–153

Mos J, Olivier B (1986) RO 15–1788 does not influence post–partum aggression in lactating female rats. Psychopharmacology 90: 278–280

Mos J, Olivier (1987) Pro–aggressive actions of benzodiazepines. In: Olivier B, Mos J, Brain PF (eds) Ethopharmacology of agonistic behaviour in animals and man. Sijthoff and Noordhoff, Dordrecht (in press)

Olivier B (1977) The ventromedial hypothalamus and aggressive behaviour in rats. Aggr Behav 3: 47–56

Olivier B (1981) Selective anti–aggressive properties of DU 27725: Ethological analysis of intermale and territorial aggression in the male rat. Pharmacol Biochem Behav 14 (S1): 61–77

Olivier B (1987) Pharmacological properties of DU 28853. Internal Duphar Report, pp 1–296

Olivier B, Van Dalen D (1982) Social behaviour in rats and mice: an ethologically based model for differentiating psychoactive drugs. Aggr Behav 8: 163–168

Olivier B, Van Aken H, Jaarsma I, Van Oorschot R, Zethof T, Bradford LD (1984a) Behavioural effects of psychoactive drugs on agonistic behaviour of male territorial rats (resident–intruder paradigm). In: Miczek KA, Kruk MR, Olivier B (eds) Ethopharmacological Aggression Research. Alan R Liss Inc, New York, pp 137–156

Olivier B, Van Oorschot R, Boschman TAC, Van der Heyden JAM, Schipper J, Mol F (1984b) Mouse killing in male TMD–S3 rats: incidence of killing and the effects of experience, castration and pharmacological manipulations Aggr Behav 10: 165–166

Olivier B, Mos J, Van Oorschot R (1985) Maternal aggression in rats: effects of chlordiazepoxide and fluprazine. Psychopharmacology 86: 68–76

Olivier B, Mos J, Van Oorschot R (1986) Maternal aggression in rats: lack of interaction between chlordiazepoxide and fluprazine. Psychopharmacology 88: 40–43

Olivier B, Mos J (1986a) A female aggression paradigm for use in psychopharmacology: maternal agonistic behaviour in rats. In: Brain PF, Ramirez JM (eds) Cross–Disciplinary Studies on Aggression. University of Seville Press, pp 73–111

Olivier B, Mos J (1986b) Serenics and aggression. Stress Med 2: 197–209

Olivier B, Van Dalen D, Hartog J (1986) A new class of psychoactive drugs: Serenics. Drugs of the Future 11: 473–499

Osborne NN (1982) Biology of Serotonergic Transmission. John Wiley and Sons, Chichester, pp 1–522

Panksepp J, Siviy S, Normansell L (1984) The psychobiology of play: theoretical and methodological perspectives. Neurosci Biobehav Rev 8: 465–492

Pazos A, Palacios JM (1985) Quantitative autoradiographic mapping of serotonin receptors in rat brain. I. Serotonin–I receptors. Brain Res 346: 205–230

Pazos A, Cortes, Palacios JM (1985) Quantitative autoradiographic mapping of serotonin receptors in the rat brain. II. Serotonin-2 receptors. Brain Res 346: 231-249

Peroutka SJ (1986) Pharmacological differentiation and characterization of 5-HT$_{1A}$, 5-HT$_{1B}$, and 5-HT$_{1C}$ binding sites in rat frontal cortex. J Neurochem 47: 529-540

Porter DB, Slusser CA (1985) Azaperone: a review of a new neuroleptic for swine. Vet Med, pp 88-92

Pradhan SN (1975) Aggression and central neurotransmitters. In: Pfeiffer CC, Smythies JM (eds) International Review of Neurobiology. Academic Press, New York, pp 213-262

Reis DJ (1974) The chemical coding of aggression in the brain. In: Myers RD, Drucker-Colin RR (eds) Neurohumoral Coding of Brain Function. Plenum Press, New York, pp 125-150

Richardson BP, Engel G (1986) The pharmacology and function of 5-HT$_3$ receptors. TINS, pp 424-428

Rossi AC (1975) The "mouse-killing" rat: ethological discussion on an experimental model of aggression. Pharmacol Res Commun 7: 199-216

Shannon NJ, Gunnet JW, Moore KE (1986) A comparison of biochemical indices of 5-hydroxytryptaminergic neuronal activity following electrical stimulation of the dorsal nucleus. J Neurochem 47: 958-965

Sheard MH, Davis M (1976) Shock-elicited fighting in rats: importance of intershock interval upon the effect of p-chlorophenylalanine (PCPA). Brain Res 111: 433-437

Sills MA, Wolfe BB, Frazer A (1984) Determination of selective and non-selective compounds for the 5-HT and 5-HT$_{1B}$ receptor subtypes in rat frontal cortex. J Pharmacol Exp Ther 231: 480-487

Smith DE, King MB, Hoebel BG (1970) Lateral hypothalamic control of killing: Evidence for a cholinoceptive mechanism. Science 167: 900-901

Soubrie P (1986) Reconciling the role of central serotonin neurons in human and animal behavior. Behav Brain Sci 9: 319-364

Steinbusch HWM (1981) Distribution of serotonin-immunoreactivity in the central nervous system of the rat - Cell bodies and terminals. Neurosci 6: 557-618

Svare BB, Mann MA (1983) Hormonal influences on maternal aggression. In: Svare B (ed) Hormones and Aggressive Behavior. Plenum Press, New York and London, pp 91-104

Symoens J, Van den Brande M (1969) Prevention and cure of aggressiveness in pigs using the sedative azaperone. Vet Rec 85: 64-77

Tedeschi RE, Tedeschi DH, Mucha A, Cook L, Mattis PA, Fellows EJ (1959) Effect of various centrally acting drugs of fighting behaviour of mice. J Pharmacol Exp Ther 125: 28–34

Timmermans PJA (1978) Social behaviour in the rat. PhD Thesis, University of Nijmegen.

Tricklebank MD (1985) The behavioural response to 5–HT receptor agonists and subtypes of the central 5–HT receptor. Trends Pharmacol Sci, pp 403–407

Valzelli L (1973) The "isolation syndrome" in mice. Psychopharmacologia 31: 305–320

Valzelli L (1981) Psychopharmacology of aggression: an overview. Int Pharmacopsychiatry 16: 39–48

Valzelli L (1984) Reflections on experimental and human pathology of aggression. Progr Neuropsychopharmacol Biol Psychiatry 8: 311–325

Valzelli L, Garattini S (1968) Behavioral changes and 5–hydroxytryptamine turnover in animals. Adv Pharmacol 6B: 249–260

Valzelli L, Garattini S, Bernasconi S, Sala A (1981) Neurochemical correlates of muricidal behavior in rats. Neuropsychobiology 7: 172–178

Van der Poel AM, Mos J, Kruk MR, Olivier B (1984) A motivational analysis of ambivalent actions in the agonistic behaviour of rats in tests used to study effects of drugs on aggression. In: Miczek KA, Kruk MR, Olivier B (eds) Ethopharmacological Aggression Research. Alan R Liss Inc, New York, pp 115–135

Vergnes M, DePaulis A, Boehrer A (1986) Parachlorophenylalanine–induced serotonin depletion increases offensive but not defensive aggression in male rats. Physiol Behav 36: 653–658

Vergnes M, Kempf E (1981). Tryptophan deprivation: Effects on mouse–killing and reactivity in the rat. Pharmacol Biochem Behav 14: 5.1: 19–23

Vergnes M, Kempf E (1982). Effect of hypothalamic injections of 5,7–dihydroxytryptamine on elicitation of mouse–killing in rats. Behav Brain Res 5: 387–397

Walsh LL (1982) Strain and sex differences in mouse killing by rats. J Comp Physiol Psychol 96: 278–283

PRO–AGGRESSIVE ACTIONS OF BENZODIAZEPINES

Jan Mos and Berend Olivier. Department of Pharmacology, Duphar B.V., P.O.Box 2, 1380 AA Weesp, The Netherlands.

INTRODUCTION

The benzodiazepines (BDZ) are known for a wide variety of pharmacological effects, among which anxiolytic, hypnotic, sedatory, muscle relaxing and anticonvulsant actions are the most prominent. Much progress has been made in the elucidation of their mechanism of action by the discovery and characterization of high affinity binding sites for BDZ in the brain. However, not all complex behavioural actions are completely explained by the concepts derived from receptor binding. One such behavioural response is the intriguing question of the pro–aggressive action of low doses of benzodiazepines.

Studies of the effects of benzodiazepines on aggression have been inconclusive with regard to pro–aggressive actions. The well–documented decrease of aggression in animals is less of a problem, since under more or less natural conditions, higher doses of BDZ lead to a marked decrease of agonistic behaviour, be it offensive or defensive in nature. In fact, one of the first papers on chlordiazepoxide (CDP) reported its taming effect at a dose of 1.0 mg/kg on monkeys (Randall 1960). The anti–aggressive actions of benzodiazepines are sometimes in a dose range in which muscle relaxing properties of the drug also become manifest, thus casting doubt upon the specificity of this anti–aggressive effect (Sofia 1969), although this is not universally true (Valzelli 1973).

Pro–aggressive actions of benzodiazepines have been reported under various testing conditions in different species. These conditions largely determine whether aggression is increased after BDZ treatment (Dantzer 1977). Reports of stimulatory effects of BDZ on human aggressive behaviour are equivocal as systematic increase in aggression by these drugs has definitely not been established. Several important questions therefore remain unanswered or unstudied.

Firstly, it is uncertain whether the pro–aggressive action found in some animal models is mediated by the benzodiazepine receptor. Secondly, the generality of the pro–aggressive action of BDZ needs to be established. Thirdly, the conditions under which pro–aggressive actions become evident need to be studied carefully. Finally, and perhaps most intriguingly is the question whether a clear relationship exists between the anxiolytic and the pro–aggressive actions of BDZ. The present paper deals with some of these aspects and initially presents a short review of the literature in animals and humans.

PRO–AGGRESSIVE ACTIONS OF BDZ IN AMIMAL STUDIES

Fox and coworkers (1970) demonstrated an increase in within–group attacks of group–housed male mice fed on a diet containing small amounts of chlordiazepoxide. Moreover, mortality was increased, presumably due to the enhanced levels of aggression. Defensive postures were not affected by the treatment. Similar findings, i.e. increased aggression and mortality have been found for diazepam (Fox and Snyder 1969), nitrazepam and flurazepam (Fox et al.

1972) and oxazepam and prazepam (Fox et al. 1974). At the same time, defence postures were either unaltered (oxazepam, prazepam, chlordiazepoxide, flurazepam) or even decreased (diazepam and nitrazepam). The chronic doses used were much smaller than those reported by other investigators demonstrating a decrease in aggression after acute oral administration. Doses in the former studies apparently caused no sedation or muscle relaxation. Zwirner et al. (1975) reported that CDP (3 mg/kg PO) given in an acute inter-group aggression test also increased aggression in mice. Control experiments revealed that this dose did not impair motor activity.

Leaf et al. (1975, 1984) in studying the muricidal response of naive rats, found that CDP (7.5 – 20 mg/kg), diazepam (1.25 – 10 mg/kg) and oxazepam (2.5 – 40 mg/kg) induced mouse killing. Other drugs used as anxiolytics such as pentobarbital and meprobamate did not induce muricidal responses in naive rats. For the induction of mouse killing, it seems essential that the experimental rats are naive. Experienced killer rats may stop killing at higher doses of chlordiazepoxide (Horovitz et al. 1966; Karli 1961), whereas similar doses facilitate mouse-killing in naive rats (Leaf et al. 1975).

A pro-aggressive action of CDP has been found by Miczek (1974) in rats employing an extinction procedure of a food-reinforced response. Aggression by the dominant rat was increased at low doses (2.5 and 5 mg/kg) whether the drug was given to the dominant or the subordinate male without any reversal of the dominance-subordination relationships. Thus besides there being direct drug effects, there may be important indirect behavioural effects of chlordiazepoxide. In experiments described by Apfelbach and Delgado (1974) and Poshivalov (1980), dominance reversal was observed after repeated treatment. Indications for the importance of environmental cues stems from experiments on mice tested in their home cage or a neutral test arena (Miczek and O'Donnell 1980). Aggression was increased by chlordiazepoxide treatment only in the neutral test arena. Since aggression levels are lower in the latter situation, it is unclear whether the pro-aggressive action is a "rate-dependent" phenomenon or due to cues from the testing environment.

A number of other studies also point to the important influence of testing conditions. Apfelbach (1978) tested the stimulatory action of CDP on prey catching behaviour of ferrets. Most notably, the size of the prey determined whether any facilitation could be observed, i.e. CDP only facilitated predation when a large prey was presented while it was without effect when the usual smaller sized prey was offered. Arnone and Dantzer (1980) only found increased aggression in pigs in an extinction procedure, when access to the lever in the operant room was inhibited. When the bar was accessible, social interactions were much less preferred, but CDP increased bar pushing.

The "response tendency" of the animal under study is similarly important as has been pointed out by Krsiak. Timid and aggressive mice react differently to a low dose of CDP (Krsiak 1975, 1979). Chlordiazepoxide increases aggression in aggressive mice, but not in "timid" mice. Moreover, Krsiak described a decrease in tail rattling after CDP treatment in aggressive mice. As tail rattling is seen as an ambivalent response representing both aggressive and flight tendencies (Krsiak 1979; Krsiak et al. 1981), this observation fits nicely with the concept that the anxiolytic action of BDZ may be important for the pro-aggressive action. Dixon (1982) drew attention to the possible olfactory component involved in diazepam's effect on social behaviour, more specifically the defensive behaviour of intruder mice.

Further systematic studies aimed at discovering the mechanism of BDZ action on aggressive behaviour are scarce. The foregoing brief resume indicates that pro-aggressive actions of BDZ in animal studies are not unique. Moreover, it is clear that experimental conditions and aggression models used, largely determine the final results of experiments, perhaps particularly those involving BDZ.

PRO–AGGRESSIVE ACTIONS OF BDZ IN HUMANS

Systematic data on pro–aggressive actions of benzodiazepines in humans are rare, certainly when viewed against the widespread use of BDZ. Early reports in the literature cautioned that BDZ may lead to increased hostility (Ingram and Timbury 1960). Reviews on the effects of BDZ have been made among others by DiMascio (1973), Azcarate (1975), Essman (1978), Bond and Lader (1979) and Sheard (1984) and most recently by Miczek (1987). It is outside the scope of this chapter to review all data of BDZ on aggressive behaviour in humans. A brief summary will be given on the best documented and well–controlled experiments on the pro–aggressive effects of BDZ.

DiMascio, Shader, Salzman and coworkers conducted an elegant series of experiments on the effects of BDZ on "low" and "high" anxious normal subjects. Basically, aggression was measured with a questionnaire method, employing the different sub–scales of the Buss–Durkee Hostility Inventory. Subjects were selected and grouped according to their scores in an anxiety test. None of the subjects, however, was clinically treated for anxiety. Treatment with CDP (15 or 30 mg daily), oxazepam (45 mg daily) or placebo for one week was followed by tests for anxiety and aggression. Chlordiazepoxide and oxazepam reduced anxiety in the medium and high anxious groups, but increased anxiety in the low anxious group (Gardos et al. 1968). Hostility scores were significantly increased in the high anxious group treated with CDP. A similar but not significant effect was noted in the medium anxious group. By contrast, oxazepam had no influence on hostility scores whatsoever, nor did placebo treatment alter hostility scores in this experiment. The importance of testing conditions was demonstrated by the same group of investigators showing that instructions of the volunteers on the expected drug effect markedly changed baseline scores and drug effects (Salzman et al. 1969). Merits of these findings and clinical consequences are briefly summarized in DiMascio et al. (1969).

Rickels and Downing (1974) found no convincing increase in aggression in an anxious outpatient population treated with CDP as measured by physician rated irritability and hostility or in self–assessed anger or hostility. Later experiments (Downing and Rickels 1984) on a group of eighty psychiatric outpatients employing the Buss Durkee Hostility Inventory revealed no increased aggression. Two important variables in these studies compared to the data of DiMascio and coworkers, are the longer period of treatment (4 weeks) and the population of subjects (psychiatric patients versus normal volunteers).

Subsequent studies using experimental aggression paradigms have not settled the issue. Wilkinson (1985), employing the Taylor competitive reaction time test reported increased aggression after a single dose (10 mg orally) of diazepam in normal male undergraduates, depending upon anxiety and provocation levels. A remarkable difference with the experiments of Gardos et al. (1968) is the increase in aggression in the low–trait anxious group under low provocation levels. Hostility was increased most markedly by CDP in the high anxious group in the Gardos experiments. Perhaps the different BDZ should be studied more carefully and systematically, not only with respect to aggression effects, but also to changes in anxiety scores in normal volunteers.

Cherek et al. (1986) have tested the effects of diazepam in an elegant experimental aggression set up. They reported no pro–aggressive actions of diazepam (2.5, 5 or 10 mg/kg) in their experimental design. Individual differences in reaction to BDZ treatment and different dose–response curves may prove to be vital in interpreting the contrasting results (see also chapter Cherek in this book).

There is no unifying hypothesis to explain these differing results but the reasons why increased aggression may not be more universally recognized are summarized by DiMascio et al. (1969). The amount of drug–released hostility, insensitivity of scales to measure aggression and a possibly beneficial effect of increased aggressiveness seem very plausible reasons. Even in well–controlled animal studies the pro–aggressive actions of BDZ are not invariably observed. The impact of subtle experimental manipulations on the eventual results in such experiments cautions against easy generalizations. The issue of increased aggression still remains of current interest, even with newer BDZ's such as alprazolam (Rosenbaum et al. 1984). It thus seems worthy to pursue this issue further, especially in outpatients employing experimental aggression paradigms to delineate which group of patients is most susceptible to adverse effects of BDZ leading to uncontrolled aggressiveness. Moreover, the differentiation between acute and chronic effects of various BDZ (preferably in the same subjects) deserves further attention.

EXPERIMENTS

Effects of different benzodiazepine agonists on maternal aggression in rats

Previous studies in our laboratory (Olivier et al. 1985; 1986) have demonstrated the pro–aggressive action of low doses of CDP. CDP essentially increased all aggressive elements in a similar dose effect curve (fig. 1). This indicated that all elements from the aggressive repertoire are potentially sensitive to the stimulatory actions of CDP, although in this particular experiment the basal levels of aggression were quite low. In subsequent experiments with other BDZ's only the number of bite attacks during the 5 minute test period were scored and no detailed analysis of the complete behaviour was performed.

Diazepam and oxazepam were tested in the maternal aggression paradigm in a similar way as described for CDP (Olivier et al. 1985; 1986). Diazepam was administered both orally (t=–60 min) and intraperitoneally (t=–30 min) and oxazepam only intraperitoneally 30 minutes before testing.

Results for the number of bite attacks, attack latency and bites per minute (i.e. number of bites corrected for the attack latency) are given in figure 2. No significant pro–aggressive action of diazepam on attacks or latency was observed after either route of administration, although there is some indication of a similar dose–effect curve as was seen for CDP (Oral administration: Friedman $X^2=5.1$, df=3, p=0.16 attacks; $X^2=5.6$, df=3, p=0.13 latency; $X^2=7.3$, df=3, p=0.06 bites/minute; IP administration: $X^2=5.8$, df=5, p=0.33 attacks; $X^2=19.0$, df=5, p=0.002 latency; $X^2=5.8$, df=5, p=0.32 bites/minute). Oxazepam, however, showed a nice biphasic dose–effect curve with pro–aggressive actions at lower doses, similar to the data obtained for CDP (Friedman $X^2=16.8$, df=5, p=0.005 attacks; $X^2=8.8$, df=5, p=0.12 latency; $X^2=14.7$, df=5, p=0.012 bites/minute).

It is difficult to explain these contrasting data. Diazepam has been reported to have pro–aggressive actions although the magnitude of these effects was not impressive (Miczek, personal communication). Studies by Friedman et al. (1986) suggest that this lack of effect cannot be attributed to unusual pharmacokinetic characteristics of diazepam metabolism in rats. Studies in humans suggest that oxazepam is less liable to result in "paradoxical aggression" (Gardos et al. 1968, Lion et al. 1975) than CDP, a finding which is not confirmed by the present animal experiments.

Brain stimulation induced attack in rats

Electrical brain stimulation–induced attack (EBS) is a model for the study of aggression with features distinct from other experimental models of animal aggression (see chapter by Kruk et al. in this volume). Electrical stimulation of specific, well localized parts of the hypothalamus in rats results in very intense forms of attack, which immediately disappear when the stimulation is discontinued (Kruk et al. 1979). The effects of CDP have been studied on this form of

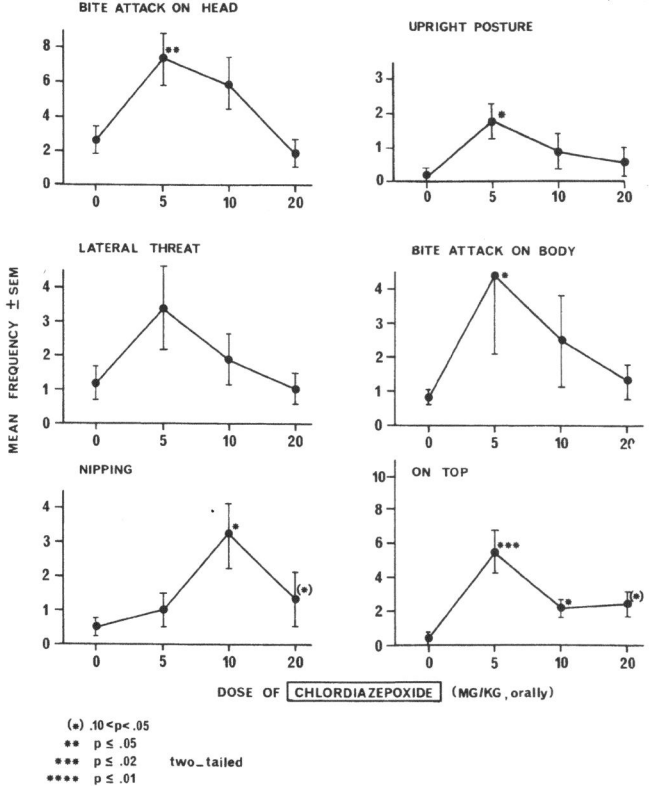

Fig. 1 Effects of chlordiazepoxide on individual elements belonging to the aggressive repertoire of maternal female rats. Chlordiazepoxide treatment of the mothers resulted in remarkably similar dose–effect curves of most aggressive elements, with significant increases at 5 mg/kg. (Reproduced with permission from Psychopharmacology 86: 68–76, 1985).

aggression, to test whether pro–aggressive actions can be detected using this paradigm.

Male CPB WE–zob rats were equipped with Pt–10% Ir electrodes, essentially similar to the techniques described previously (Kruk et al. 1979). Threshold currents needed to induce attacks were determined with an up–and–down design according to a modified procedure of Wetherill (1966). Inhibition of aggression was reflected in an increased current threshold necessary to induce aggression, while pro–aggressive actions of drugs should lower the current needed to induce attack on male conspecifics.

Seven animals were tested with 3 doses of CDP (5, 10 and 20 mg/kg, 60 min before testing, PO) and vehicle administered in a random sequence with at least a two day interval between treatments. Figure 3A summarizes the changes in current thresholds for aggression after CDP treatment. No significant overall effects were observed (Friedman $X^2 = 2.6$, df=3, p=0.46), but a tendency to inhibit aggression was evident at the highest dose tested. Lowered current thresholds at 5 and 10 mg/kg, indicative of pro–aggressive actions of CDP, were not observed. Similar

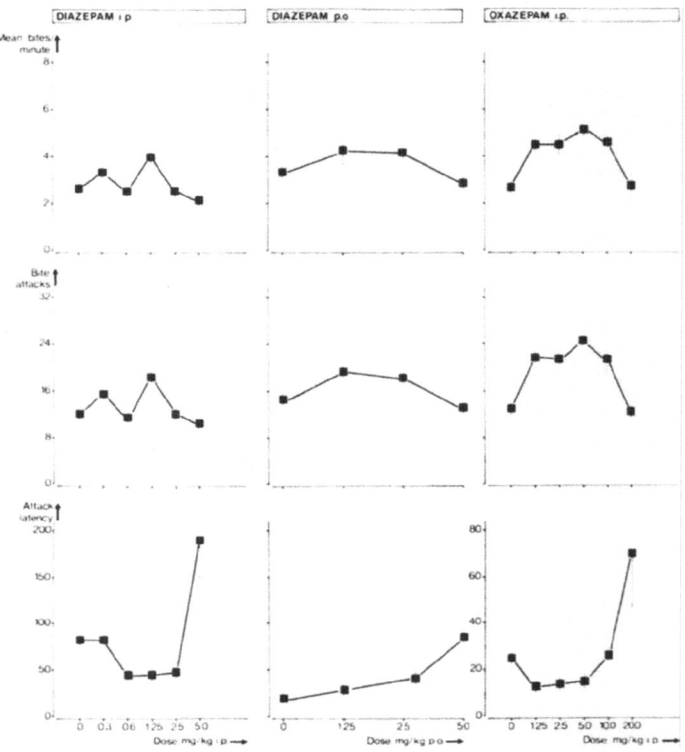

Fig. 2   Mean bites per minute, number of bite attacks and latency to first attack
(± SEM) of maternal aggression after treatment with diazepam (IP and PO)
or oxazepam (IP). Significant increases at low doses were only observed
with oxazepam. Although a slight increase was found for some doses of
diazepam this never reached statistical significance.

results have been found by Kruk et al. (see his chapter in this book) testing female
rats

These experiments seem to add to the already confusing data on the
pro-aggressive actions of CDP. In this particular model for aggression, however,
there were no clear indications of concurrent inhibitory processes. The rat is not
aggressive when the current is off, there is a sudden episode of violence when the
stimulation is on, and during this period, male as well as female opponents, are
attacked. Introductory social behaviour, usually seen during territorial aggression is
largely absent, while threat behaviour also seldom occurs. Under many more
naturalistic testing environments both flight and fight tendencies are invariably
present (Van der Poel et al. 1984).

If there are no concurrent inhibitory processes affecting EBS–induced attacks, a
facilitation by BDZ might not be expected. An alternative explanation may be that
threshold current stimulation induces an almost maximal aggressive response which

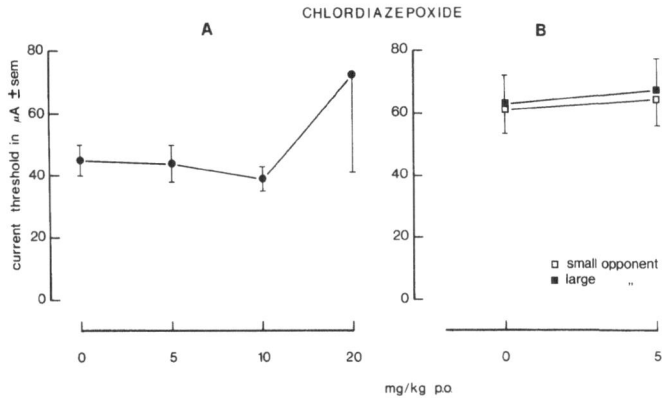

Fig. 3 A. Current thresholds for aggression induced by electrical stimulation in the lateral hypothalamus of male WE–zob rats (± SEM). Chlordiazepoxide did not significantly alter the current thresholds for attack. At the highest dose a tendency to increased thresholds (i.e. lower aggression) is noted, probably due to nonspecific sedative effects.
   B. Using large and small opponents, current thresholds were determined in male S3 rats. In contrast to maternal aggression, neither an effect of opponent size nor of a single selected 5 mg/kg dose CDP was observed,

can hardly be facilitated because of the "ceiling effects". In such a case, the lack of pro–aggressive action can simply be ascribed to rate–dependency.

Influence of opponent size on pro–aggressive actions of CDP

Experiments with BDZ's have demonstrated the importance of testing conditions. It is tempting to speculate that environmental situations contribute to the inhibition of aggression and that BDZ in turn result in increased aggression by disinhibitory actions. Opponent size has been shown to be an important factor in maternal aggression in rats (Flannelly et al. 1986; Mos et al. 1987) as well as in prey catching in ferrets (Apfelbach 1978). Large sized opponents or prey reduce aggression. Inhibition of aggression by varying the size of the opponents is an attractive possibility to study the pro–aggressive effects of BDZ, although such a procedure may introduce rate–dependency phenomena. In two experiments the role opponents may play in revealing possible pro–aggressive actions of CDP were assessed.

On the basis of previous experiments a single dose of 5 mg/kg (PO) was selected because it reliably increases aggression (Olivier et al. 1985; 1986). Maternal aggression was tested in S3 females (250–300 gram) with high basal levels of aggression which were confronted with small (ca 180 g) and large (ca 420 g) male opponents. Testing procedures were similar to those described previously (Olivier et al. 1985). The behaviour of the lactating females was recorded during 5 minute tests under vehicle or CDP conditions. Detailed ethological analysis revealed a significant effect of the opponent's size and drug treatment.

Large opponents in their own might change the strategy used by maternal females who show more lateral threat and upright posture, but who are also less successful in bite attacks and gaining the on top position (Fig. 4). The frequency of

194

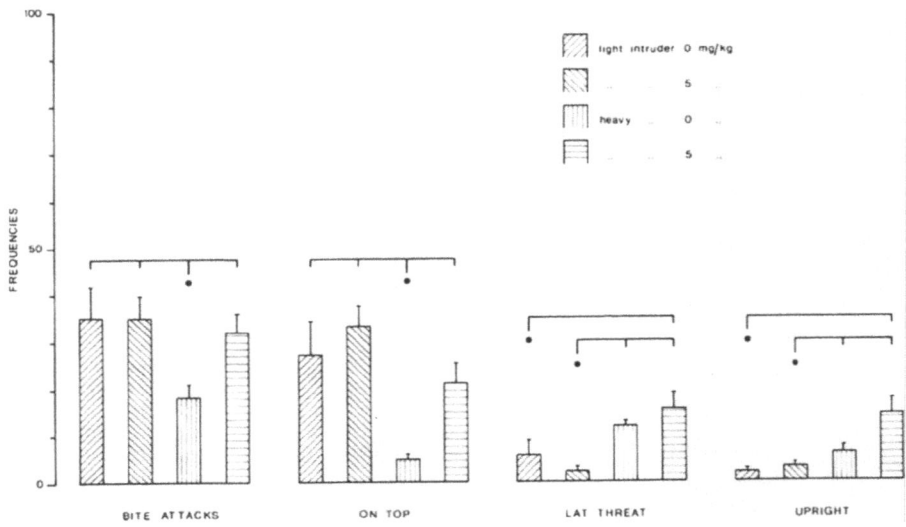

Fig. 4. Maternal aggression towards different sized opponents under control or 5 mg/kg PO chlordiazepoxide treatment. Large opponents reduce the frequency of some aggressive elements, but increase the occurrence of others. CDP treatment of the females increased the duration of On Top in the small opponent condition (not shown) leaving most other aggressive acts uninfluenced. When confronted with heavy intruders, the CDP treated females responded with increased aggression, most notably Bite Attacks (X^2=13.0, df=3, p=0.0005) and On Top (X^2=19.5, df=3, p=0.0002). Significantly different groups are connected and indicated by an asterisk. (Reproduced with permission from Pharmacol Biochem Behav 26: 577–584, 1987).

aggressive acts was increased by CDP treatment when large opponents were used, but not when small sized opponents were used. The time spent on aggression was, however, significantly increased under both intruder conditions (data not shown). Inspection of the individual aggressive elements such as on top, lateral threat, bite attacks and upright posture reveals the different strategies used by CDP treated animals. With large opponents, CDP treatment specifically increases bite attacks and on top, (the more intense aggressive acts).

From these experiments one can conclude that increasing the opponent's size inhibits maternal aggression. Under those "inhibited aggression" conditions, the pro-aggressive actions of CDP become more evident. These data, however, do not allow definite conclusions about the mechanism of action of CDP. The inhibition of aggression and rate-dependency phenomena are intermingled. One can conclude however, that experimental conditions play an important role: If aggression levels are high, stimulatory effects are hard to detect, a finding that may explain the lack of pro-aggressive actions in several experiments.

Aggression induced by electrical stimulation (see earlier) is not facilitated by low doses of BDZ. In a similar fashion as employed in maternal aggression, an attempt was made to study if under heavy opponent conditions the pro-aggressive action of CDP became evident. Figure 3B summarizes the results: a selected dose of 5 mg/kg CDP did not facilitate aggression. In fact, large opponents were attacked at similar threshold currents (Friedman $X^2=5.2$, df=3, p=0.16). Although the sample size is small (n=6), the animals served as their own control, and the consequences of similar treatments in maternal aggression justified the expectation that larger effects were possible.

In the first experiment, the high level of aggression of the maternal females, virtually precluded any possible pro-aggressive actions of CDP. By analogy, it can be hypothetized that the EBS attacks are so intense that facilitation by CDP is impossible because of ceiling effects. By contrast to maternal aggression, EBS induced attacks were not reduced/inhibited by large opponents. This finding suggest that behavioural inhibition plays a less important role in EBS-induced attacks than in other aggression models. Also, if it is true that BDZ act by disinhibiting suppressed behaviour, there is no reason to expect pro-aggressive actions in this aggression model, because inhibitions are not of paramount importance in EBS-induced attacks.

Colony aggression in rats

Low doses of BDZ's may increase aggression in several animals under varying test conditions (Essman 1978). There are a few drugs that increase aggression in a rather naturalistic way, in contrast to hyperreactivity and defence-like reactions sometimes observed following drug withdrawal. The mechanisms by which BDZ stimulate aggression have not yet been unravelled. Some explanations suggest a drug-related release of suppressed or inhibited behaviours (Miczek and O'Donnell 1980).

In groups of rats usually one rat becomes dominant (Blanchard and Blanchard 1977). This alpha male is the most aggressive individual, both within the group and when a strange rat is placed into the colony. All other colony males and females attack at a much lower rate or show no aggression at all. Apparently, the dominant male suppresses the attack by other, subordinate animals on an intruder; the latter either take no initiative, or are chased away by the alpha male.

A slightly modified colony situation was used to assess the pro-aggressive effects of low doses of CDP on a dominant and subordinate male rat towards an intruding male. Two male and one ovariectomized female S3 (Tryon Maze Dull) rat were permanently housed in a 80x60x60 cm cage. Cages were kept in a temperature ($22\pm1°C$) and humidity (60–65%) controlled room under a reversed day and night schedule (lights off 07:00 hours, light on 19:00 hours local time). Aggression tests were conducted during the first part of the morning under dim red lighting. Ten minute aggression tests were performed weekly by placing an intruder into the cage after the female had been removed. The hierarchy between the two males had already been established on the basis of previous baseline observations. Time spent on aggression by the dominant and the subordinate was recorded. Aggression varied from relatively mild skin pulling, upright posture, sideways threat and teeth chattering to highly threatening or damaging acts such as biting the head and body, clinch fights, chasing, jump attacks and "on top or full aggressive posture". Aggression towards the intruder as well as aggression between the cage-mates was recorded. Differences in times spent on aggression were tested by Friedman analysis of variance followed by multiple comparison tests between the doses. The changes in aggression after drug treatment with respect to control condition of the dominant and subordinate rats was tested by Wilcoxons matched pairs comparison.

CDP (5, 10 and 20 mg/kg) was suspended in 1% tragacanth and administered orally 60 minutes before the test. 1% tragacanth served as the control injection. Drugs were given in a randomized design to fourteen pairs of rats; each member of the pair receiving the same treatment.

Results are summarized in fig. 5. Under control conditions, the dominant males spent more time on aggression towards the intruders than the subordinates (222 vs

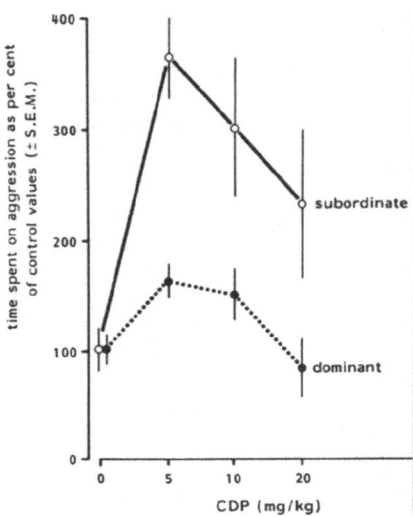

Fig. 5 Mean time spent on aggression (as % of control) by the dominant and the subordinate rat in a small colony situation towards a male intruder rat. Both the dominant and subordinate resident were treated with 0, 5, 10 and 20 mg/kg CDP (PO) 30 minutes before testing. The increase in aggression after CDP treatment was significantly higher in the subordinate than in the dominant male.

78 seconds). To compare the change in aggression after CDP treatment, data are expressed as percentage of vehicle condition. CDP treatment significantly increased aggression towards the intruder both in the dominant and subordinate at 5 and 10 mg/kg. However, the increase in aggression was much higher in the subordinate than the dominant ($p<0.05$, $p=0.07$, $p<0.05$ at 5, 10 and 20 mg/kg respectively). At the same time, aggression between the cage mates is unchanged.

These data confirm previous observations of a pro–aggressive effect of low doses of CDP (Miczek and O'Donnell 1980; Olivier et al. 1985). Moreover, they show that the dominant and subordinate react differentially towards CDP. The dominant rat still attacked more frequently after CDP, which indicates that aggression suppressing mechanisms are active even in readily attacking males. Subordinate males, whose aggression is clearly suppressed by the dominant male and most likely also by the mechanisms referred to above, are more sensitive to the pro–aggressive effects of CDP. The differential increases in aggression of subordinate animals demonstrates that CDP counters the inhibition or suppression of behaviour caused by social interactions as well as that evident after punishment such as after the use of foot shocks. There are a multitude of ways to bring about an inhibition or suppression of behaviour. For a further insight into the mechanisms of action of the benzodiazepines it seems necessary to study the neural substrates involved in the inhibition of aggresive behaviour in more detail.

Muricidal responses by naive rats

A wide variety of factors inhibit aggression but it is not always clear whether they are related to anxiety or fear. Novelty is a factor that may inhibit behaviour and is sensitive to BDZ treatment (File 1980; Gardner and Guy 1984). It was decided to study one of these "novelty-inhibited" behaviours in more detail.

Leaf et al. (1975, 1984) have demonstrated an increase in the percentage of killer rats after BDZ treatment of naive rats. If inhibition of mouse-killing by the novel experience (a naive rat confronted for the very first time with a mouse) is mainly due to novelty, than CDP treatment should be most effective during the first session and show little effect later on. When there is a direct pro-aggressive action of CDP, one might expect facilitation of mouse killing also at a later time of testing, i.e. when novelty-induced inhibition has disappeared due to repeated testing.

Naive female rats were treated with a selected dose of CDP (5 mg/kg PO) and tested for muricide. Two separate experiments were conducted. The first consisted of a 3 trial cross-over design in which 24 animals were treated three times with CDP (t=-60 min), followed by three tests with vehicle (1% tragacanth). A second group of 24 female rats were subjected to the reverse regimen. Tests were conducted at weekly intervals.

Table I   Proportion of female rats killing a mouse in a 30 minute test period (median killing latency in minutes between brackets). Mann Whitney U tests revealed significant shorter killing latencies in the CDP group (p<0.001). Moreover, a comparison between the latencies of the third and fourth tests showed a carry-over effect (p=0.0003).

	treatment	1	2	3	treat-ment	4	5	6	test number and sequence
group I	CDP	17/24 (8)	18/24 (1)	19/24 (1)	vehicle	17/24 (1)	19/24 (1)	19/24 (1)	
group II	vehicle	5/24 (>30)	5/24 (>30)	5/24 (>30)	CDP	7/24 (>30)	6/24 (>30)	7/24 (>30)	

Results of the first experiment are summarized in table 1. CDP treatment resulted in a significantly higher percentage of female rats displaying mouse killing. Repeated testing did not dramatically change the percentage of killers in either group, but killing performance, as indicated by the shortening killing latencies, improved. After the cross-over, the first group of animals continued to kill, while CDP treatment was unable to facilitate mouse killing in (now) experienced non-killers. The significant carry over effect reveals two possible aspects of the mechanism of action of BDZ with regard to pro-aggressive effects. First, the inhibition to kill a mouse can only be overcome by CDP if it is related to novelty (fear), or lack of experience with the response. CDP is no longer "needed" after some positive experiences (learning). At the same time, however, it seems that the opposite also holds true. Repetitive negative experience cannot be changed by CDP treatment. Although mouse-killing is not in all respects representative for other forms of aggressive behaviour, these findings indicate that (negative) experience makes some animals insensitive to possible pro-aggressive actions of BDZ. The results fit more into an anxiolytic mechanism of action than a direct pro-aggressive action, since, in the latter case, a stimulatory effect in the second group whould be expected.

198

Table II Proportion of female rats killing a mouse in a 30 minute test period
(median killing latency in minutes between brackets).
A single trial cross–over design was employed to test whether one dose
of CDP during the first test (when the female is naive) resulted in a
carry over effect in the second test. In the first test, CDP significantly
decreased the killing latency (p<0.0001); after the first test a small but
significant carry–over effect was observed (p=0.014).

	treatment	1	treatment	2	test number
group I	CDP	17/25 (6)	vehicle	12/25 (>30)	
group II	vehicle	1/25 (>30)	CDP	11/25 (>30)	

In order to test the time course of carry–over effects of CDP, the experiments
were repeated with another group of naive female S3 rats in a single cross–over
design Table 2 summarizes the results. Again, CDP had a marked effect on mouse
killing during the first test. After the single cross–over, however, CDP was able to
stimulate mouse killing in the group of rats with only one session of negative
experience. For the group of rats first treated with CDP a carry–over effect was
found at the second test. In this test situation, the first experience is apparently of
great consequence for future reactions. This simple experiment shows the
subtleties in experimental design that may lead to important behavioural
differences. Incidentally, the inability of CDP to induce mouse killing in rats
already tested three times, is in accordance with earlier data by Karli (1956) who
reported that rats that failed to kill three consecutive times never killed
afterwards. Moreover, in experienced male killer rats it was also observed that
CDP was unable to facilitate killing (unpublished results).

Play–fighting
Play–fighting in young rats has recently received attention as an interesting
paradigm to study the effects of psychotropic agents on social, affiliative
behaviour. The psychopharmacology of play–fighting has been reviewed by
Panksepp et al. (see chapter in this volume). Although much has to be learned about
the nature and function of play–fighting, it is clear that this juvenile precursor of
adult agonistic behaviour has a number of behavioural elements (e.g. pinning, on
top) which are very similar to those displayed during adult agonistic behaviour.
Panksepp et al. (see chapter in this book) found that CDP had no stimulatory action
on the incidence of pinning. However, they used a design in which the animals were
repeatedly tested. In the present experiment, naive juvenile rats were tested for
play–fighting in a novel environment. Briefly, male S3 pups were weaned at day
21–23 and subsequently housed in isolation. After 14 days of isolation, pups from
different litters were placed into a test cage (40x30x30 cm) in dyads which had
received the same treatment. The behaviour of these rats was recorded and the
number of pins and the latency to the first pin were measured. Animals were
treated with 0, 1.25, 2.5, 5 and 10 mg/kg CDP suspended in gelatin–mannitol (t=–30
min, IP. Each group consisted of 10 pairs of animals. Each animal was used only
once.
Results are summarized in fig. 6. A one–way analysis of variance
(Kruskal–Wallis) revealed a significant overall effect in the latency (H^2=17.2,
df=4, p=0.0017) and number (H^2=21.3, df=4, p=0.0003) of pins. Mann Whitney–U

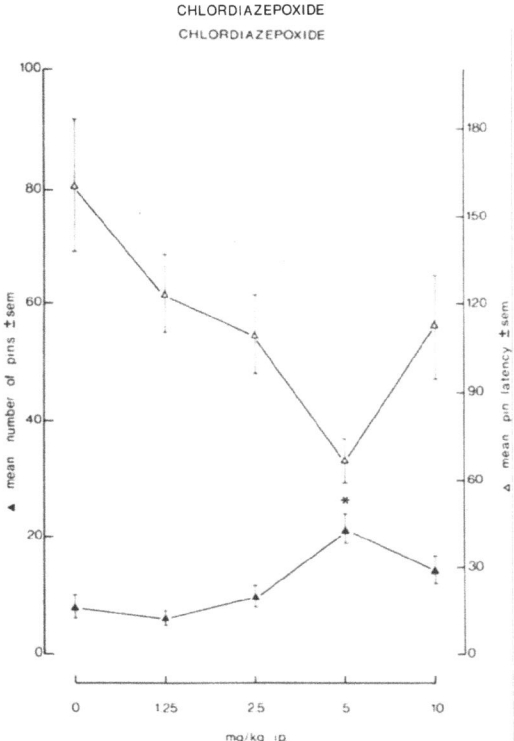

CHLORDIAZEPOXIDE

Fig. 6   Number of pins (mean ± SEM) and latency to the first pin in a test for play-fighting in male rat pups. CDP increased the number of pins and decreased the latency significantly at 5 mg/kg IP. The dose effect curve was biphasic, similar to maternal aggression (n=10 pairs of animals for each dose).

tests for independent samples revealed significant differences between control and 5 mg/kg conditions, both for latencies and number of pinnings.

These data are consistent with previous findings in maternal aggression and mouse-killing behaviour in which 5 mg/kg CDP was the optimal dose for its pro-aggressive actions. Although CDP induced a statistically significant effect, the behaviour displayed by the young rats lends support to the idea that aggression in more experienced rats can not be facilitated by CDP. Animals given CDP freeze less and therefore engage more readily in social contact. Repeated testing reduces the period of freezing in the new cage (personal communication; D.Einon) and the animals may start play-fighting more quickly. Direct pro-aggressive actions of CDP should result in some increase in experienced play-fighters too. A reasonable hypothesis may be that the anxiolytic action of CDP reduces freezing, thus facilitating the dominant behavioural response (i.e. social contact, play-fighting). A more detailed analysis, not limited to pinning alone, in both experienced and naive rats is necessary to elucidate the behavioural mechanism of action.

Benzodiazepine receptor antagonists and maternal aggression

It is of interest to determine whether the pro-aggressive actions of BDZ are caused by direct interaction with the BDZ receptor. This is especially important as a w.de variety of drugs which vary from full agonists through neutral antagonists to full or partial inverse agonists have been identified. Our own evidence (unpublished data) suggests that the anxiogenic inverse agonist carboline β-CCE only has inhibitory effects on maternal aggression but Beck and Cooper (1986) reported dose-related decreases in agonistic behaviour between pair-housed male rats by the β-carboline FG7142 (another partial inverse benzodiazepine agonist). Total social interaction was unaffected. The purported neutral BDZ antagonist RO 15-1788 has been studied in several aggression paradigms. Rodgers and Waters (1984) described differential effects of RO 15-1788 on social behaviour of · male resident and intruder mice. Offensive threat was increased in residents, but not in a consistent dose-related way. Olfactory investigation was reduced. In intruders, behavioural effects were only observed at 1.25 mg/kg, suggesting a reduced defensiveness. It thus appears likely that the behavioural effects of RO 15-1788 are subtle and much dependent upon the status of the animal. Skolnick et al. (1985) studied the antagonism of diazepam effects on isolation-induced aggression in mice. Diazepam reduced aggression in their paradigm, while RO 15-1788 (10 ng/kg) counteracted this reduction. RO 15-1788 itself was virtually without effect. Moreover, they identified anti-aggressive properties of β-CCE in this model; a reduction which was also antagonized by RO 15-1788. Sulcova and Krsiak (1984) discriminated between aggressive and timid isolated mice. Diazepam reduced tail rattling and aggressive unrest significantly in the aggressive mice and increased sociable activities, all effects which were counteracted by RO 15-1788 (20 ng/kg). In timid mice, diazepam reduced defensive postures and escapes, effects which were also antagonized by RO 15-1788. RO 15-1788 itself increased alert postures and locomotion in timid mice. Our own experiments with RO 15-1788 in maternal aggression (Mos et al. 1986) revealed no significant effects of RO 15-1788 on aggression. Summarizing the evidence reported thusfar in the literature, the anti-aggressive actions of BDZ can be antagonized by RO 15-1788, but pro-aggressive actions of BDZ have not been shown to be antagonized.

The purpose of the present experiment was to study the potential antagonism of CDP induced increases in aggression by RO 15-1788, the most neutral BDZ receptor antagonist presently available. Experimental procedures were similar to those described in Olivier et al. (1985) and Mos and Olivier (1986). Eighteen maternal females were tested for five minutes against naive male intruders. Treatment with 0 or 5 mg/kg CDP (PO, t=-60 min) was followed by RO 15-1788 (0, 1.25 or 10 mg/kg, IP, t=-30 min). Drugs were given according to a randomized block design, each female serving as her own control. The number of bite attacks and the latency to the first attack were recorded. The number of attacks was corrected for the latency and the resultant mean bites/minute are plotted in figure 7. Analysis of variance revealed a significant pro-aggressive effect of CDP (p=0.02), no effects of RO 15-1788 (p=0.26) and no interaction between RO 15-1788 and CDP (p=0.70). The absence of antagonism was contrary to expectations and the complete behaviour was therefore analysed (Mos and Olivier; in press). This more sophisticated analysis did not change the basic result, which was that pro-aggressive actions of CDP were not antagonized by RO 15-1788.

Since most behaviours elicited by agonistic actions of benzodiazepines – among these, the anxiolytic effects – are effectively counteracted by RO 15-1788, the possibility must be taken seriously that the pro-aggressive actions of benzodiazepines are not directly caused by interactions with the BDZ receptor. Of course, more BDZ should be tested (e.g. oxazepam) before any definite conclusion can be drawn. The possibility also exists that higher doses of RO 15-1788 are needed, but this seems unlikely in view of the many studies that demonstrate antagonistic activity at doses lower than 10 mg/kg (Bonetti et al. 1982). Moreover,

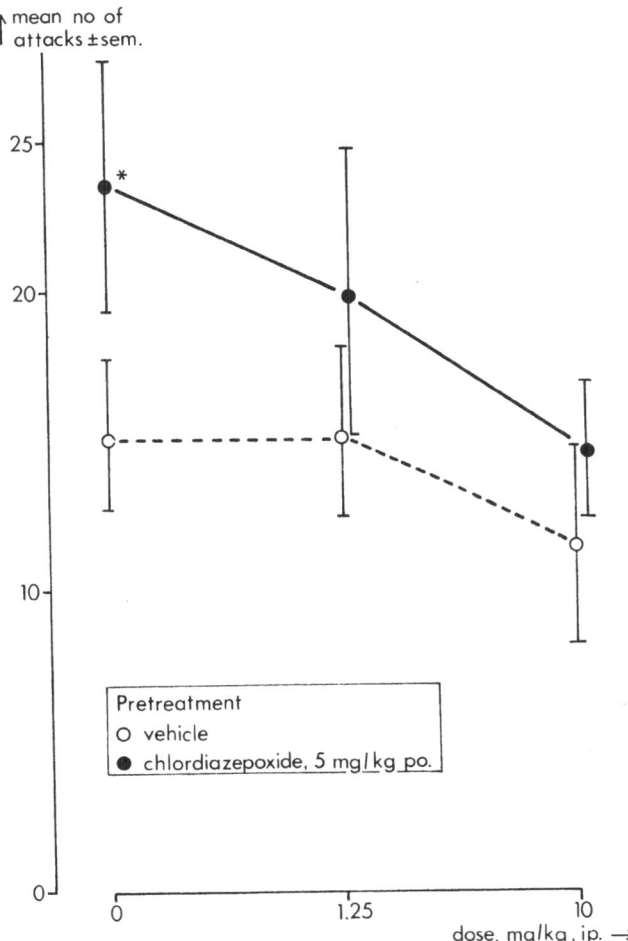

RO 15-1788

Figure 7. Interaction of CDP and RO 15-1788. Maternal aggression of female rats tested with standard opponents is expressed as mean bites per minute, corrected for attack latency. 5 mg/kg CDP (PO) increased aggression, but RO 15-1788 (IP) failed to antagonize this pro-aggressive effect. RO 15-1788 itself was without significant effects on aggression in the dose range tested.

the pharmacokinetics of RO 15-1788 are peculiar, with a short half life (±17 minutes; Lister et al. 1984), suggesting that the injection of RO 15-1788 thirty minutes before testing may be inappropriate. However, similar RO 15-1788 administration antagonized CDP-induced feeding (unpublished results). On the other hand, complex interactions between agonists of the BDZ receptor and RO 15-1788 have been described, which cannot simply be explained by pharmacokinetic factors (Lister and File 1986).

These results thus lead to the suggestion that pro–aggressive actions of BDZ are not mediated by the BDZ receptor, but are side–effects of the drugs acting at as yet unknown sites. That is not to say that anxiolytic actions do not, or do never contribute to increased aggression. However, it seems reasonable to suspect that the pro–aggressive actions of (some ?) BDZ arise from the summation of a number of smaller effects, each of which may not be of sufficient magnitude to be recognized and tested appropriately.

CONCLUSIONS AND FUTURE PROSPECTS

The experiments described in this chapter indicate once more the difficulties in establishing the actions of BDZ on aggression. The question posed by DiMascio in 1973: "The effects of benzodiazepines on aggression: reduced or increased ?" is still unanswered. It seems clear that higher doses of BDZ decrease aggression, perhaps with doubtful specificity. Pro–aggressive actions seem to depend on the basal level of aggression, the testing environment, experience (both with fighting and the test conditions), status, and the measures with which aggression is quantified (behavioural elements, time and frequency) and perhaps many other variables.

Assuming that disinhibition is one of the main factors in the anxiolytic actions of BDZ, the literature shows a bewildering variety of inhibitory factors. Inhibition may occur in many forms and different situations. This may depend on familiarity of test environment (Miczek and O'Donnell 1980), lighting conditions (File 1980), novel experience (e.g. muricide in naive rats; Leaf et al. 1975, 1984), brain stimulation induced suppression (Kozak et al. 1984), animal traits (Krsiak; timid and aggressive mice), prey size (Apfelbach 1978) and social status (Apfelbach and Delgado 1974). The wide variety of factors causing inhibition is not always clearly related to anxiety or increased fear. Dantzer (1977) has characterised the actions of BDZ as being mainly "a response perseveration effect or the strengthening of the prevailing behavioural output" (p.80). Whether this is exclusive to or partially overlapping with the disinhibition is outside the scope of this chapter and perhaps more a matter of definitions and semantics. The idea is attractive because it intuitively "explains" what is observed. However, it may not always be evident what the prevailing behavioural tendency is, especially in non–operant test situations. Moreover, if the external conditions vary, the behavioural tendency may change. When the prevailing behavioural tendency cannot unequivocally be established and/or may change during the observation period, the elegant description of the mechanism of action cannot be falsified.

To reconcile all the data within the framework of the two afore–mentioned hypotheses seems an impossible task. A summarizing description, without any pretention to this being an alternative hypothesis, would be as follows. BDZ enhance aggression when the basal levels are low to moderate. Highly aggressive animals are not stimulated, either because inhibitions play no significant role, or because the maximal behavioural output has been attained and no further increase is possible due to physical limitations. Why basal levels of aggression are low in some animals and some test situations is a question that cannot be simply answered. If anxiety reduces aggression, the anxiolytic actions of BDZ may result in a disinhibition of (anxiety) suppressed aggression. Our data do not support the idea that the pro–aggressive effect is exclusively BDZ receptor mediated. Perhaps BDZ exert other pharmacological actions that cannot be neglected as an interpretation of the pro–aggressive effects? Studies with other anxiolytic drugs on the one hand and basic experiments on the factors promoting and inhibiting aggression on the other, may provide avenues to elucidate the mechanisms of the pro–aggressive actions of BDZ.

Acknowledgements: We like to thank Ruud van Oorschot, Theo Zethof and Albert Ramakers for technical support and Marijke Mulder for typing the manuscript.

REFERENCES

Apfelbach R (1978) Instinctive predatory behavior of the ferret (Putorius putorius furo L.) modified by chlordiazepoxide hydrochloride (Librium). Psychopharmacology 59: 179–182

Apfelbach R, Delgado JMR (1974) Social hierarchy in monkeys (Macaca mulatta) modified by chlordiazepoxide hydrochloride. Neuropharmacology 13: 11–20

Arnone M, Dantzer R (1980) Effects of diazepam on extinction induced aggression in pigs. Pharmacol Biochem Behav 13: 27–30

Azcarate CL (1975) Minor tranquilizers in the treatment of aggression. J Nerv Ment Disease 160: 100–107

Beck CHM, Cooper SJ (1986) The effect of the β–carboline FG 7142 on the behaviour of male rats in a living cage: An ethological analysis of social and nonsocial behaviour. Psychopharmacology 80: 203–207

Blanchard RJ, Blanchard DC (1977) Aggressive behavior in the rat. Behav Biol 21: 197–224

Bond A, Lader M (1979) Benzodiazepines and aggression. In: Sandler M (ed), Psychopharmacology of Aggression. Raven Press, New York, pp 173–182

Bonetti EP, Pieri L, Cumon R, Schaffner R, Pieri M, Gamzu ER, Muller RKM, Haefely W (1982) Benzodiazepine antagonist RO 15–1788: Neurological and behavioral effects. Psychopharmacology 78: 8–18

Cherek DR, Kelly TH, Steinberg JL, Friedman TT (1986) Effects of acute alcohol or diazepam administration on human aggressive and non–aggressive responding. In: Shagass C et al. (eds) Biological Psychiatry, Elsevier, New York, pp 497–499

Dantzer R (1977) Behavioral effects of benzodiazepines: a review. Biobehav Rev 1: 71–86

DiMascio A (1973) The effects of benzodiazepines on aggression: reduced or increased? Psychopharmacologia 30: 95–102

DiMascio A, Shader RI, Harmatz J (1969) Psychotropic drugs and induced hostility. Psychosomatics 10: 46–47

Dixon AK (1982) A possible olfactory component in the effects of diazepam on social behavior of mice. Psychopharmacology 77: 246–252

Downing RW, Rickels K (1981) Hostility conflict and the effect of chlordiazepoxide on change in hostility levels. Compr Psychiatry 22: 362–367

Essman WB (1978) Benzodiazepines and aggressive behaviour. In: Modern Problems in Pharmacopsychiatry. Valzelli L (ed), Psychopharmacology of Aggression. Karger, Basel, pp 13–28

File SE (1980) The use of social interaction as a method for detecting anxiolytic activity of chlordiazepoxide–like drugs. J Neurosci Meth 2: 219–238

Flannelly KJ, Flannelly L (1985) Opponents' size influences maternal aggression. Psychol Rep 57: 883–886

Fox KA, Guerriero FJ, Zanghi DC (1974) Oxazepam, prazepam, and aggression in mice. Pharmacol Res Comm 6: 301–306

Fox KA, Snyder RL (1969) Effect of sustained low doses of diazepam on aggression and mortality in grouped male mice. J Comp Physiol Psychol 69: 663–666

Fox KA, Tuckosh JR, Wilcox AH (1970) Increased aggression among grouped male mice fed chlordiazepoxide. Eur J Pharmacol 11: 119–121

Fox KA, Webster JC, Guerriero FJ (1972) Increased aggression among grouped male mice fed nitrazepam and flurazepam. Pharmacol Res Comm 4: 157–162

Friedman H, Abernethy DR, Greenblatt DJ, Shader RI (1986) The pharmacokinetics of diazepam and desmethyldiazepam in rat brain and plasma. Psychopharmacology 88: 267–270

Gardner CR, Guy AP (1984) A social interaction model of anxiety sensitive to acutely administered benzodiazepines. Drug Devl Res 4: 207–216

Gardos G, DiMascio A, Salzman C, Shader RI (1968) Differential actions of chlordiazepoxide and oxazepam on hostility. Arch Gen Psychiatry 18: 757–760

Horovitz ZP, Piala JJ, High JP, Burke JC, Leaf RC (1966) Effects of drugs on the mouse–killing (muricide) test and its relationship to amygdaloid function. Int J Neuropharmacol 5: 405–411

Ingram IM, Timbury GC (1960) Side–effects of librium. Lancet 2: 766

Karli P (1956) The Norway rat's killing response to the white mouse: an experimental analysis. Behaviour 10: 80–103

Karli P (1961) Action du methaminodiazepoxide (Librium) sur l'agressivité interspécifique Rat–Souris. C R Soc Biol 625–627

Kozak W, Valzelli L, Garattini S (1984) Anxiolytic activity on locus coeruleus–mediated suppression of muricidal aggression. Eur J Pharmacol 105: 323–326

Krsiak M (1975) Timid singly–housed mice: their value in prediction of psychotropic activity of drugs. Br J Pharmacol 55: 141–150

Krsiak M (1979) Effects of drugs on behavior of aggressive mice. Br J Pharmacol 65: 525–533

Krsiak M, Sulcova A, Tomasikova Z, Dlohozkova N, Kosar E, Masek K (1981) Drug effects on attack, defense and escape in mice. Pharmacol Biochem Behav 14 (suppl 1): 47–52

Kruk MR, Van der Poel AM, De Vos–Frerichs TP (1979) The induction of aggressive behaviour by electrical stimulation in the hypothalamus of male rats. Behaviour 70: 292–322

Leaf RC, Wnek DJ, Gay PE, Corcia RM, Lamon S (1975) Chlordiazepoxide and diazepam induced mouse killing by rats. Psychopharmacologia 44: 23–28

Leaf RC, Wnek DJ, Lamon S (1984) Oxazepam induced mouse killing by rats. Pharmacol Biochem Behav 20: 311–313

Lion JR, Azcarate CL, Koepke HH (1975) "Paradoxical rage reactions" during psychotropic medication. Dis Nerv System 36: 557–558

Lister RG, File SE (1986) A late appearing benzodiazepine–induced hypoactivity that is not reversed by a receptor antagonist. Psychopharmacology 88: 520–524

Lister RG, Greenblatt DJ, Abernethy DR, File SE (1984) Pharmacokinetic studies on RO 15–1788, a benzodiazepine receptor ligand, in the brain of the rat. Brain Res 290: 183–186

Miczek KA (1974) Intraspecies aggression in rats: Effects of d–amphetamine and chlordiazepoxide. Psychopharmacology 39: 275–301

Miczek KA (1987) The psychopharmacology of aggression. In: Iversen LL, Iversen S, Snyder SH (Eds) Handbook of Psychopharmacology. Vol. 19: Behavioural Pharmacology, Plenum Press, New York, pp 183–328

Miczek KA, O'Donnell JM (1980) Alcohol and chlordiazepoxide increase suppressed aggression in mice. Psychopharmacology 69: 39–44

Mos J, Olivier B (1986) RO 15–1788 does not influence maternal aggression in rats. Psychopharmacology 90: 278–280

Mos J, Olivier B, Van Oorschot R (1987) Maternal aggression towards different sized male opponents: effect of chlordiazepoxide treatment of the mothers and d–amphetamine treatment of the intruders. Pharmacol Biochem Behav 26: 577–584

Olivier B, Mos J, Van Oorschot R (1985) Maternal aggression in rats: effects of chlordiazepoxide and fluprazine. Psychopharmacology 86: 68–76

Olivier B, Mos J, Van Oorschot R (1986) Maternal aggression in rats: lack of interaction between chlordiazepoxide and fluprazine. Psychopharmacology 88: 40–43

Poshivalov VP (1980) The integrity of the social hierarchy in mice following administration of psychotropic drugs. Br J Pharmacol 70: 367–373

Randall LO (1960) Pharmacology of methaminodiazepoxide. Dis Nerv System 21: 7–10

Rickels K, Downing RW (1974) Chlordiazepoxide and hostility in anxious out-patients. Am J Psychiatry 131: 442–444

Rodgers RJ, Waters AJ (1984) Effects of the benzodiazepine antagonist RO 15–1788 on social and agonistic behaviour in male albino mice. Physiol Behav 33: 401–409

Rosenbaum JF, Woods SW, Groves JE, Klerman GL (1984) Emergence of hostility during alprazolam treatment. Am J Psychiatry 141: 792–793

Salzman C, DiMascio A, Shader RI, Harmatz JS (1969) Chlordiazepoxide, expectation and hostility. Psychopharmacologia 14: 38–45

Sheard MH (1984) Clinical pharmacology of aggressive behaviour. Clin Neuropharmacol 7: 173–183

Skolnick P, Reed GF, Paul SM (1985) Benzodiazepine–receptor mediated inhibition of isolation–induced aggression in mice. Pharmacol Biochem Behav 23: 17–20

Sofia RD (1969) Effects of centrally active drugs on four models of experimentally–induced aggression in rodents. Life Sci 8: 705–716

Sulcova A, Krsiak M (1984) The benzodiazepine–receptor antagonist RO 15–1788 antagonizes effects of diazepam on aggressive and timid behaviour in mice. Activ Nerv Sup (Praha) 26: 255–256

Valzelli L (1973) Activity of benzodiazepines on aggressive behavior in rats and mice. In: Garattini S, Mussini E, Randall LO (eds) The Benzodiazepines. Raven Press, New York, pp 405–417

Van der Poel AM, Mos J, Kruk MR, Olivier B (1984) A motivational analysis of ambivalent actions in the agonistic behaviour of rats in tests used to study effects of drugs on aggression. In: Miczek KA, Kruk MR, Olivier B (eds) Ethopharmacological aggression research, Alan R Liss, New York, pp 115–136

Wetherill GB (1966) Sequential estimation of points on quantal response curves. In: Wetherill GB (ed) Sequential methods in statistics, London: Methuen and Co, Ltd

Wilkinson CJ (1985) Effects of diazepam (Valium) and trait anxiety on human physical aggression and emotional state. J Behav Med 8: 101–114

Zwirner PP, Porsolt RD, Loew DM (1975) Inter–group aggression in mice. Psychopharmacologia 45: 133–138

SEROTONIN, SOCIAL BEHAVIOUR, AND AGGRESSION IN VERVET MONKEYS

Michael T. McGuire[1,2] and Michael J. Raleigh [1,2,3]

1) Department of Psychiatry–Biobehavioral Sciences School of Medicine, University of California at Los Angeles, Los Angeles, California 90024, USA

2) Nonhuman Primate Laboratory, Sepulveda Veterans Administration Medical Center, Sepulveda, California 91343, USA

3) Neurobiochemistry Laboratory, Brentwood Veterans Administration Medical Center, Los Angeles, California 90073, USA

INTRODUCTION

In our view, studies of the psychopharmacology of aggression are greatly enhanced by investigating animals or humans living in social groups. For example, differences in the form and intensity of social status relationships, the age–sex composition, and the subject's tenure in their groups can all affect an individual's behavioural options and degree of social inhibition. These factors are known to be associated with different physiological states. Different physiological states, are associated with different probabilities of specific behaviours (including aggression), different rates of drug metabolism, and different behavioural responses to drugs (McGuire et al. 1982). In effect, social behaviour, physiology, and pharmacology may be viewed as three corners of a triangle, each corner interacting with and exerting influence on the other two. Pharmacological studies designed within this paradigm result in new types of data and interpretations, different from findings and conclusions developed using isolated or paired animals.

With the exception of results shown in Table 1, this paper reviews studies illustrating points mentioned in the preceding paragraph. The following will provide focuses; behavioural conditions that alter the function of serotonin systems; relationships between these alterations and aggressive behaviour; and, the effects of drugs altering serotonin function and aggressive behaviour. Findings are largely taken from studies of Cercopithecus aethiops sabaeus, the vervet or West African green monkey. The present state of animal and human research will also be summarized, the questions remaining to be answered delineated, selected findings reviewed and, tentative conclusions reached about serotonergic functions and aggressive behaviour.

PRESENT STATE OF ANIMAL AND HUMAN STUDIES

This selective review of the present state of animal and human research on aggression underscores the facts that there are a variety of approaches to the study of aggression and, our understanding of aggression is far from complete.

Several (not necessarily mutually exclusive) approaches to the study of aggression namely evolutionary, taxonomic, and social, as well as interactions between different systems will be discussed.

A recent evolutionary biological overview of the phenomena of aggression has been developed by McKenna (1983; see also Blanchard and Blanchard 1984). From an evolutionary perspective, aggression is a phylogenetically old behaviour and a favoured product of selection. Its ultimate causes appear to lie in events that favour positive outcomes to attack and defensive behaviour, such as self-defence, protection of kin and resources, and establishing competitive advantages. Developmentally, aggressive capacities require the integration of many skills (see, for example, Adamec et al. 1980a, b, c).

For most higher species, skills are acquired and rehearsed at different points during development and not fully integrated until adulthood. Thus, aggression develops in a mosaic fashion. The functions of aggression are multiple and include those listed above, as well as frightening or destroying potentially dangerous precators. Proximal mechanisms also appear to be multiple. Aggression is sometimes instantaneous as, for example, when one is unexpectedly attacked. At other times it seems a response to repeated minor "frustrations" and appears to be "calculated". Such differences arise from different interactions between physiological and cognitive systems.

To say that evolution has favoured aggression does not imply selection for uncontrolled or indiscriminate aggression. Rather, evolution has and is likely to continue to favour the selective use of aggression under circumstances in which there is a reasonable chance that it will alter events to the benefit of the animal that is aggressing. One implication of the idea of selective use is that cognition (e.g., evaluating circumstances and their implications) is a critical element in understanding aggressive behaviour (see, for example, McGuire and Troisi, in press). This point is especially relevant to Old World monkeys, apes, and humans who live in environments that frequently change and are characterized by unpredictable events. A second implication is that studies which focus on event sequences leading to aggression (e.g., shock > irritation > attack another animal) may be less instructive than studies in which animals or persons are living in natural settings and the characteristics of the settings are closely monitored. Different social characteristics alter the probability of aggression. A third implication is that aggression will be associated with different conditions among different species. Such differences reflect diverse evolutionary histories.

A number of taxonomies of aggression have been developed. Perhaps the most historically influential is that of Moyer (1968), who subdivided aggressive behaviour into the following categories: predator, intermale, fear–induced, irritable, territorial defence, maternal, and instrumental. Revisions of Moyer's categories have been undertaken (see, for example, Averill 1982; Blanchard and Blanchard 1984; Brain 1984) not only to deal with the implications of mutual exclusiveness and independent causation that are implied in Moyer's categories, but also to devise categories that may have greater relevance to humans. In part for reasons suggested above (e.g., interactions between social conditions and an animal's or person's physiological state), as well as those discussed below, any existing taxonomy of aggression is unlikely to be completely satisfactory or achieve general acceptance. Individual investigators thus may find it necessary to develop their own taxonomies which accurately reflect the purposes and conditions of their studies.

Studies of aggression have covered the spectrum from field research to highly controlled laboratory investigations. Clearly, different types of studies have varied advantages and limitations. For example, in laboratory studies conditions can be controlled and replicability is possible. However, findings from studies using animals living alone, in pairs, or animals that are the product of special kinds of breeding may not be easily generalizable. The latter point is particularly relevant to studies involving the standard laboratory rat which, compared to the feral rat, is a docile and relatively unaggressive animal. Physiological mechanisms identified in laboratory rats may be generalizable, but functional generalizations are as likely to be wrong as right. For example, extreme conditions often are required to produce aggression in laboratory–bred rats whereas field living rats will frequently fight with less provocation. Studies of feral–raised animals living in natural settings have the advantage of increasing our knowledge of when and how animals behave aggressively in settings akin to those in which they evolved. A disadvantage of such studies is that contextual variables are difficult to control.

Often, however, contextual variables can be used to a research advantage. For example, in an ongoing study by the authors, administration of a specific drug to an animal living in a stable social group results in a significant decrease in the frequency of aggressive behaviour as well as ataxia. The same drug, at the same dose, given to the same animal in the same social group, with the added element that an unknown male is placed nearby (a frequent occurrence in natural settings) results in a similar decrease in the frequency of aggression but no evidence of ataxia. In this study, animals do not engage in physical contact. Thus, a cognitive mechanism is implicated: for example, the recognition of an unknown conspecific may initiate specific physiological changes (e.g., arousal) that offset the ataxic effects of the drug.

In attempting to delineate what system and system interactions are most frequently implicated in aggression, one is faced with a mass of perplexing data. For example, within–individual and cross–species studies (see, for example, Anderson and Chamove 1985; Chamove et al. 1985; Huck et al. 1985; Thierry 1985) point both to similarities in the conditions under which aggressive behaviour is likely to occur as well as similarities in physiological measures (e.g., elevated testosterone levels among dominant animals). Such findings would be expected among related species for phylogenetically old behaviour. Moreover, cross–species similarities suggest that the number of involved systems is limited. The degree to which such findings generalize to other species, particularly humans, is, however, unclear as yet.

For example, Averill (1982) has identified possible causes of both anger and aggression in humans, as including: disappointment in the behaviour of others, frustration with ongoing activities, loss of self-esteem, property damage, and injury and pain. Although these are phenomenological categories, and therefore have limitations, the noted differences in the causes suggest that many physiological or anatomical systems may be involved. Injury and pain information, for example, are carried along different neural pathways and affect different brain areas compared to those usually implicated in disappointment or frustration. Moreover, recent studies of nonhuman primates (see, for example, McGuire et al. 1986; Steklis et al. 1985) suggest that the generally accepted physiological correlates of aggression (e.g., elevated testosterone) are seen only under specific conditions.

Other studies indicate that many possibly interrelated systems are likely to change in association with aggression, e.g., blood pressure, pulse rate, testosterone levels, adrenal function, epinephrine function, glucose metabolism, and autonomic nervous system activity (see Brain 1984 for a recent review). Neurotransmitter systems implicated in these responses include (at least) norepinephrine, epinephrine, dopamine, and serotonin (see McGuire and Troisi, in press). Neuropeptide contributions to aggression, which seem probable, have not as yet been adequately studied. The diverse components of neuroanatomical systems implicated in aggressive behaviour include, the hypothalamus, amygdala and midbrain central grey (see Adams 1979; Albert and Walsh 1984). Indeed, a review of the literature reveals that so many systems have been implicated that it may be worthwhile asking which systems are not involved.

Table 1 Behaviour physiology interactions

This table summarizes findings from studies designed to assess behaviour–physiology interactions in semi–natural settings. A variety of modalities are implicated in physiological changes including: behaviour, olfaction, visual communication, and cognition. The table is further discussed in text.

SERUM CORTISOL
1. (Squirrel monkeys): levels higher in dominant than in subordinate males. (Coe et al. 1979)
2. (Squirrel monkeys): levels increase in both dominant and subordinate males during mating season (Coe et al. 1983)
3. (Talapoin monkeys): levels higher in subordinate than in dominant males. Levels increase in both subordinate and dominant males when females are made more attractive by oestradiol treatment (Eberhart et al. 1983)
4. (Squirrel monkeys): levels higher in dominant males than in subordinate males during group formation using males only. Differentiation increased when females added to groups (Mendoza et al. 1978)
5. (Vervet monkeys): levels higher in and, differentiate those animals who become dominant during competition for dominant status among subordinate males (McGuire et al. 1986)

NEOSTRIATAL DOPAMINE
(Pheasants): levels higher in dominant males than in subordinate males (McIntyre and Chew 1983)

SERUM LUTEINIZING HORMONE
(Talapoin monkeys): levels higher in both dominant and subordinate males when caged with females than when caged alone (Eberhart and Keverne 1979)

SERUM PROLACTIN
1. (Mangabey and patas monkeys): levels increase in response to ambiguity in social and challenge situations (Aidara et al. 1981)
2. (Talapoin monkeys): levels decrease in males in the presence of females who are made more attractive by oestradiol treatment (Eberhart et al. 1983)
3. (Talapoin monkeys): levels higher in subordinate than in dominant males (Hansen et al. 1979)

WHOLE BLOOD SEROTONIN
1. (Vervet monkeys): levels higher in dominant than in subordinate males (Raleigh et al. 1984)
2. (Vervet monkeys): levels decrease in dominant males when subordinate males are removed from social group (McGuire et al. 1983b)
3. (Squirrel monkeys): levels higher in dominant than in subordinate males (Steklis et al. in press)
4. (Humans): levels higher in high–status males compared to lower status males (McGuire et al. 1983b)
5. (Humans): levels higher in males who engage in type–A behaviour compared to males who engage in type–B behaviour (Madson and McGuire 1984)

Table 1 continued

SERUM TESTOSTERONE
1.   (Rhesus monkeys): levels increase in males introduced to females both in and out of breeding season. Levels greater in breeding season (Bernstein et al. 1977)
2.   (Squirrel monkeys): levels higher in dominant than subordinate males (Coe et al. 1979)
3.   (Talapoin monkeys): levels increase in both dominant and subordinate males when housed with females compared to being housed alone (Eberhart and Keverne 1979)
4.   (Talapoin monkeys): levels higher in dominant than in subordinate monkeys (Hansen et al. 1979)
5.   (Humans): levels increase in winners but not in losers of sports competitions (Mazur and Lamb 1980)
6.   (Squirrel monkeys): levels higher in dominant males than in subordinate males during periods of group formation (Mendoza et al. 1978)
7.   (Rhesus monkeys): levels and aggression positively correlate. Levels higher in dominant than in subordinate males (Rose et al. 1971)
8.   (Vervet monkeys): levels increase following aggression (Steklis et al. 1985)

CEREBROSPINAL FLUID 5-HYDROXYINDOLEACETIC ACID
(Talapoin monkeys): levels increase in males and females who become subordinate (Yodyingyuad et al. unpublished data)

While efforts to develop detailed taxonomies of aggression have not been entirely successful, there is little doubt that behaviours classifiable in broad categories, such as attack vs. defensive aggression, can be identified and agreed upon across investigators. A number of the preceding comments, as well as recent research reports, suggest that different anatomical and physiological systems may be associated with these two categories (see, for example, Adams 1979 and Albert and Walsh 1984). Further, the underlying physiological systems and the degree of their contributions to the elicitation or propogation of aggression appear to differ across species. To the extent that different systems and species are involved in studies, pharmacological investigations of aggression will be complex and formidable. Moreover, there is little likelihood that the complexity can be easily reduced. This conclusion is reinforced by the findings of Smith and Byrd (1983) who have documented the behavioural effects of different drugs in a variety of different group–living nonhuman primates: different group situations and/or species studied have significant influences on drug effects and, therefore, on attempts to use pharmacological probes to identify contributing systems.

REMAINING QUESTIONS TO BE ANSWERED

The preceding review hints at the kinds of questions remaining to be answered in studies of aggression. For example: "What are the contributions of and interactions between cognitive, neurotransmitter, neuropeptide, and discrete anatomical structures to aggressive behaviour?" "In what ways do different kinds of developmental experiences influence these interactions and contributions?" And, "What are the different functions of aggression?" A more complete understanding of interactions between contextual–behavioural variables and physiological function also appears to be an essential prerequisite for understanding psychopharmacological effects on aggression. The latter point is underscored by the findings shown in Table 1, which provides a selected review of behaviour–physiology interactions among animals living in social groups.

A number of points important to the theme of this paper are implicit in Table 1. For example: there are clear sex differences in the physiological responses to specific stimuli or group composition changes; alterations of contextual variables often are associated with robust physiological changes; different neurotransmitter and hormone changes are associated with different behaviours; and, there are status–linked differences in cognition that influence behaviour and physiology. To this list should be added the point that changes in different physiological systems occur over different time frames, e.g., peripheral serotonin function changes slowly while cortisol function changes rapidly. What Table 1 does not reveal, and what cannot be determined from studies like those cited in the table, are the exact contributions of specific neurotransmitters and hormones to specific behaviours. It is in clarifying such contributions that pharmacological studies have particular value. Table 1 also does not reveal a second and equally important point: namely, animals who are otherwise similar but in different physiological states will differ in both their metabolism of and behavioural responses to drugs (see below).

REVIEW OF FINDINGS

Characteristics of aggression in vervet monkeys

Morphologically and behaviourally, vervets are a generalized species. They eat nearly any type of food and, next to man, live in the largest variety of habitats of any primate. Reproductively, they are the second most successful primate (humans are the first), and they are not endangered. Compared to rhesus monkeys, vervets are only moderately aggressive (McGuire et al. 1983b). Generally they live in multi–male, multi–female groups although single–male (age–graded male groups) are sometimes observed. Multi–male groups nearly always have an identifiable dominant male.

Subordinate males differ in their relationships to each other and to dominant males (Fairbanks and McGuire 1979; Fairbanks et al. 1978). In most instances there is a recognizable linear hierarchy among adult males. When males reach adulthood, they usually transfer out of their natal group, often to a group in which they have a male relative. Females, on the other hand, tend to remain in their natal group. The frequency and type of aggression observed among males and females differs. With rare exceptions, aggression is confined to the same sex. In established groups, both males and females will develop alliances and aggress against other animals of the same sex. Alliances are more common among females than males (McGuire et al. 1983a).

We will focus on male–male aggression. The study of female aggression is still in its infancy, largely because females assimilate into newly formed groups at a significantly slower rate than males. In using the term aggression, we refer to those situations in which there is actual contact between animals and an injury occurs or the intent to injure another animal appears likely (as in situations of an intense and prolonged chase accompanied by specific threats, vocalizations, and attack behaviour). Among adult male vervets, aggressive interactions are most easily classified as attack or defence. More subtle taxonomic distinctions often are difficult to make.

In nearly all instances, aggression is preceded by threats and displays. Vervets seldom attack each other without a warning, although the period between the warning and an actual attack may be short. There are instances of aggression in which there is no observable behavioural precursor: one animal will simply threaten another animal and quickly escalate the encounter to the point of attack. Such behaviour suggests that animals are settling 'old scores' or that repeated small irritations have added up to the point that an animal attacks. It is possible, however, that behaviour cues exist and they have not been identified. Unless they are repeatedly challenged, dominant males usually engage in fewer aggressive interactions with any given subordinate male than the reverse. However, when there are many subordinate males in a group, the total number of agonistic or aggressive interactions for the dominant male will exceed the total for any subordinate male (McGuire, unpublished data).

Among captive groups living in semi-natural settings, part of the variance in the frequency and type of aggressive behaviour is related to an animal's tenure within a group. When male vervets which are unknown to each other are placed together to form a new social group, there is relatively little aggressive behaviour for a period of several weeks. During this period, animals develop affiliative alliances. Competition for dominance then begins and the number of agonistic and aggressive encounters increases rapidly until status relationships are established, a period usually lasting 4 to 6 weeks. In newly formed groups, behavioural frequencies among males become stable between 16–20 weeks. Once stability is achieved, the frequency of aggressive behaviour is generally lower than it was during the period in which dominance relationships were contested. Thus, dominance and aggression are not correlated on a 1-to-1 basis. There are exceptions, however. Some newly formed groups do not stabilize rapidly and the frequency of aggressive interactions and injuries remains high and clearly differentiated status relationships do not emerge.

An established dominant male generally wins greater than 80% of his agonistic encounters and is injured about one–quarter as frequently as an average subordinate male. While subordinate males are injured more, injuries are seldom debilitating and literally never lethal. Moreover, the eliciting factors, the form, and the apparent function of aggression differs as a function of social status. Aggression tends to occur in the following conditions: when a dominant male is frequently challenged by a subordinate male (during such periods subordinate males threaten and display but seldom engage in attack aggression); when a subordinate male is frequently challenged by a dominant male; when a subordinate male fails to submit to or be displaced from a favoured position by a dominant male; when a subordinate male copulates in the presence of a dominant male; and, when males compete for dominant status.

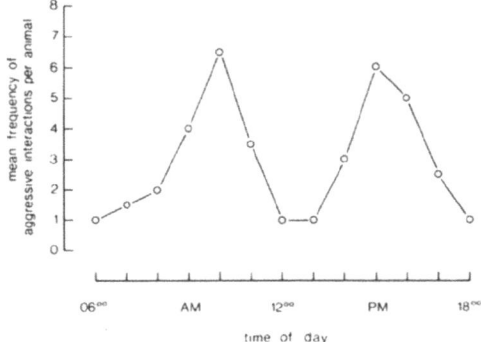

Fig. 1 Plots the frequency of threats and displays and aggressive bouts (animal contact) among adult male vervet monkeys living in socially stable groups. The line depicts the mean number of agonistic bouts per hour for groups (n=6) composed of one dominant and two subordinate males. Results represent findings from 32 observation days. The figure is further discussed in text (C.Johnson, unpublished data).

Among captive stable groups, the frequency of aggressive behaviours changes over the course of the day. These differences are shown in fig. 1. Figure 1 depicts two peak periods of aggressive behaviour, mid–morning and mid–afternoon. During those periods in which there are few threats, displays, or aggressive interactions, it is difficult to identify which animal in a group is dominant or subordinate. However, during the periods of peak agonistic activity, an inexperienced observer has little difficulty identifying dominant and subordinate males. In addition, the peak frequencies shown in the figure may be in part an artifact of enclosure living. In field situations, at comparable times, subordinate animals usually have dispersed themselves and often are several hundred meters from the dominant male of their group (McGuire 1974). This finding suggests that subordinate animals may actually avoid dominant males at specific times in order to reduce the rate of receiving attacks. The frequency of aggression in the field, therefore, is usually less than that seen in enclosure settings, even though enclosures in which animals have been studied to obtain the data for fig. 1 were at least 100 square meters and contained not more than three adult males and three adult females.

The findings shown in fig. 1 have important implications for pharmacological studies. For example, in vervets, short-acting drugs designed to reduce the frequency of aggression, if given outside the peak aggression periods, are likely to show minimal effects because of the low baseline of behaviour. Drugs given during the peaks may show differing effects as a function of physiological changes occurring during the periods of increased aggressive activity. These points also have clear implications for the treatment of humans and underscore the need of knowing the times in which target behaviours are likely to occur, the effective period of a drug, and the subject's drug taking practices.

Serotonin change

Compared to norepinephrine, epinephrine, and cortisol, peripheral levels of serotonin change relatively slowly. Studies among animals living in stable social groups show that there is less than 10% $\pm$ variance for repeated samples over 24 hour periods (Raleigh, unpublished data). Recall that aggressive behaviour often is rapid in onset and that it may occur without obvious preparation. Thus, contributing physiological systems are likely to be those that undergo rapid functional change.

This requirement suggests that peripheral serotonin function is not a major contributing factor in any specific episode of rapid–onset aggression. CNS serotonin function may be another matter, however. Elevated CNS serotonin function may serve to ameliorate the effects of a rapid increase in norepinephrine or dopamine function. Alternatively, increases in catecholamine function may be less effectively modulated when serotonin function is low or reduced.

Whole blood serotonin levels and behaviour differ in dominant and subordinate males

Studies of adult male vervets living in multi–male groups have repeatedly shown that whole blood serotonin (WBS) levels are higher in dominant compared to subordinate males (Raleigh et al. 1984). Dominant males have levels in the 1000 ng/ml range, while subordinate males have levels in the 600 ng/ml range. Pharmacological studies have demonstrated that these differences correlate positively with central serotonin function differences (Raleigh et al. 1985). Similar WBS differences are seen when comparing high and low–status squirrel monkeys (Steklis et al. unpublished data) and among high and lower–status human males (McGuire et al. 1983a,b).

WBS differences appear to depend on male–male interactions. Dominant males who are isolated from their groups show a decline in serotonin levels over a 14–21 day period (Raleigh et al. 1984). When subordinate males are removed from their groups, leaving only dominant males, females and their offspring, serotonin levels also decline over the same time period (McGuire et al. 1983b). One–way mirror studies (dominant males are placed behind one–way mirrors where they can see members of their group but can not be seen or heard) also point to the necessity of specific behaviours to maintain elevated serotonin levels in dominant males. In such situations, dominant males threaten and display towards subordinate males but are not responded to with submissive behaviours. Serotonin levels decline over a 14–21 day period (McGuire et al. 1983b). These studies suggest that display behaviour and/or physical activity in themselves are not sufficient to keep WBS levels elevated in dominant males. Submissive behaviours on the part of subordinate males also are required. It is of course possible that factors other than those noted above (e.g., olfaction) are the basis for observed differences. However, repeated attempts to demonstrate alternative bases for WBS differences have produced negative results.

Possible mechanisms mediating the physiological effects of subordinate behaviour are discussed in McGuire and Troisi (in press) and emphasize the importance of visual communication and cognitive function, primarily because many agonistic encounters do not result in physical contact. A further and as yet unexplained finding is the effect of isolation on WBS levels. When group–living males are socially isolated so that they cannot see other animals but can hear and smell them, WBS levels stabilize at approximately 600 ng/ml irrespective of their pre–isolation levels. Levels in the 600 ng/ml range are observed for approximately 60–70 days, at which time levels begin to rise and, at approximately 100 days, levels average 1100 ng/ml, (McGuire 1983b).

Dominant and subordinate males differ in their response to a variety of individual behavioural tests. These tests are conducted when animals are temporarily removed from their social group in order to reduce the effects of social inhibition on behaviour and to determine if behavioural differences observed in social settings extend outside of such settings. Compared to subordinate males, dominant males explore mazes more calmly, are more likely to aggress against unknown conspecifics, and are less likely to approach novel objects (McGuire et al. 1984). Such studies lend support to the view that physiological differences have important behavioural consequences. As discussed below, these differences correlated with cross–status drug response and drug metabolism differences.

Dominant and subordinate males differ in their response to tryptophan loads

Dominant and subordinate males differ in their responses to tryptophan loads (see table 2).

Table 2   Whole blood serotonin response to tryptophan load

	Dominant N=7	Nondominant N=9
Basal level (ng/ml)	1136 ± 40	668 ± 26
Tryptophan +60 min (20 mg/kg)	2428 ± 88	1029 ± 48
Absolute difference	1292 ± 70	361 ± 36
Relative difference	2.14 ± 0.07	1.54 ± 0.06

(Data are mean $\pm$ SEM)

The findings shown in Table 2 raise a number of interesting interpretative questions. For example, pre-tryptophan load measures of peripheral tryptophan do not differentiate dominant from subordinate males. This finding suggests that the total amount of circulating tryptophan is not the basis for either dominant–subordinate WBS differences or response differences to tryptophan loads shown in Table 2. When diets are held constant, WBS differences persist. Thus, food intake is not a differentiating variable. Differences in gut activity might also explain WBS differences because most peripheral serotonin is manufactured in the gut. However, repeated (daily for two weeks) IP injections of tryptophan do not chronically elevate WBS levels in subordinate animals. Thus, gut-related explanations are difficult to support. One cannot rule out the possibility that animals with low WBS levels also have a higher percentage of bound tryptophan, and thus less available tryptophan for conversion to brain serotonin. The implication of the findings shown in Table 2 for drug-related research seems clear: animals in different physiological states will metabolize drugs differently. Such differences may explain much of the between subject variance noted in pharmacological studies.

The effects of pharmacologically induced rapid reductions in serotonin levels

Before discussing the effects of rapidly reducing serotonin levels in adult males, it is worth reemphasizing that in behaviourally and physiologically stable groups, WBS levels do not reliably predict the frequency of aggression. They do, however, generally correlate with the type and degree of success of aggression i.e. dominant males engage most often in attack aggression and subordinate males engage most often in defence aggression. The absence of a relationship between WBS and the rate of aggression exists because the frequency of aggression changes as a function of within-group dynamics. For example, there are times when subordinate males repeatedly challenge dominant males and the latter appear relatively tolerant of such behaviour.

The relationships between serotonin function and aggressive behaviour were investigated. Under certain natural circumstances, elevations in WBS are associated with increases in aggressive behaviour. This situation occurs when a dominant animal is removed from a stable social group. Remaining subordinate males all compete for dominance status (McGuire et al. 1983a) and the animal that will become dominant shows a rise in WBS beginning at about 4 days and attaining the levels characteristic of dominant animals in about 14 days. During the period of the rapid rise, there may be a five fold increase in the frequency of aggressive behaviour of both the emerging dominant and other subordinate males. The preceding data strongly suggest that serotonin function is likely to be only one amongst many probable factors (e.g., neuropeptides, catecholamine function, and social behaviour options) that contribute to aggression.

Rapid changes in serotonin function are likely to result in temporarily physiologically destabilized animals (e.g., compensatory changes have not had time to occur) and are associated with changes in behavioural frequencies. For example, parachlorophenylalinine (PCPA), chronic fenfluramine, and cyproheptadine all decrease serotonin function (Raleigh and McGuire 1980; Raleigh et al. 1985). Behaviourally, these drugs produce irritable animals who are more likely to engage in aggressive behaviour than untreated counterparts (Raleigh and McGuire 1980; Raleigh et al. in press). Chronic fenfluramine treatment, for example, results in an eight-fold increase in the frequency of aggression towards conspecific males and a two-fold increase towards humans (Raleigh et al. in press). This increased aggressive behaviour is associated with concomitant reductions in WBS and CSF 5-HIAA, an observation which further implicates reduced central serotonergic function in aggression.

However, it has not been conclusively shown that the above mentioned effects are due specifically to a decline in serotonin function. It is possible that rapid declines in serotonin function are primarily associated with irritability. The increase in aggression would be a secondary effect, due to the alteration of other physiological systems. Another alternative is that rapidly reduced serotonin function is associated with a relative increase in catecholamine function. The latter interpretation (which can be tested by administering drugs that increase catecholamine function in serotonin depleted animals) is currently favoured in this laboratory.

Secondary effects of rapid declines in serotonin function

The secondary effects of pharmacologically induced falls in serotonin function not only are interesting but serve to illustrate two important points: the social consequences of drug treatment of single animals in social groups; and, the need to consider the social context in which measures are made. For example, the effects of PCPA treatment of a single animal in a group can be assessed on <u>untreated</u> subordinate animals. Results from studies of this kind are shown in fig. 2.

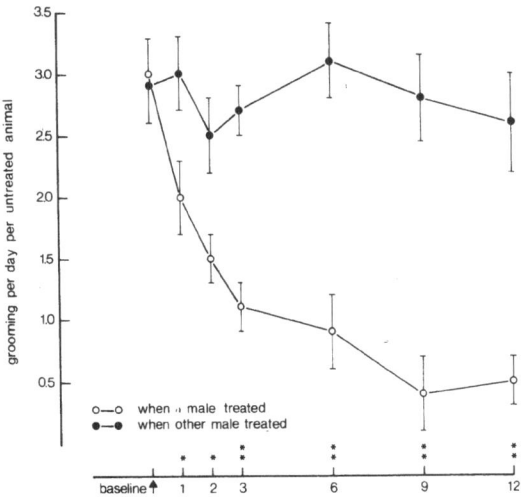

Fig. 2   This figure depicts grooming bouts per day per untreated animal on the last day of a baseline period (saline) and on day 1, 2, 3, 6, 9 and 12, of a 14 day treatment period. Animals received 80 mg/kg of dl–PCPA IP two hours prior to the onset of observations during the treatment period. In six groups dominant alpha males were treated and in six groups other males were treated. Data are in mean ± SEM. * p<0.05; ** p<0.005 (Adapted from Raleigh and McGuire 1980).

Animals treated with PCPA become irritable and are less predictable in their responses to the behaviour of other animals. If only treated animals are observed, the effects of PCPA treatment are similar in both dominant and subordinate males. However, as Figure 2 shows, the consequences of the irritability on untreated subordinate animals differ. Subordinate animals are significantly more likely to reduce the frequency of the social grooming if the dominant animal has received PCPA than if a subordinate animal has received PCPA (Raleigh and McGuire 1980). In our view, the potential impact of Figure 2, especially for studies of humans, has not been fully appreciated. It emphasizes that changes in the behaviour of one member of a social group may significantly alter the activities of other members.

The effects of pharmacologically induced increases in serotonin function

Pharmacologically induced rapid increases in serotonin function following administration of small doses of tryptophan, fluoxetine, or quipazine result in an increase in the frequency of 'approach', 'groom', 'rest', and 'eat', and, a decrease in the frequency of 'locomotion', 'avoid', and 'be-vigilant'. In these studies we did not observe significant decreases in the behaviours 'aggress' or 'be-aggressed' (Raleigh et al. 1985), although such changes may not occur because of overriding dominance relationships or because the drugs used did not influence a particular serotonin sub-system. Selected findings from these studies are shown in Table 3.

Table 3 Quantitative effects of fluoxetine, quipazine and tryptophan treatment. The table summarizes qualitatively the behavioural consequences of fluoxetine (FL), quipazine (QP) and tryptophan (TR) treatment. Each drug was given to a separate set of 5 dominant and 10 subordinate males at vehicle, low, moderate, and high doses. A plus sign in the FL, QP or TR columns indicates that a drug treatment increased the rate of a behaviour and a minus sign indicates that the treatment decreased the rate of a behaviour. The status interaction column indicates whether there was a significant status x dose interaction. An asterisk by a behaviour indicates that previous pharmacological and physiological data imply that it is serotonergically-influenced (Adapted from Raleigh et al. 1985).

| Behaviour | Behavioural effect | | | Status interaction | |
	FL	QP	TR	FL	QP
Approach*	+	+	+	Yes	Yes
Groom*	+	+	+	Yes	Yes
Rest*	+	+	+	No	No
Eat*	+	+	+	No	No
Locomote*	–	–	–	Yes	Yes
Avoid*	–	–	–	Yes	Yes
Be vigilant*	–	–	–	Yes	Yes
Be solitary*	–	–	–	Yes	Yes
Huddle	+	0	0	No	No
Aggress	0	0	0	No	No
Be aggressed	0	0	0	No	No
Submit	0	–	0	No	No
Sex	–	0	0	Yes	No

As a precursor, receptor agonist, and reuptake inhibitor result in essentially the same behavioural changes (see Table 3), this strongly implicates serotonin as a contributing factor in the observed changes. These findings lend further support to the idea that serotonin function differs among dominant-high WBS and subordinate low-WBS animals.

CONCLUSION

We have reviewed studies dealing with serotonin function and aggression in group-living adult male vervet monkeys. When animals live in stable social groups, dominant and subordinate animals differ in their levels of peripheral serotonin. Indirect (pharmacological) evidence suggests that status-related differences in CNS serotonin function also exist. There are clear and consistent differences in the types of aggression and agonistic behaviours among animals of different social status: dominant males tend to engage in attack aggression while subordinate animals tend to engage in defensive responses. These differences correlated with WBS differences but it is not known whether there is a direct causal relationship. When animals change social status, they change the type of aggression in which they most frequently engage. Pharmacologically–induced decreases in serotonin levels and function are associated with increases in irritable and aggressive behaviour, irrespective of the treated animal's status. This is perhaps the strongest finding that we have developed, indicating a close relationship between serotonin function and aggression. Pharmacologically induced increases in serotonin function (within physiological limits) do not uniformly result in either increases of decreases in aggressive behaviour. In certain situations (e.g., dominance competition) increases in serotonin function may be associated with an increase in aggression.

Acknowledgements:The authors wish to thank G.Brammer, R.Schuster and J.Dillon for their helpful comments on this chapter and D.Bolden for her assistance in manuscript preparation. The studies were in part funded by a Veterans Administration Merit Review, by the Giles and Elise Mead Foundation, and the University of California.

REFERENCES

Adamec RE, Stark–Adamec C, Livingston KE (1980a) The development of predatory aggression and defence in the domestic cat (Felis catus). I. Effects of early experience on adult patterns of aggression and defense. Behav Neural Biol 30: 389–409

Adamec RE, Stark–Adamec C, Livingston KE (1980b) The development of predatory aggression and defense in the domestic cat (Felis catus). II. Development of aggression and defense in the first 164 days of life. Behav Neural Biol 30: 410–434

Adamec RE, Stark–Adamec C, Livingston KE (1980c) The development of predatory aggression and defense in the domestic cat (Felis catus). III. Effects on development of hunger between 180 and 365 days of age. Behav Neural Biol 30: 435–447

Adams DB (1979) Brain mechanisms for offense, defense, and submission. Behav Brain Sci 2: 201–241

Aidara D, Tahiri–Zagret C, Robyn C (1981) Serum prolactin concentrations in mangabey (Cercocebus atys lunulatus) and patas (Erythrocebus patas) monkeys in response to stress, ketamine, TRH, sulpiride, and levodopa. J Reprod Fertil 62: 165–172

Albert DJ, Walsh ML (1984) Neural systems and the inhibitory modulation of agonistic behavior: a comparison of mammalian species. Neurosci Biobehav Rev 8: 5–24

Anderson JR, Chamove AS (1985) Early social experience and the development of self–aggression in monkeys. Biol Behav 10: 147–157

Averill JR (1982) Anger and Aggression. An Essay on Emotion. New York, Springer–Verlag

Bernstein IS, Rose RM, Gordon TP (1977) Behavioural and hormonal responses of male rhesus monkeys introduced to females in breeding and non-breeding seasons. Anim Beh 25: 609-614

Blanchard DC, Blanchard RJ (1984) Affect and aggression: an animal model applied to human behavior. In: Blanchard RJ, Blanchard DC (eds) Advances in the Study of Aggression, Vol 1, Academic Press, New York, pp 1-62

Brain PF (1984) Biological explanations of human aggression and the resulting therapies offered by such approaches: a critical evaluation. In: Blanchard RJ, Blanchard DC (eds) Advances in the Study of Aggression, Vol 1, Academic Press, New York, pp 63-102

Chamove AS, Bayart F, Nash VJ, Anderson JR (1985) Dominance, physiology and self-aggression in monkeys. Aggr Behav 11: 17-26

Coe CL, Mendoza SP, Levine S (1979) Social status constrains the stress response in the squirrel monkey. Physiol Behav 23: 633-638

Coe CL, Smith ER, Mendoza SP et al. (1983) Varying influence of social status on hormone levels in male squirrel monkeys. In: Steklis HD, Kling AS (eds) Hormone, Drugs and Social Behavior in Primates, SP Medical and Scientific Books, New York, pp 7-32

Eberhardt JA, Keverne EB (1979) Influences of the dominance hierarchy on lutenizing hormone, testosterone, and prolactin in male talapoin monkeys. J Endocrinol 83: 42-43

Eberhardt JA, Keverne EB, Meller RE (1983) Social influences on circulating levels of cortisol and prolactin in male talapoin monkeys. Physiol Behav 30: 361-369

Fairbanks LA, McGuire MT (1979) Inhibition of control role behaviors in captive vervet monkeys (Cercopithecus aethiops sabaeus). Behav Processes 4: 145-153

Fairbanks LA, McGuire MT, Page N (1978) Social roles in captive vervet monkeys (Cercopithecus aethiops sabaeus). Behav Processes 3: 335-352

Hansen S, Keverne EB, Martensz ND et al. (1980) Behavioral and neuroendocrine factors regulating prolactin in monkeys. In: Anandkumar TC (ed) Nonhuman Primate Models For Study Of Human Reproduction, Basel, Karger, pp 148-158

Huck UW, Banks EM, Wang SC (1985) Behavioral and physiological correlates of aggressive dominance in male brown lemmings (lemmus sibricus) Aggr Behav 12: 139-148

Madsen D, McGuire MT (1984) Whole blood serotonin and type A behavior pattern. Psychosomatic Med 46: 546-548

Mazur A, Lamb TA (1980) Testosterone, status, and mood in human males. Horm Behav 14: 236-246

McGuire MT (1974) The St. Kitts Vervet. Basel, Karger

McGuire MT, Raleigh MJ, Brammer GL (1982) Sociopharmacology. Ann Rev Pharmacol Toxicol 22: 643–661

McGuire MT, Raleigh MJ, Johnson C (1983a) Social dominance in adult male vervet monkeys: behavioral–biochemical relationships. Social Sci Inf 22: 311–328

McGuire MT, Raleigh MJ, Johnson C (1983b) Social dominance in adult male vervet monkeys: general considerations. Social Sci Inf 22: 89–123

McGuire MT, Raleigh MJ, Brammer GL (1984) Adaptation, selection, and benefit–cost analysis: implications from behavior–physiology and studies of social dominance in vervet monkeys. Ethol Sociobiol 5: 269–277

McGuire MT, Brammer GL, Raleigh MJ (1986) Basal cortisol and the emergence of dominant male status in vervet monkeys. Horm Behav 20: 106–117

McGuire MT, Troisi A (in press) Anger: An evolutionary interpretation. In: Plutchik R, Kellerman H (eds) Emotions. Academic Press, New York

McIntyre DC, Chew GL (1983) Relation between social rank, submissive behavior, and brain catecholamine levels in ring-necked pheasants (Phasians colchicus). Behav Neurosci 97: 595–601

McKenna JJ (1983) Primate aggression and evolution: an overview of sociobiological and anthropological perspectives. Bull Am Acad Psychiat Law 11: 105–130

Mendoza S, Coe CL, Lowe EL, Levine S (1978) The physiological response to group formation in adult male squirrel monkeys. Psychoneuroendocrinology 3: 221–229

Moyer KE (1968) Kinds of aggression and their physiological basis. Commun Behav Biol 2: 65–87

Raleigh MJ, McGuire MT (1980) Biosocialpharmacology. Maclean Hosp J 5: 73–86

Raleigh MJ, McGuire MT, Brammer GL, Yuwiler A (1984) Social and environmental influences on blood serotonin concentrations in monkeys. Arch Gen Psychiat 41: 405–410

Raleigh MJ, Brammer GL, McGuire MT, Yuwiler A (1985) Dominant social status facilitates the behavioral effects of serotonergic agonists. Brain Res 348: 274–282

Raleigh MJ, Brammer GL, Ritvo ER, Geller E, McGuire MT, Yuwiler A (in press) Effects of chronic fenfluramine on blood serotonin, cerebrospinal fluid metabolites, and behavior in monkeys.

Rose RM, Holaday JW, Bernstein IS (1971) Plasma testosterone, dominance rank and aggressive behavior in male rhesus monkeys. Nature 231: 366–368

Smith EO, Byrd LD (1983) Studying the behavioral effects of drugs in group–living nonhuman primates. In: Miczek KA (ed) Ethopharmacology. Primate Models of Neuropsychiatric Disorders. Alan R. Liss, New York, pp 1–31

Steklis HD, Brammer GL, Raleigh MJ, McGuire MT (1985) Serum testosterone, male dominance, and aggression in captive groups of vervet monkeys (Cercopithecus aethiops sabaeus). Horm Behav 19: 154–163

Thierry B (1985) Patterns of agonistic interactions in three species of macaque (Macaca mulatta, M fasicularis, M tonkeana) Aggr Behav 11: 223–233

ALCOHOL EFFECTS ON THE AGGRESSIVE BEHAVIOUR OF SQUIRREL MONKEYS AND MICE ARE MODULATED BY TESTOSTERONE

James T. Winslow, Joseph F. DeBold and Klaus A. Miczek. Department of Psychology, Tufts University, Medford, Massachusetts 02155, U.S.A.

A number of recent reviews of the literature have noted and critically examined the correlational statistics that link alcohol, more than any other drug, to a high incidence of violence and aggression (e.g. Brain 1986; Miczek 1987). The acute or chronic consumption of alcohol is associated with high rates of homicides, suicides, physical assaults toward others, especially family members and associates, and sexual assaults, particularly rape (e.g. Abel and Zeidenberg 1985; Corenblum 1983; Frankel et al. 1976; Johnson et al. 1978; Leonard et al. 1985; Rada 1975; Tinklenberg et al. 1974). These statistics focus attention on an extremely serious problem for which the environmental and social determinants have only begun to be studied and no neurobiological mechanisms have been identified.

EPIDEMIOLOGICAL STUDIES

The effects of alcohol on aggressive behaviour vary enormously among individuals, not only quantitatively, but also qualitatively (e.g. Boyatzis 1974; Cherek et al. 1984, 1985). The sources for this variability may be found in a host of situational, social and personality factors, and these have been the topic of numerous investigations. For example, the linkage between the genetic component in abusing alcohol and in the development of antisocial personalities have been explored (e.g. Bohman 1978; Lewis et al. 1985; Stabenau 1984). However, alcohol abuse and antisocial personalities co-vary weakly, and often the interaction between alcoholism and violence is not statistically significant. Heightened aggressive behaviour is only one of many symptoms in antisocial personalities (e.g. Cloninger 1983), and it is unclear whether the behavioural problems are induced by alcohol abuse, are concomitant with alcohol abuse, or actually result in alcohol abuse. Another related line of inquiry found levels of acid metabolites of brain monoamines, especially the serotonin metabolite 5-hydroxyindolacetic acid, in the cerebrospinal fluid (CSF) of violent offenders with explosive or antisocial personalities who also abused alcohol that were significantly lower than those in passive aggressive patients (Linnoila et al. 1983). Earlier observations suggested that low CSF levels of 5-HIAA appear to characterize subgroups of depressed patients with strong suicidal tendencies (Asberg et al. 1976; Van Praag 1982). To what degree this particular biochemical marker identifies poor impulse control and how it is related to the propensity to abuse alcohol or to certain types of alcohol effects remains to be further investigated.

ALCOHOL AND ANIMAL AGGRESSION

The results of many studies on alcohol and aggression in animals have been limited to the demonstration of the sedative and ataxic properties of alcohol (e.g.

Berry and Smoothy 1986). Recent reviews have summarized the repeated findings of alcohol's anti-aggressive effects that are usually obtained in the range of sedative and ataxic doses (e.g. Miczek 1987; Miczek and Winslow 1987). Acute alcohol doses decrease (1) aggressive behaviour in isolated mice (Bertilson et al. 1977; Lagerspetz and Ekqvist 1978; Smoothy et al. 1982; Smoothy and Berry 1983), (2) defensive bites and postures in mice, rats and Squirrel monkeys that are evoked by painful, noxious stimulation (Bammer and Eichelman 1983; Emley and Hutchinson 1983; Irwin et al. 1971; Smoothy and Berry 1984), and (3) mouse killing in rats (Bammer and Eichelman 1983).

In contrast to these systematic, but rather disappointing observations, several reports during the past 30 years have identified aggression-enhancing effects of alcohol at selected doses in certain animal species and under specific conditions. The earliest and most convincing demonstration of the aggression-heightening action of alcohol came from studies in Siamese fighting fish and cichlids. When exposed to low alcohol concentrations resident fish will threaten a mirror image or attack an intruder with higher frequency and longer duration, whereas at higher alcohol concentrations these aggressive responses are suppressed (Ellman et al. 1972; Figler and Peeke 1978; Peeke et al. 1973, 1975, 1981; Peeke and Figler 1981; Raynes et al. 1968; Raynes and Ryback 1970).

In mammalian species, aggression-heightening effects of acute alcohol doses have been difficult to demonstrate in a systematic, dose-related manner. Yet, during the 1970's several reports on increased aggressive behaviour after a selected low alcohol dose in mice, rats, dogs, cats and Macaque monkeys under varied conditions suggested species-generality to the biphasic effects of alcohol (e.g. Chamove and Harlow 1970; Chance et al. 1973; MacDonnell et al. 1971; Pettijohn 1979; Weitz 1974). Quantitative analysis of the effects of alcohol on several salient behavioural acts, movements and postures that are exhibited during agonistic confrontations between pairs of mice and rats revealed the primary "behavioural sites of action" for alcohol (e.g. Krsiak 1975, 1976; Miczek and Barry 1977). Low doses of alcohol increased the frequency of attacks and threat displays by isolated timid mice (0.4 and 0.8 g/kg, PO) and dominant rats (0.5 g/kg, IP), 2 to 3 times higher doses decreased various aggressive acts and postures (>1.6 g/kg, PO in mice; >1.5 g/kg, IP in rats), and a further doubling in alcohol dose level resulted in sedation and ataxia. Similar biphasic alcohol dose-effect relationships for aggressive behaviour have been seen in resident mice confronting an intruder into their home cage (Yoshimura and Ogawa 1983). The aggression-enhancing effects of low alcohol doses are particularly prominent when aggressive behaviour is suppressed by removing a resident mouse from his home cage and exposing him to unfamiliar surroundings for the encounter with an opponent. Under these conditions, low alcohol doses (0.15, 0.3 g/kg, PO) increase attack and threat behaviour 2 to 3 fold (Miczek and O'Donnell 1980).

ALCOHOL AND GONADAL HORMONES

The synthesis, release and metabolism of steroid hormones, particularly testosterone are altered by alcohol abuse (Cicero et al. 1981, 1982; Mendelson et al. 1977; Van Thiel et al. 1975). For example, the clinical profile of male alcoholics can include testicular atrophy, reduced plasma testosterone, feminization, but normal LH and plasma cortisol (Cicero 1983; Mendelson and Mello 1974; Mendelson et al. 1978). Chronic administration of alcohol to rats also significantly reduces plasma testosterone levels, and causes testicular atrophy (van Thiel et al. 1975). Since testosterone has effects on aggressive behaviour in many species (Svare 1983), it is possible that alcohol might alter aggression through its effects on testosterone. However, unequivocal studies on the physiological effects of chronic alcohol administration are complicated by the problem of having to supply calorically and nutritionally matched controls (Cicero and Badger 1977). In addition, the decrease in testosterone during alcohol abuse is generally associated with reduced aggression and this appears to argue against a link between the effects of chronic alcohol on steroidal mechanisms and increased aggression in alcoholics (Coid 1982).

Studies of the effects of acute alcohol administration have provided important insights into the actions of alcohol on steroid hormones. Badr and Bartke (1974) describe a dose–dependent decrease in plasma testosterone levels of male mice following administration of alcohol (0.155–1.24 g/kg, PO). Doses of alcohol (1–3 g/kg) reduce testicular steroidogenesis in the rat, and this effect is blocked by administration of the dehydrogenase inhibitor pyrazole (Cicero et al. 1981). Thus, oxidation of alcohol to acetaldehyde appears to be required for this reduction in testosterone synthesis. Ellingboe and Varanelli (1979) have demonstrated in vitro competition for the oxidizing agent NADH by alcohol dehydrogenase and enzymatic conversion of pregnenolone to progesterone in the Leydig cells of rat testes. They proposed that increased metabolism of alcohol reduces NADH availability for testosterone synthesis. Murono and Fisher–Simpson (1985a,b) have also demonstrated enhanced metabolism of dihydrotestosterone to androstandiols by reductase enzymes in rat Leydig cells. This is a rate–limiting step in testosterone catabolism and it may also be affected by changes in the NADH/NAD+ ratio produced by increased alcohol metabolism in these cells. Cicero et al. (1982) studied acute alcohol (1–3 g/kg) effects on naloxone–induced release of luteinizing hormone–releasing hormone from the hypothalamus. They compared this effect to the effects of alcohol on pituitary release of LH and concluded that a dose related decrease in LH produced by alcohol is mediated by changes in hypothalamic secretion of LHRH. Alcohol had little direct effect on LH release from the pituitary gland.

IMPLICATIONS FOR CURRENT STUDIES

The currently available evidence provides important clues about some of the critical determinants of alcohol effects on aggressive behaviour. These appear to include: (1) the environmental antecedents and consequences of the aggressive behaviour, (2) the behavioural history of the individual with attack or, alternatively, with defensive, submissive or flight behaviour, (3) the past and present pharmacological conditions to which the individual is subjected. In addition, the biphasic effect of alcohol on aggressive behaviour suggests at least two separate mechanisms for the ascending and the descending portions of the dose–effect curve.

It appears that a source for individual variation in alcohol's effects on aggressive behaviour may be found in certain biochemical processes involving gonadal steroids, opioid peptides or brain monoamines. It will be important to learn how genetic dispositions, social experiences and environmental variables influence these biochemical processes that are the targets for alcohol action and ultimately mediate the changes in aggressive behaviour.

Primates provide an excellent opportunity to study the role of social factors in determining the effects of drugs, as well as providing direct access to physiological and endocrinological mechanisms. We have chosen to study Squirrel monkeys (Saimiri sciureus) as an example of an infra–human primate species. Squirrel monkeys, a widely studied New World species, form reliable social relationships and the resulting monkey groups may be studied for the role of complex social contingencies on physiological and pharmacological processes. In addition, this species exhibits large natural variations in gonadal hormone levels, providing a further opportunity to study possible alcohol interactions with different endocrine states. The goal of maintaining stable, social groups of primates limits, however, their use in invasive techniques to study brain and physiological mechanisms controlling behaviour. Rodents may exhibit high levels of aggressive behaviour under naturalistic testing conditions which provide a sensitive measure of drug and endocrine manipulations. Reliability of occurrence of aggression in rodents and the relative simplicity of the social conditions permit a more rapid analysis of underlying brain mechanisms. In the following sections a series of experiments designed to clarify the role of gonadal hormones in mediating the effects of alcohol on the aggressive behaviour of Squirrel monkeys and of mice are described.

PRIMATE STUDIES

The effects of alcohol on the behaviour of Squirrel monkeys (Saimiri sciureus) living in captive, social groups were investigated. Our objective was to identify environmental and physiological factors controlling the effects of alcohol, particularly those on agonistic behaviour. In this account we will (1) focus on the agonistic interactions between members of the same group, (2) describe the pattern of social organization these interactions reveal, (3) describe features of the unique behavioural and physiological profiles associated with the principle characteristic of this social organization – dominance relationships, and (4) summarize experiments designed to describe possible mechanisms mediating the status–dependent effects of alcohol on agonistic behaviour.

Four established groups of Peruvian and Columbian Squirrel monkeys were composed of 2–3 adult males, 2–3 adult females and infants/juveniles of both sex. Each group lived in separate rooms (2.4 m x 2.2 m x 2.4 m) with a one–way vision window, controlled climate (27°C, 50% relative humidity), fixed light cycle (12h light/dark), free access to water and chow. Approximately half the animals were feral. Captive groups of Squirrel monkeys maintain a stable social organization which can be quantified by measuring the frequency and direction of agonistic initiatives by group members (e.g. Alvarez 1975).

Figure 1 portrays a cluster analysis of group structure based on agonistic interactions. Each circle represents an individual member of the group and the size of the circle is proportionate to the number of agonistic behaviours initiated compared to those received. The arrows indicate the directions of dyadic agonistic interactions, and the thickness of the line represents the proportion of all agonistic encounters in a group accounted for by these interactions. Stable dominance–subordination relationships emerged from this analysis which persisted for three years of the current experiments, and reliably predicted unique, status–related behavioural and endocrine profiles.

Time budget analysis is a useful method of portraying the behavioural differences associated with dominance and subordination. Using an exhaustive and mutually exclusive behavioural catalogue it is possible to portray the proportion of time spent in a particular class of behaviour as degrees of a circle. Such 'pie charts' have several useful features. Firstly, they demonstrate the finite and interactive nature of an organism's behavioural repertoire: changes in the amount of time allocated to one class of behaviour are necessarily compensated for by changes in other classes. Secondly, they provide a simple, lucid means of identifying quantitative patterns unique to specific circumstances such as social status or drug treatment.

Figure 2 portrays the proportion of time spent walking, sitting with curled tail, in a stationary alert posture, feeding, marking, sending agonistic behaviours, and exhibiting associative or submissive behaviour for four dominant and four subordinate animals. Dominant monkeys spend more time walking and engaging in social and agonistic behaviour, and less time in the stationary alert posture than subordinate monkeys. Agonistic behaviour accounts for less than 10% of the total

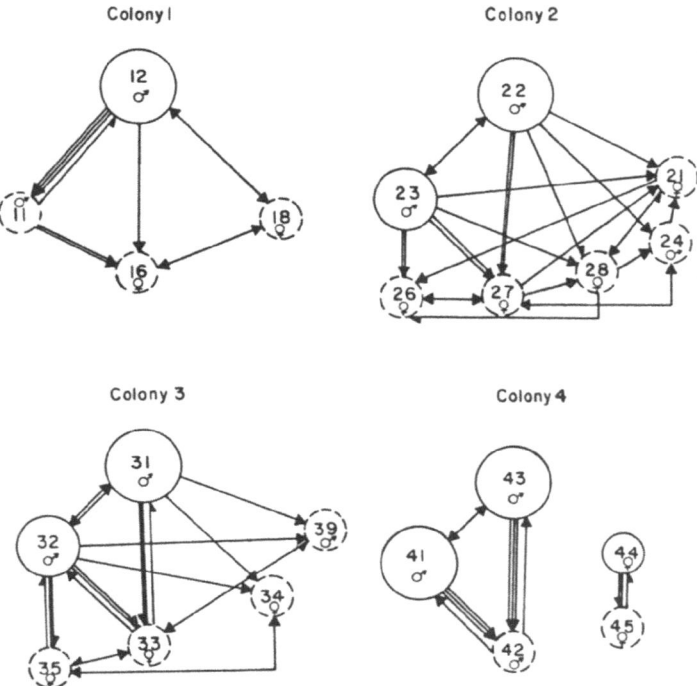

Fig. 1   Sociograms of four colonies of monkeys observed in 1985. The size of the circle represents the number of agonistic behaviours which an animal initiated compared to those received. The arrows indicate the direction of agonistic interactions: the number of lines represents the frequency of these interactions between specific pairs.

activity in both high and low status animals. This stands in striking contrast to the pervasive influence of status on other features of the animal's internal and external environment. Social status (determined as a function of the frequency, direction and success of aggressive behaviour) is associated with unique differences in the pattern of non–aggressive social and motor behaviour.

Social structure in Squirrel monkeys is also reflected in significant status–related differences in gonadal and glucocorticoid hormones (Coe et al. 1983, 1985; Leshner and Candland 1972; Mendoza et al. 1978). Dominant male Squirrel monkeys show dramatic changes in blood levels of testosterone with concentrations increasing from 42.7 ng/ml during the non–mating season to over 211 ng/ml during the mating season. Subordinate male monkeys also show seasonal variation in testosterone levels but at significantly lower levels than dominant monkeys (Figure 3).

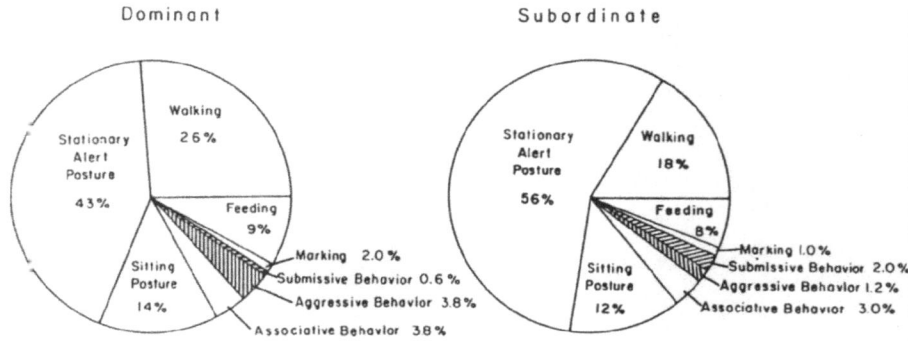

Fig. 2 The proportion of time spent walking, sitting, in stationary alert posture, feeding, marking and sending aggressive, associative, or submissive behaviours. The data represent the means of four 40-min observations each for five dominant and three subordinate Squirrel monkeys (from Winslow and Miczek 1985).

The seasonal changes in testosterone described here are consistent with data reported by Coe et al. (1985), and are accompanied by increases in body weight, and intensification of sexual and aggressive behaviours among dominant male monkeys. Figure 4 portrays the mean monthly body weight of four reproductively active male Squirrel monkeys. These data were collected between October 1983 and February 1986 and provide evidence for the relatively stable annual mating cycle of this species in our laboratory. Estimates of the blood levels of testosterone (ng/ml) likely to occur during this interval are also portrayed as adapted from Coe et al. (1985). Actual plasma testosterone values collected from the four dominant males in our colonies are represented by the shaded columns. A reliable mating cycle was predicted by large increases in body weight among dominant male monkeys between December and February, and was associated with large increases in plasma testosterone. Our studies of alcohol effects on social behaviour were scheduled to coincide with the seasonal mating cycle during the course of a three-year interval. The period of the cycle was identified by the characteristic weight gain measured in reproductively active male Squirrel monkeys ("the fatted male syndrome").

Significant status-related differences in the sensitivity of Squirrel monkeys to alcohol have been previously reported (Winslow and Miczek 1985). After adapting monkeys to the drug administration process, dominant and subordinate Squirrel monkeys exhibited stable baselines of aggressive behaviour. Low to moderate doses (0.1–0.6 g/kg, PO) of alcohol reliably increased the frequency of aggressive behaviour exhibited by dominant monkeys during the first hour after oral administration (Fig. 5A). The highest dose (1.0 g/kg) reliably decreased the frequency of aggressive behaviour. In contrast, such behaviour was unchanged in subordinate monkeys by any of these doses. Figure 5B portrays the time course for enhancement of aggression at the 0.6 g/kg dose in dominant monkeys. Aggression was increased during the first 20–40 minutes then returned to baseline levels for the remainder of the 2 hr session (Winslow and Miczek 1985).

Plasma Testosterone

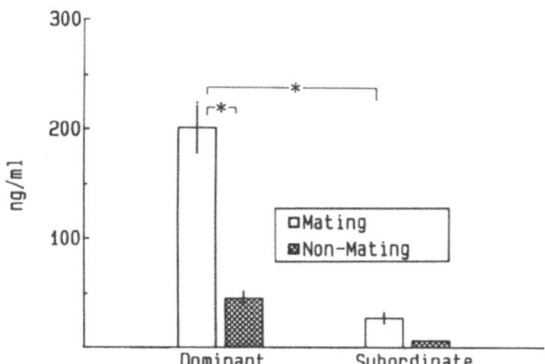

Fig. 3   Plasma testosterone (ng/ml) measured during the mating and non–mating seasons of 1985. Data represent the mean of four dominant and four subordinate Squirrel monkeys. Vertical lines at each data point represent $\pm$ 1 SEM. Asterisks indicate $p<0.05$.

Seasonal Reproductive Rhythm

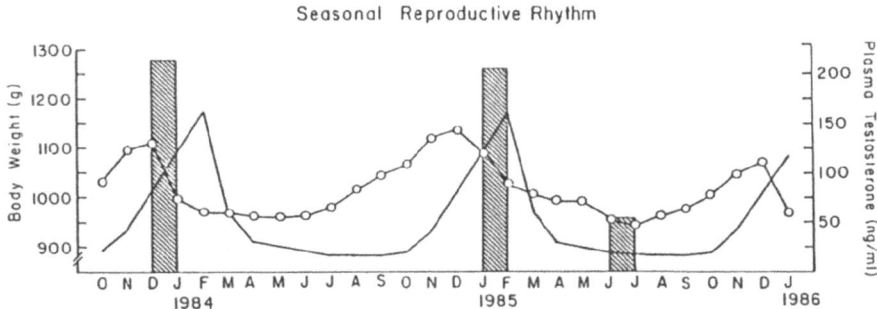

Fig. 4   The mean monthly body weight of four dominant male Squirrel monkeys collected between October 1983 and February 1986 (open circles). Also given are estimates of the plasma values of testosterone likely to occur during the annual mating cycle (solid line) which were adapted from Coe et al. (1985). Actual mean plasma testosterone concentrations measured in four dominant male monkeys are portrayed in the shaded columns.

230

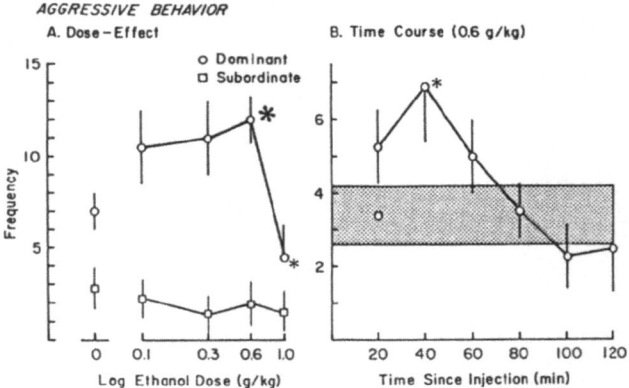

Fig. 5 (A) The frequency of aggressive behaviour (grasps, display, displacements) during the 40–min period beginning 5 min after administration of alcohol (0.0, 0.1, 0.3, 0.6, 1.0 g/kg, PO) to dominant (n=5) and subordinate (n=6) members of groups of captive, free–ranging Squirrel monkeys. Vertical lines at each data point represent ± 1 SEM. Asterisks indicate p<0.05 versus 0 g/kg. (B) The frequency of aggressive behaviours measured in consecutive 20–min segments of a 2–h observation. The data represent the effects of 0.6 g/kg alcohol on the aggressive behaviour of dominant Squirrel monkeys (n=5). The shaded area represents the mean ± 1 SEM of five water control tests for each of the five dominant monkeys (from Winslow and Miczek 1985).

During the next mating season, these alcohol effects were replicated with low and moderate doses of alcohol producing significant increases in the frequency of agonistic behaviours exhibited by dominant monkeys. However, during the non–mating season, this alcohol effect on agonistic behaviour of dominant monkeys was significantly reduced. The aggressive behaviour of subordinate monkeys was unaffected by alcohol during both mating and non–mating periods (Fig. 6).

The seasonal and status–dependent effects of alcohol coincided with differences in circulating testosterone. In an effort to describe the role of testosterone in modulating the seasonal changes in alcohol sensitivity of dominant monkeys, the plasma testosterone levels of subordinate monkeys were elevated with daily, subcutaneous injections of testosterone propionate (TP, 25 mg/kg). This treatment produces increased blood testosterone levels in the subordinates in excess of those seen in dominant males during the mating season. Hormone was administered daily at 06:30, 2 hours before a one hour observation of the treated monkey's behaviour in the group. A two week investigation of the effects of daily TP administration on behaviour and plasma testosterone concentrations was followed by a study of the effects of alcohol on the TP–treated subordinate monkeys. Two experiments were conducted, both during the non–mating season and separated by 4 months. In the first experiment, the composition of the groups remained unaltered. In the second study, the dominant male monkey from each colony was removed and housed separately.

Aggressive Behaviour (Grasp, Threat, Displace)

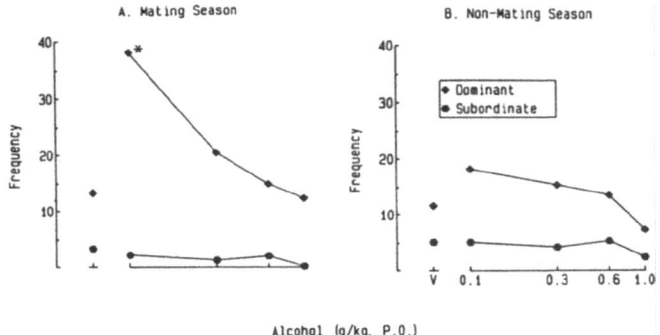

Alcohol (g/kg. P.O.)

Fig. 6 (Left) The frequency of aggressive behaviour during the 60–min period beginning 5 min after administration of alcohol (vehicle, 0.1, 0.3, 0.6, 1.0 g/kg, PO) to dominant (n=4) and subordinate (n=4) monkeys during the mating season of 1985. (Right) The frequency of aggressive behaviour during the 60–min period beginning 5 min after administration of alcohol (vehicle, 0.1, 0.3, 0.6, 1.0 g/kg, PO) to dominant (n=4) and subordinate (n=4) monkeys during the non–mating season of 1985. Asterisks indicate $p < 0.05$ versus vehicle.

Marking Behavior of Subordinates

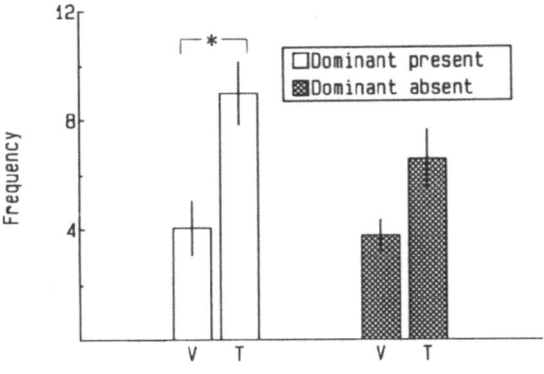

Testosterone Propionate (25 mg/kg, s.c.)

Fig. 7 The frequency of marking behaviour exhibited by subordinate Squirrel monkeys 2–h after daily SC injections of sesame oil vehicle (V) or 25 mg/kg testosterone propionate (T). Data represent the mean of four observations collected with the dominant monkey in each colony present (open columns), or housed separately (hatched columns). Vertical lines at each column represent ± 1 SEM. The asterisk indicates $p < 0.05$.

The social behaviour of subordinate monkeys was largely unaffected by daily injections of TP even though plasma testosterone actually increased to levels even greater than those measured in dominant monkeys during the mating season (934 ± 43 ng/ml). A significant increase in the frequency of marking behaviour was evident compared to measures collected during pre-treatment observations (Figure 7) but the frequency of marking behaviour exhibited by testosterone-treated subordinates remained significantly lower than that of dominant monkeys during the mating season.

Figure 8 portrays the effects of alcohol on the aggressive behaviour of testosterone treated subordinate monkeys in the presence and absence of the dominant monkey. Low and moderate doses of alcohol (0.1, 0.3 g/kg, PO) significantly increased the frequency and duration of aggressive behaviour exhibited by TP-treated subordinates in the presence of the dominant monkey, but not when the dominant monkey was removed. The measures of aggressive behaviour were not increased to the absolute levels exhibited by dominant monkeys, but the pattern of alcohol effects on this group of activities was similar to that seen in dominant monkeys during the mating season. These data suggest that testosterone may modulate the effects of alcohol on Squirrel monkey social behaviour, but that the social history and environment of the treated monkey continues to exert a significant influence. These results add to the evidence that CNS mechanisms that control gonadal hormones and gonadally-dependent behaviour may be the targets of alcohol.

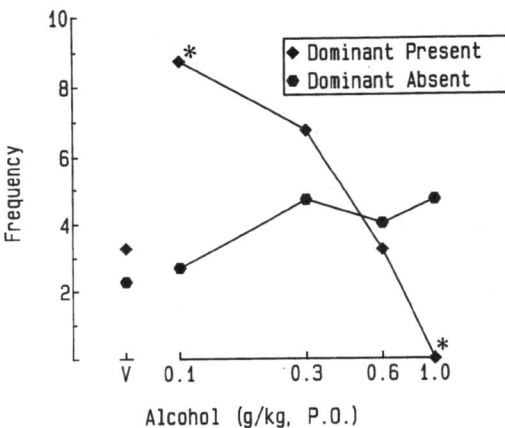

Fig. 8   The frequency of aggressive behaviour of testosterone treated subordinate monkeys during the 60-min period beginning 5 min after administration of alcohol (0.0, 0.1, 0.3, 0.6, 1.0 g/kg, PO). Data represent behaviour exhibited with the dominant monkey in each colony present (diamonds), or housed separately (hexagons). Asterisks indicate p<0.05 versus vehicle.

MOUSE STUDIES

In comparison with the limited research on alcohol and aggression in monkeys, a more extensive literature exists on the effect of alcohol on aggression in rodent species (see reviews by Brain 1986; Miczek 1987; Miczek and DeBold 1983; Miczek and Krsiak 1979). In rodents the control of aggressive behaviour by testosterone and reproductive state is even more pronounced than in monkeys (Svare 1983), and thus it seemed possible that the potential for an interaction of alcohol and testosterone on aggression might be even greater in rodents. The influence of sex hormones on aggressive behaviour was investigated and an attempt was made to determine whether these gonadal hormones can alter some of the behavioural effects of alcohol in rats and mice. In an initial series of three experiments, the effects of acutely administered alcohol on the aggressive behaviour of male mice and the role of testosterone in determining such responses were determined (DeBold and Miczek 1985). In these experiments, pairs of male and female CFW mice were housed together for a few weeks, the female and any offspring were temporarily removed and a group–housed male mouse introduced into the resident male's home cage. This procedure engenders a reliable, high level of attack by the resident male on the "intruder" (Miczek and O'Donnell 1978). After the baseline level of aggression toward intruders was determined, male resident mice were given 0.1, 0.3, 1.0, 1.7 or 3.0 g/kg of ethanol or the water vehicle (IP). Fifteen minutes later, a group–housed male intruder was placed into the home cage and the aggressive, social and locomotor behaviours of the mice were measured with quantitative ethological methods (Miczek 1982).

A small aggression–enhancing effect of alcohol at the 1.0 g/kg dose was evident but the only statistically significant effect of ethanol was the suppression of male resident's aggression by the 3.0 g/kg dose (Fig. 9). This suppressive effect of 3.0 g/kg alcohol on male mouse aggression could be seen within 5 min of alcohol administration and lasted at least 60 min (DeBold and Miczek 1985). This effect of alcohol was specific to aggressive behaviours as no significant effects on measures of non–aggressive behaviours such as the time the male resident spent walking or rearing were seen.

To examine the role of testosterone in the effects of alcohol on aggression, another set of male mice were castrated and then implanted with a 7.5 mm or a 2.5 mm long silastic capsule filled with testosterone or cholesterol as a control. These capsules slowly and steadily release their contents into the blood stream to produce tonic plasma levels of steroid (Smith et al. 1977). The 2.5 mm testosterone capsule produces plasma testosterone levels that are significantly lower than gonadally intact values and the 7.5 mm capsule produces testosterone levels somewhat higher than intact levels. The release of testosterone from these capsules is apparently unaltered by alcohol, in contrast to testicular release of androgen (Badr and Bartke 1978; Ellingboe and Varanelli 1979). The capsule–implanted, castrated mice were examined for their response to alcohol in much the same fashion as the intact males.

The castrated mice with 2.5 mm testosterone capsules and the castrated males with cholesterol implants had lower baseline levels of aggression than intact mice or castrated males with 7.5 mm testosterone capsules, but they exhibited an alcohol dose–response pattern that was similar to that of the intact males (compare fig. 9 and 10). The castrated males with 7.5 mm testosterone capsules showed a distinctly shifted dose–response curve for alcohol's effects on aggression: first, moderate alcohol doses (1.0 and 1.7 g/kg) significantly enhanced the frequency of attack bites and threat behaviours; second, the 3.0 g/kg dose of alcohol (which suppressed aggression in all other treatment groups) had no effect on the 7.5 mm testosterone males, instead a very high dose of alcohol (5.6 g/kg) was required to suppress aggression (fig. 10). This effect of testosterone does not appear to reflect a general reduction in sensitivity to alcohol. Castrated males with 7.5 mm testosterone capsules actually appear to be somewhat more sensitive to the sleep–inducing effects of alcohol. They take longer to regain their righting reflex after IP alcohol than intact males. This suggests that the reduced sensitivity to 3.0 g/kg alcohol is not due to an increase in alcohol clearance. It appears more

234

likely that the interaction between testosterone and alcohol on aggression occurs in brain sites where testosterone and alcohol effects overlap rather than on peripherally determined alcohol metabolism.

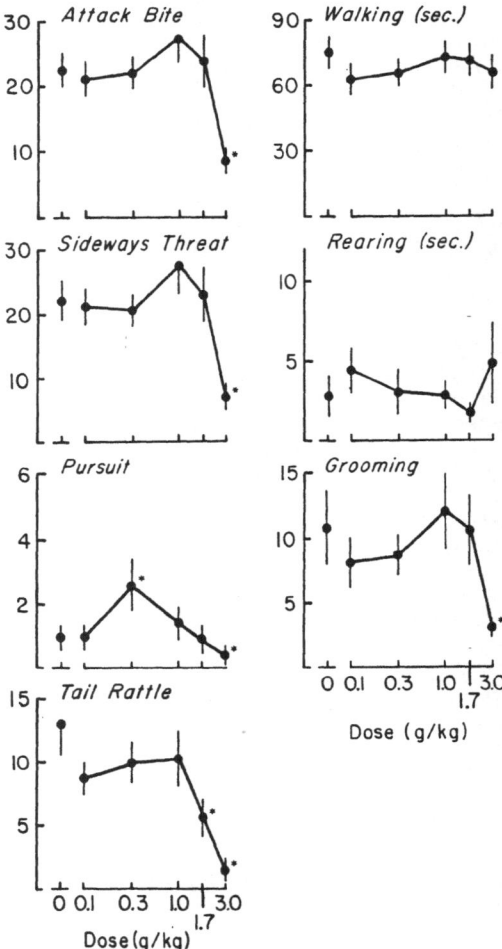

Fig. 9 The left panel shows the frequency of attacks with bites (top), sideways threat (center), pursuit and tail rattling (bottom) behaviour by gonadally intact male mice as a function of dose (g/kg, IP, log scale) of ethanol.
The right panel shows the non-aggressive behaviours of the same animals and contains the means ± 1 SEM for the duration of all bouts of locomotion, and rearing, and the frequency of grooming as a function of ethanol dose. Statistically significant differences (p<0.05) from the control tests (0 g/kg) are indicated by asterisks (from DeBold and Miczek 1985).

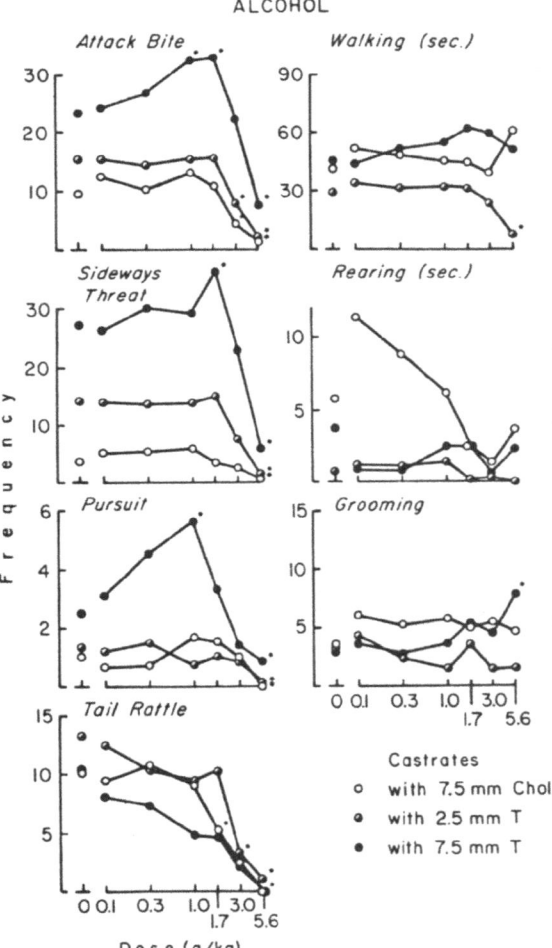

ALCOHOL

Fig. 10 The left panel shows the aggressive behaviours of castrated male mice as a function of ethanol dose and testosterone replacement. The right panel shows the non-aggressive behaviours of the same mice. Open symbols represent castrated male mice with 7.5-mm silastic capsules of cholesterol, half-filled symbols represent castrates with 2.5-mm capsules of testosterone, and solid symbols represent castrates with 7.5-mm capsules of testosterone. Asterisks indicate p<0.05 versus 0 g/kg (from DeBold and Miczek 1985).

Female rodents can also be aggressive toward intruders into their home cage. However, their aggressive behaviour is under a different form of hormonal control.

Testosterone Treated Female Mice
Attack Bites

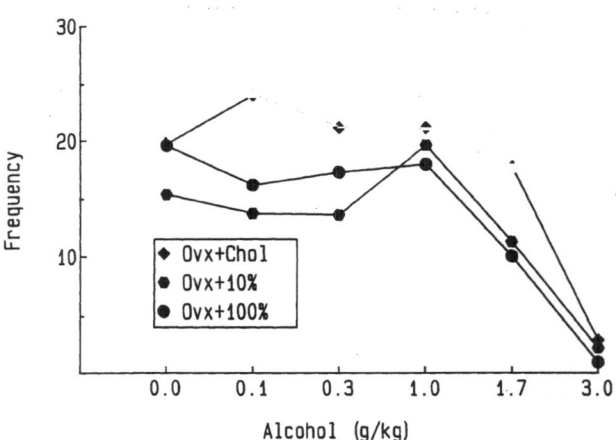

Alcohol (g/kg)

Fig. 11 The frequency of attack bites of ovariectomized female mice as function
of alcohol dose and testosterone treatment. <u>Diamond</u> symbols represent
ovariectomized female mice implanted with 7.5-mm silastic capsules of
cholesterol, <u>hexagonal</u> symbols represent 10% testosterone capsules, and
<u>circles</u> represent 100% testosterone capsules.

Female mice and rats are most aggressive when they are lactating (Svare and
Marn 1981). Ovariectomized females are less aggressive than lactating females,
but they may still attack and threaten intruders (DeBold and Miczek 1983). Their
aggression obviously does not depend on the presence of their gonads. In addition,
gonadectomized females, unlike castrated males, are relatively unaffected by
testosterone. We have been studying the effect of alcohol on aggressive behaviour
shown by female rodents characterised by a number of different endocrine
conditions.

Lactating female mice and rats show baseline levels of attack toward female
intruders that are comparable in frequency to those of male residents toward male
intruders (DeBold and Miczek 1983; DeBold et al. 1986). When given alcohol,
lactating females show a dose-response curve for aggressive behaviour that is
similar to that of intact males. However, significant suppression of aggression
occurs in lactating females at a somewhat lower dose (1.7 g/kg ethanol in females
vs 3.0 in males). This variation does not appear to be related to the sex difference
in testosterone levels seen in adults. When female mice are ovariectomized and
given silastic capsules of testosterone or cholesterol, testosterone had no effect on
aggression in the presence or absence of alcohol (fig. 11). Female mice have been
previously reported to be much less sensitive to the augmenting effects of
testosterone on male-typical aggression and copulation (Edwards 1969; Floody
1983). We can now add that they are also less sensitive than males to the
interaction between testosterone and alcohol on aggressive behaviour.

Sex differences in behavioural sensitivity to testosterone are generally
attributed to developmental processes occurring during sexual differentiation (Goy
and McEwen 1980). This can be assessed by altering sexual differentiation by
adding or removing testosterone during a sensitive period of early development. In
mice this sensitive period is just before and just after birth (vom Saal 1983).

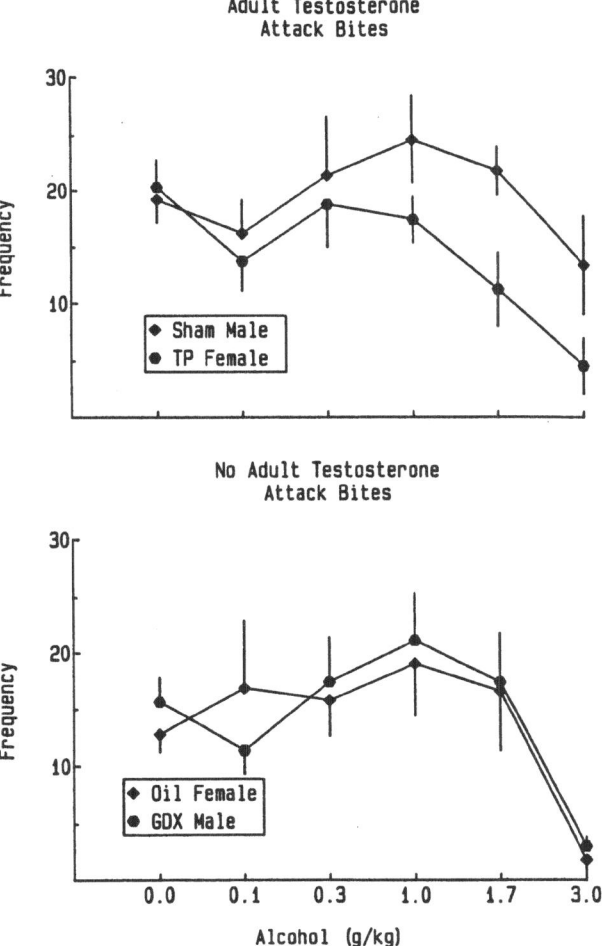

Fig. 12 (Top) The effects of adult testosterone administration and doses of alcohol (0.0, 0.1, 0.3, 1.0, 1.7, 3.0 g/kg, PO) on the frequency of attack biting by sham-treated male mice (diamonds), and neonatally androgenized female mice (hexagons). (Bottom) The effects of no adult testosterone treatment and doses of alcohol on the frequency of attack biting by sham-treated females (diamonds) and neonatally castrated males (hexagons). Asterisks indicate p<0.05 versus 0 g/kg (from Lisciotto et al. 1987).

Newborn female mice were injected with testosterone and males castrated at birth to test whether manipulating part of sexual differentation would change the sex specificity of the interaction between testosterone and alcohol seen in adult mice (Lisciotto et al. 1987). As can be seen in fig. 12, neonatally androgenized female mice responded somewhat like males to adult treatment with testosterone and alcohol while neonatally castrated males appeared to be more female-like. These data demonstrate that the testosterone – alcohol interaction is in part a function of sexual differentation and the physiological basis for this interaction can be modified by exposure to testosterone during early development.

DISCUSSION

It is important to emphasize that the high level of aggressive behaviour and the elevated testosterone titre of dominant monkeys changed significantly according to the annual mating cycle, while the endocrine and behavioural profile of subordinate monkeys remained relatively constant. The sensitivity of Squirrel monkeys to alcohol thus depends on social status and on seasonal factors. During the mating season, the behaviour of dominant monkeys was increased by low, and reduced or unaffected by higher doses of alcohol. The behaviour of subordinate monkeys was unaffected at the same doses of alcohol.

Previous studies of the behavioural pharmacology of alcohol in Squirrel monkeys have focussed on changes in conditioned behaviour (Barrett 1985). These studies demonstrated that the so-called "disinhibiting" effects of alcohol may be more accurately and efficiently described in terms of baseline levels of behaviour. For example, Glowa and Barrett (1976) found that alcohol increased low rate, and decreased high rate behaviour, independent of whether the activity was suppressed by some environmental contingency. The data reported here provide further support for the observation that disinhibition is not a sufficient explanation for alcohol's effects on social behaviour. In particular, the behaviour of subordinate male monkeys in social groups appears suppressed by the threat and attack behaviour of dominant male monkeys. Alcohol did not increase social initiatives by subordinate monkeys, and actually decreased them at high doses. In contrast, the already high level of aggressive behaviour of dominant monkeys could be increased by low doses of alcohol. Also, the low rate of aggressive behaviour by castrated mice failed to be "disinhibited" by alcohol. The effects depended on the endocrine status of the monkeys or mice, and unlike operant behaviour, were independent of baseline levels.

Treatment of subordinate monkeys with a large daily dose of testosterone did not trigger an increase in social or aggressive behaviour, but did modify the sensitivity of these animals to the effects of alcohol on aggressive behaviour. As in dominant monkeys during the mating season, the frequency of aggressive behaviour exhibited by testosterone-treated subordinate animals was increased by a low and unchanged by higher doses of alcohol.

Testosterone treatment also altered the effect of alcohol on mouse aggressive behaviour in the resident-intruder paradigm. Alcohol produced a dose-related decrease in such behaviour in intact male mice, castrated male mice, and castrated mice administered continuous low doses of testosterone. An aggression-enhancing effect of alcohol was only measured in castrated male mice administered continuous, high concentrations of testosterone. The aggression increasing effects depend on the presence of high plasma testosterone concentrations. It would be interesting to examine whether pulsatile augmentation would also be sufficient to obtain these alcohol effects (Brain et al. 1983).

Several potential sites of interaction have been identified for alcohol and testosterone. Alcohol dose-related decreases of plasma testosterone may depend on competition for essential coenzymes at rate-limiting steps of both synthesis and metabolism of testosterone (Cicero 1983; Ellingboe and Varanelli 1979). However, reduction of testosterone is associated with a decline of aggressive behaviour, thus providing an unlikely mechanism for the aggression-enhancing effect described here. Conversely, varying concentrations of testosterone may be associated with different rates of alcohol metabolism, resulting in modifications of time course, absorption and dose effect. A common condition necessary for the rate–increasing effects of alcohol on both dominant monkey and mouse aggression is a high concentration of plasma testosterone. Treatment of subordinate monkeys with testosterone propionate demonstrates that elevated testosterone concentration is not a sufficient condition for modifying the treated monkey's aggressive behaviour, but does permit the aggression increasing effects of alcohol. These results suggest that testosterone may modulate neural systems controlling aggressive behaviour which are also sensitive to alcohol. High levels of plasma testosterone represent a necessary condition for alcohol induced increases in aggression associated with the mating season in dominant monkeys, aggression by testosterone treated subordinates, and increased attacks by male mice on male intruders.

Evidence from studies of prenatal treatment of female mouse pups with testosterone, and castrated male pups suggest that the substrate for alcohol–testosterone interaction is organized during a critical period shortly before and after birth, and is activated by high levels of testosterone in adults (vom Saal 1983). Studies of the interaction of stress, adrenal response and alcohol have demonstrated that alcohol effects may both modify and be modified by the hormonal status of an animal (Pohorecky 1981). The results of the current studies suggest that a similarly complex interaction exists between alcohol and gonadal hormones. A complete analysis of alcohol's effect on behaviour and the physiological substrate must include consideration of the endocrine status of the treated animal.

Acknowledgements: Preparation of this contribution and research was supported by United States Public Health Service grants AA 05122 and DA 2632. We thank Dr.J.Ellingboe, Carol Paronis, Eugene Brandon, and Chris Lisciotto for assistance and comments.

REFERENCES

Abel EL,Zeidenberg P (1985) Age, alcohol and violent death: A postmortem study. J Studies Alc 46: 228–231

Alvarez F (1975) Social hierarchy under different criteria in groups of Squirrel monkeys, Saimiri sciureus. Primates 16: 437–455

Asberg M, Thoren P, Traskman L, Bertilsson L, Ringberger V (1976) "Serotonin depression" – A biochemical subgroup within the affective disorders? Science 191: 478–480

Badr FM, Bartke A (1974) Effect of ethyl alcohol on plasma testosterone level in mice. Steroids 23: 921–928

Bammer G, Eichelman B (1983) Ethanol effects on shock–induced fighting and muricide by rats. Aggr Behav 9: 175–181

Barrett JE (1985) Behavioral pharmacology of the Squirrel monkey. In: Rosenblum LA, Coe CL (eds) Handbook of Squirrel Monkey Research. Plenum Press, New York, pp 315–348

Berry MS, Smoothy R (1986) A critical evaluation of claimed relationships between alcohol intake and aggression in infra–human animals. In: Brain PF (ed) Alcohol and Aggression. Croom Helm, Dover, NH, pp 84–137

Bertilson HS, Mead JD, Morgret MK, Dengerimk HA (1977) Measurement of mouse squeals for 23 hours as evidence of long-term effects of alcohol on aggression in pairs of mice. Psychol Rep 41: 247–250

Bohman M (1978) Some genetic aspects of alcoholism and criminality. Arch Gen Psychiatry 35: 269–276

Boyatzis RE (1974) The effect of alcohol consumption on the aggressive behavior of men. Quart J Studies Alc 35: 959–972

Brain PF, Haug M, Kamis A (1983) Hormones and different tests for aggression with particular reference to the effects of testosterone metabolites. In: Balthazart J, Pröve E, Gilles R (eds) Hormones and behaviour in higher vertebrates. Springer Verlag, Heidelberg, pp 290–304

Brain PF (ed) (1986) Alcohol and aggression. Croom Helm, London

Chamove AS, Harlow HF (1970) Exaggeration of self–aggression following alcohol ingestion in rhesus monkeys. J Abn Psychol 75: 207–209

Chance MRA, Mackintosh JH, Dixon AK (1973) The effects of ethyl alcohol on social encounters between mice. J Alcoholism 8: 90–93

Cherek DR, Steinberg JL, Manno BR (1985) Effects of alcohol on human aggressive behavior. J Studies Alc 46: 321–328

Cherek DR, Steinberg JL, Vines RV (1984) Low doses of alcohol affect human aggressive responses. Biol Psychiatry 19: 263–267

Cicero TJ (1983) Behavioral significance of drug–induced alterations in reproductive endocrinology in the male. In: Gottheil E, Drury KA, Skoloda TE, Waxman HM (eds) Alcohol, Drug Abuse, and Aggression. Charles C Thomas, Springfield, Ill, pp 203–227

Cicero TJ, Badger TM (1977) Effects of alcohol on the hypothalamic–pituitary–gonadal axis in the male rat. J Pharmacol Exp Ther 201: 427–433

Cicero TJ, Newman KS, Gerrity M, Schmoeker PF, Bell RD (1982) Ethanol inhibits the naloxone–induced release of luteinizing hormone–releasing hormone from the hypothalamus of the male rat. Life Sci 31: 1587–1596

Cicero TJ, Newman KS, Meyer ER (1981) Ethanol–induced inhibitions of testicular steroidogenesis in the male rat: Mechanisms of actions. Life Sci 28: 871–877

Cloninger CR (1983) Antisocial behavior. In: Hippius H, Winokur G (eds) Psychopharmacology 1, Elsevier, Amsterdam, pp 353–370

Coe CL, Smith ER, Levine S (1985) The endocrine system of the Squirrel monkey. In: Rosenblum LA, Coe CL (eds) Handbook of Squirrel Monkey Research. Plenum Press, New York, pp 191–218

Coe CL, Smith ER, Mendoza SP, Levine S (1983) Varying influence of social status on hormone levels in male Squirrel monkeys. In: Kling AS, Steklis HD (eds) Hormones, drugs and social behavior. Spectrum, New York, pp 7–32

Coid J (1982) Alcoholism and violence. Drug Alc Dep 9: 1–13

Corenblum B (1983) Reactions to alcohol–related marital violence: Effects of one's own abuse experience and alcohol problems on causal attributions. J Studies Alc 44: 665–674

DeBold JF, Miczek KA (1981) Sexual dimorphisms in the control of aggressive behavior of rats. Pharmacol Biochem Behav 14 (S1): 89–94

DeBold JF, Miczek KA (1983) Testosterone alters alcohol effects on aggression in male but not female mice. Fed Proc 42: 887

DeBold JF, Miczek KA (1985) Testosterone modulates the effects of ethanol on male mouse aggression. Psychopharmacology 86: 286–290

DeBold JF, Haney M, Miczek KA (1986) Alcohol and maternal behavior in female rats: Differentiation between aggression and pup retrieval. Soc Neurosci Abstr 12, part 1, p 281

Edwards DA (1969) Early androgen stimulation and aggressive behavior in male and female mice. Physiol Behav 4: 333–338

Ellingboe J, Varanelli CC (1979) Ethanol inhibits testosterone synthesis by direct action on Leydig cells. Res Comm Chem Path Pharmacol 24: 87–102

Ellman GL, Herz MJ, Peeke HVS (1972) Ethanol in a cichlid fish: Blood levels and aggressive behavior. Proc Western Pharmacol Soc 15: 92–95

Emley GS, Hutchinson RR (1983) Unique influences of ten drugs upon post–shock biting attack and pre–shock manual responding. Pharmacol Biochem Behav 19: 5–12

Figler MH, Peeke HVS (1978) Alcohol and the prior residence effect in male convict cichlids (Cichlasoma nigrofasciatum). Aggr Behav 4: 125–132

Floody OR (1983) Hormones and aggression in female mammals. In: Svare BB (ed) Hormones and Aggressive Behavior. Plenum Press, New York, pp 39–90

Frankel BG, Ferrence RG, Johnson FG, Whitehead PC (1976) Drinking and self–injury: Toward untangling the dynamics. Br J Add 71: 299–306

Glowa JR, Barrett JE (1976) Effects of alcohol on punished and unpunished responding of Squirrel monkeys. Pharmacol Biochem Behav 4: 169–173

Goy RW, McEwen BS (1980) Sexual Differentiation of the Brain. MIT Press, Cambridge, Mass

Irwin S, Kinohi R, Van Sloten M, Workman MP (1971) Drug effects on distress–evoked behavior in mice: Methodology and drug class comparisons. Psychopharmacologia 20: 172–185

Johnson SD, Gibson L, Linden R (1978) Alcohol and rape in Winnipeg, 1966–1975. J Studies Alc 39: 1887–1894

Krsiak M (1975) Timid singly–housed mice: Their value in prediction of psychotropic activity of drugs. Br J Pharmacol 55: 141–150

Krsiak M (1976) Effect of ethanol on aggression and timidity in mice. Psychopharmacology 51: 75–80

Lagerspetz KMJ, Ekqvist K (1978) Failure to induce aggression in inhibited and in genetically non–aggressive mice through injections of ethyl alcohol. Aggr Behav 4: 105–113

Leonard KE, Bromet EJ, Parkinson DK, Day NL, Ryan CM (1985) Patterns of alcohol use and physically aggressive behavior in men. J Studies Alc 46: 279–282

Leshner AI, Candland DK (1972) Endocrine effects of grouping and dominance rank in Squirrel monkeys. Physiol Behav 8: 441–445

Lewis CE, Robins L, Rice J (1985) Association of alcoholism with antisocial personality in urban men. J Nerv Mental Dis 173: 166–174

Linnoila M, Virkkunen M, Scheinin M, Nuutila A, Rimon R, Goodwin FK (1983) Low cerebrospinal fluid 5–hydroxyindoleacetic acid concentration differentiates impulsive from nonimpulsive violent behavior. Life Sci 33: 2609–2614

Lisciotto CA, DeBold JF, Miczek KA (1987) Sexual differentiation and the effects of alcohol on aggression in mice. Pharmacol Biochem Behav (in press)

MacDonnell MF, Fessock L, Brown SH (1971) Ethanol and the neural substrate for affective defense in the cat. Quart J Studies Alc 32: 406–419

Mendelson JH, Mello NK (1974) Alcohol, aggression and androgens. In: Frazier SH (ed) Aggression. Research in Nervous Mental Disorders, Vol 52, Williams and Wilkins, Baltimore, pp 225–247

Mendelson JH, Mello NK, Ellingboe J (1977) Effects of acute alcohol intake on pituitary–gonadal hormones in normal human males. J Pharmacol Exp Ther 202: 676–632

Mendelson JH, Mello NK, Ellingboe J (1978) Effects of alcohol on pituitary–gonadal hormones, sexual function, and aggression in human males. In: Liptor MA, DiMascia A, Killam KF (eds) Psychopharmacology: A generation of progress. Raven Press, New York, pp 1677–1692

Mendoza SP, Lowe EL, Resko JA, Levine S (1978) Seasonal variations in gonadal hormones and social behavior in Squirrel monkeys. Physiol Behav 20: 515–522

Miczek KA (1982) Ethological analysis of drug action on aggression, defense and defeat. In: Spiegelstein MY, Levy A (eds) Behavioral Models and the Analysis of Drug Action. Elsevier, Amsterdam, pp 225–239

Miczek KA (1987) The psychopharmacology of aggression. In: Iversen LL, Iversen SD, Snyder SH (eds) Handbook of Psychopharmacology, Vol 19, Behavioral Pharmacology. Plenum Press, New York, pp 183–328

Miczek KA, Barry H III (1977) Effects of alcohol on attack and defensive–submissive reactions in rats. Psychopharmacology 52: 231–237

Miczek KA, DeBold JF (1983) Hormone–drug interactions and their influence on aggressive behavior. In: Svare BB (ed) Hormones and Aggressive Behavior. Plenum Press, New York, pp 313–347

Miczek KA, Krsiak M (1979) Drug effects on agonistic behavior. In: Thompson T, Dews PB (eds) Advances in Behavioral Pharmacology. Academic Press Inc, New York, pp 87–162

Miczek KA, O'Donnell JM (1978) Intruder–evoked aggression in isolated and nonisolated mice: Effects of psychomotor stimulants and l–dopa. Psychopharmacology 57: 47–55

Miczek KA, O'Donnell JM (1980) Alcohol and chlordiazepoxide increase suppressed aggression in mice. Psychopharmacology 69: 39–44

Miczek KA, Winslow JT (1987) Psychopharmacological research on aggressive behavior. In: Greenshaw A, Dourish CT (eds) Experimental Psychopharmacology. Humana Press, Clifton, New Jersey, pp 27–113

Murono EP, Fisher–Simpson V (1985a) Ethanol directly stimulated dihydrotestosterone conversion to 5–alpha–androstan–3–alpha,17–beta–diol and 5–alpha–androstan–3–beta,17–beta–diol in rat liver. Life Sci 36: 1117–1124

Murono EP, Fisher–Simpson V (1985b) Ethanol directly increases dihydrotestosterone conversion primarily to 5 alpha–androstan–3 beta, 17 beta–diol in rat leydig cells. Life Sci 36: 1381–1387

Peeke HVS, Figler MH (1981) Modulation of aggressive behavior in fish by alcohol and cogeners. Pharmacol Biochem Behav 14 (supl 1) 79–84

Peeke HVS, Cutler L, Ellman G, Figler M, Gordon D, Peeke SC (1981) Effects of alcohol, cogeners, and acetaldehyde of aggressive behavior of the convict cichlid. Psychopharmacology 75: 245–247

Peeke HVS, Ellman GE, Herz MJ (1973) Dose dependent alcohol effects on the aggressive behavior of the convict cichlid (Cichlasoma nigrofaciatum). Behav Biol 8: 115–122

Peeke HVS, Peeke SC, Avis HH, Ellman G (1975) Alcohol, habituation and the patterning of aggressive responses in a cichlid fish. Pharmacol Biochem Behav 3: 1031–1036

Pettijohn TF (1979) The effects of alcohol on agonistic behavior in the Telomian dog. Psychopharmacology 60: 295–301

Pohorecky LA (1981) The interaction of alcohol and stress, a review. Neurosci Biobehav Rev 5: 209–229

Rada RT (1975) Alcoholism and forcible rape. Am J Psychiatry 132:444–446

Raynes A, Ryback R, Ingle D (1968) The effect of alcohol on aggression in Betta splendens. Commun Behav Biol 2: 141–146

Raynes A, Ryback RS (1970) Effect of alcohol and congeners on aggressive response in Betta splendens. Quart J Studies Alc 5: 130–135

Smith ER, Damassa DA, Davidson JM (1977) Hormone administration: Peripheral and intracranial implants. In: Myers RD (ed) Methods in Psychobiology. Academic Press, New York, pp 259–279

Smoothy R, Berry MS (1983) Effects of ethanol on behavior of aggressive mice from two different strains: A comparison of simple and complex behavioral assessments. Pharmacol Biochem Behav 19: 645–653

Smoothy R, Berry MS (1984) Effects of ethanol on murine aggression assessed by biting of an inanimate target. Psychopharmacology 83: 268–271

Smoothy R, Bowden NJ, Berry MS (1982) Ethanol and social behaviour in naive Swiss mice. Aggr Behav 8: 204–207

Stabenau JR (1984) Implications of family history of alcoholism, antisocial personality, and sex differences in alcohol dependence. Am J Psychiatry 141: 1178–1182

Svare BB (ed) (1983) Hormones and aggressive behavior. Plenum Press. New York

Svare BB, Mann MA (1983) Hormonal influences on maternal aggression. In: Svare BB (ed) Hormones and Aggressive Behavior. Plenum Press, New York, pp 91–104

Tinklenberg JR, Murphy PL, Murphy P, Darley CF, Roth CF, Kopell BS (1974) Drug involvement in criminal assaults by adolescents. Arch Gen Psychiatry 30: 685–689

Van Thiel DH, Gavaler JS, Lester R, Goodman MD (1975) Alcohol induced testicular atrophy: An experimental model for hypogonadism occuring in chronic alcoholic men. Gastroenterology 69: 326–332

Van Praag HM (1982) Depression, suicide and the metabolism of serotonin in the brain. J Aff Dis 4: 275–290

Vom Saal FS (1983) Models of early hormonal effects on intrasex aggression in mice. In: Svare BB (ed) Hormones and Aggressive Behavior. Plenum Press, New York, pp 197–222

Weitz MK (1974) Effects of ethanol on shock–elicited fighting behavior in rats Quart J Studies Alc 35: 953–958

Winslow JT, Miczek KA (1985) Social status as determinant of alcohol effects on aggressive behavior in Squirrel monkeys (Saimiri sciureus). Psychopharmacology 85: 167–172

Yoshimura H, Ogawa N (1983) Pharmaco–ethological analysis of agonistic behavior between resident and intruder mice: Effects of ethylalcohol. Folia Pharmacol Jap 81: 135–141

PSYCHOPHARMACOLOGY OF HUMAN AGGRESSION: LABORATORY STUDIES

Don R. Cherek and Joel L. Steinberg. Department of Psychiatry, Louisiana State University, School of Medicine, Shreveport, U.S.A.

INTRODUCTION

Human aggressive behaviour occurring in the natural environment has frequently been related to prior drug ingestion (e.g. Cherek and Steinberg 1986; Tinklenberg and Stillman 1970). The association between alcohol consumption and subsequent human aggressive behaviour has been frequently reported and is generally accepted as a causative relationship. Such naturalistic observations obviously focus upon the administration of drugs which precede increased probabilities of aggressive behaviour. One can expect, however, that the probability of human aggressive behaviour can be reduced following administration of other drugs or the same drugs in different environmental situations. A more precise delineation of the effects of particular drugs on human aggressive behaviour can be advanced only by studies conducted in the laboratory under more controlled conditions.

Before any laboratory studies could be initiated, the problem of establishing a definition of aggression had to be resolved. Historically, definitions of human aggressive behaviour have distinguished between verbal and physical aggressive responses, and have defined the latter as responses which produced or are intended to produce physical injury or pain in the victim. The lack of a consensus regarding the types of responses which should be referred to as aggressive has impeded research on human aggressive behaviour.

Two basic problems had to be addressed prior to an initition of the laboratory study of human aggressive behaviour: (1) establishment of a precise definition of the aggressive response, and (2) development of an objective measure of the aggressive response which would not involve any physical injury. Buss (1961) operationally defined aggressive responses as responses which result in the delivery of a noxious, aversive stimulus to another person. The terms noxious, and aversive signify that subjects would attempt to escape or avoid the presentation of such stimuli. The aversive stimulus which Buss employed was electric shock, the presentation of which in the laboratory seems directly analogous to the presentation of the types of aversive stimuli, e.g. physical blows, which occur in the natural environment. The presentation of the aversive stimulus to another person requires the subject to push a button, which facilitates the objective, automatic recording of aggressive responses.

Three methodologies have been developed to study human aggressive behaviour in the laboratory. All three employ the same operational definition of the aggressive response, and record responses by measuring button presses which actually or ostensibly deliver noxious stimuli to another person. In the Buss (1961) procedure, the research subject is informed that the effects of punishment on learning are being studied. He/she is allowed to ostensively administer an electric shock to another person when they make an 'error' on a learning task. The research subject chooses the intensity of electric shock to be directed to the other person following each error, by pressing one of a series of buttons representing increasing shock intensities. The measures of aggressive responses recorded are: (1) the shock intensity selected, i.e., which button in the series was depressed, and (2) the shock duration, i.e., the amount of time the button was depressed.

Taylor (1967) developed a second procedure (the Taylor Competitive Reaction Time test) also involving the ostensive presentation of electric shock to another person. Subjects were informed that they were competing in a reaction time task with another subject. Prior to each reaction time trial, the subject selects a shock intensity which is potentially delivered to a fictitious opponent. Following each trial, the person with the slowest reaction time is said to receive the electric shock. As with the Buss procedure, the measure of aggression is the intensity of shock selected by the subject for presentation to his/her opponent.

A third procedure (Cherek 1981), provides both aggressive and non-aggressive response options to research subjects. The non-aggressive response is button pressing maintained by the accumulation of points on a counter which are exchanged for money at the end of daily sessions. The aggressive response available to the research subject is the ostensive subtraction of similar points from a fictitious subject. Aggressive responding is stimulated by substracting points from the research subject which are attributed to the other subject. The measure of aggression is the number of times the subject chooses to respond aggressively during each session.

Thus, all three methodologies may be used in the determination of drug effects.

EFFECTS OF DRUGS ON HUMAN AGGRESSIVE RESPONDING IN THE LABORATORY

The use of the above types of study in assessing the effects of alcohol (which provides the bulk of the data), stimulant and sedative drugs is briefly reviewed below.

1. Alcohol

Bennett et al. (1969) conducted the first laboratory investigation of the effects of alcohol on human aggressive behaviour utilizing the Buss procedure. Administration of low doses of alcohol (0.24 and 0.5 g/kg) resulted in slight non-significant increases in the shock intensity selected for delivery to another person, the measure of aggression. Zeichner and Pihl (1979, 1980) modified the basic Buss procedure to include a counter-aggressive or retaliatory response. When subjects ostensively administered electric shock to the other person, they were able to present an auditory tone of varying intensity through headphones to the subject. With this modified Buss procedure, the investigators observed significant increases in the intensity and duration of shocks presented by subjects administered the high alcohol dose (1.2 g/kg) compared to subjects given a placebo drink or no beverage.

The first laboratory study utilizing the Taylor's paradigm (Shuntich and Taylor 1972) reported increased aggressive responses following the administration of a similar alcohol dose (approximately 0.5 g/kg). Subjects given this alcohol dose set significantly higher shock intensities for their opponents than subjects given a placebo beverage or no beverage. Subsequent studies comparing the effect of higher (1.2 g/kg) and the same alcohol dose indicated that subjects given the higher dose set higher shock intensities during the first trial before they received information regarding the shock intensity ostensively selected by their opponent (Taylor and Gammon 1975). In addition, subjects given the higher alcohol dose were much more likely to select the highest available shock intensity for their opponent. Subjects given this high alcohol dose were also more likely to set an extreme shock intensity added to the usual shock intensity panel which was described as very painful and potentially injurious (Taylor et al. 1979).

The Cherek procedure utilizes a repeated measures single subject design, in which each subject is administered placebo and different doses of the drug over successive sessions. Using this procedure, Cherek et al. (1985) found that alcohol administered in doses of 0, 0.12, 0.23 and 0.46 g/kg produced dose–dependent increases in the number of aggressive responses per session.

Thus, all three procedures employed to study human aggressive behaviour in the laboratory have found that alcohol administration may increase the frequency or intensity of aggressive responses. As there is inadequate space to extensively review the diverse studies which have been conducted on alcohol's effects on aggressive behaviour and their interaction with other non–pharmacological variables, the reader is referred to Taylor and Leonard (1983) for a good account.

2. Drugs with stimulant properties

Nicotine

The two studies reported involve manipulations of tobacco smoking but the effects can be attributed to nicotine which is generally regarded as the behaviourally active pharmacological agent in this substance. Schechter and Rand (1974) used the Buss procedure to assess and compare aggressive responding in cigarette smokers and non–smoking subjects. The latter did not change the duration or intensity of shocks ostensibly administered to another person over sessions. In contrast, smoking subjects increased the intensity and/or duration of shocks presented, during sessions in which they were not allowed to smoke. Thus, tobacco deprivation was associated with increased aggressive responding as assessed by the Buss procedure.

Cherek (1981) determined the effects within subjects of not smoking or smoking high or low nicotine cigarettes on aggressive responding using a repeated measures design. Compared to non–smoking conditions, subjects emitted fewer aggressive responses during sessions conducted after smoking experimental cigarettes. The decreases in aggressive responding were dose–related in that greater decreases were observed following smoking high rather than low nicotine cigarettes. It seems apparent that nicotine decreases the probability of aggressive responding but these studies involved regular tobacco smokers. The non–smoking conditions in these studies may have resulted in increased aggressive responses as part of a tobacco withdrawal syndrome. The apparent decrease in aggressive responding following smoking may represent a suppression of tobacco withdrawal symptoms. It would consequently be most interesting to assess the effect of nicotine on aggressive responding in non–smokers.

Caffeine

Two studies have been conducted assessing the effects of caffeine and coffee on aggressive responding using the Cherek Procedure. The first study (Cherek et al. 1983) investigated the effects of the oral administration of placebo and three doses of caffeine (1, 2 and 4 mg/kg) on aggressive responses. Compared to placebo, the number of aggressive responses was decreased following caffeine administration. The effects of caffeine were dose related with the greatest decrease in aggressive responding observed following the highest caffeine dose. A second study (Cherek et al. 1984) compared the effects of drinking regular and decaffeinated coffee on aggressive responding using the same procedure. Each subject drank two cups of coffee prior to each session which contained either four teaspoons of regular coffee, two teaspoons of regular and decaffeinated coffee, or four teaspoons of decaffeinated coffee. Compared to the decaffeinated coffee, the consumption of regular coffee decreased aggressive responding.

The amount of caffeine that subjects normally consume influences the suppression of aggressive responses following drinking regular coffee. Those subjects consuming more than 200 mg of caffeine per day had larger decreases in aggressive responding after drinking regular coffee, than counterparts consuming less than 100 mg of caffeine per day. As with nicotine studies, it is possible that some subjects experienced caffeine withdrawal symptoms which could be countered by the administration of the highest caffeine dose (4 mg/kg) or regular coffee, decreasing aggressive responding. The fact that the highest consumers of caffeine were most sensitive to caffeine and regular coffee is consistent with this interpretation. Studies of aggressive responding involving the administration of caffeine or regular coffee to very infrequent caffeine consumers or abstainers could provide information on the possible role of caffeine withdrawal.

d-Amphetamine

The Cherek procedure has been employed to assess the effects of placebo and three doses of d-amphetamine (5, 10 and 20 mg per 70 kg of body weight) on aggressive responding (Cherek et al. 1986d). D-Amphetamine produced increases in aggressive responding at the 5 and 10 mg per 70 kg doses. At the highest dose (20 mg per 70 kg), aggressive responding decreased (relative to other d-amphetamine doses) to the level of or slightly below frequencies observed during placebo sessions. Biphasic or inverted-U-shaped dose-response curves for aggressive responses observed in this investigation are similar to those reported in studies of aggressive behaviour in animals (e.g. Smith and Byrd 1984).

Drugs with stimulant properties generally decrease aggressive responding. As most of these studies have been conducted with the Cherek procedure, it is possible that stimulant drugs increase the probability of continuing to select the most likely response (earning points) and diminish the probability that the subject will select a less probable aggressive response (ostensively subtracting points from another subject). If this is true, such effects would represent suppression of low probability behaviour rather than a specific effect upon aggressive responding.

3. Drugs with sedative properties

Delta-9-tetrahydrocannabinol (THC)

There has been considerable controversy in the scientific literature regarding the relationship between marihuana use and aggressive behaviour (e.g. Abel 1977). Two studies employing the Taylor procedure have involved the oral administration of the active principle of marihuana, THC. In a first study (Taylor et al. 1976), subjects were assigned to a high (0.3 mg/kg) or a low (0.1 mg/kg) THC dose. Subjects given the high THC dose set lower shock intensities for their fictitious opponents, than subjects given the low THC dose. Shock intensities set for the opponent during the first trial represents a measure of aggressive responding in the absence of provocation, since the subject is informed of the shock intensity ostensively selected by his opponent after each trial. During this first trial, subjects given the high THC dose selected lower shock intensities than subjects given the low dose.

Myerscough and Taylor (1985), examined the effects of oral administration of three doses of THC (0.1, 0.25 and 0.4 mg/kg) using the same procedure. The level of provocation provided by the opponent was increased compared to the earlier study. In the first study (Taylor et al. 1976), the opponent typically ostensively set gradually increasing shock intensities for the subject over trials. In the latter study, the opponent set high shock intensities for the subject throughout the experimental session. The mean shock intensities set by subjects for their opponents did not differ in the three THC dose conditions but there was a tendency for more intense shock settings over blocks of trials among subjects receiving the low THC dose. In addition to the typical ten shock intensities, a very intense potentially injurious shock intensity was also available to both opponents and subjects. After two trials, the subject was told that the opponent had selected this very intense shock for presentation to him/her (shock was not actually presented).

Following this intense provocation, subjects administered the low THC dose set significantly higher shock intensities and were more likely to retaliate by setting this very intense shock for their opponent. This response to intense provocation was not observed in subjects given the medium and high doses of THC. Thus, both studies indicate a tendency for THC to decrease the intensity of aggressive responses and to diminish the retaliatory response to intense provocation.

Diazepam

Two very recent studies have examined the effects of diazepam (Valium) on aggressive responding in the laboratory. The first study (Wilkinson 1985) utilized Taylor's paradigm and subjects were assigned to placebo or 10 mg diazepam treatment. As with previous experiments employing the Taylor procedure, the fictitious opponent selected more intense shocks for the subject across the 21 trials of the experimental session, thus exposing subjects to increasing provocation over time. During the initial (low provocation) block of trials, where the opponent set low shock intensities for the subjects, subjects given diazepam selected significantly higher shock intensities for opponents than subjects given placebo. During the second and third block of trials when the opponent set higher shock intensities for the subjects, both placebo and diazepam subjects increased the shock intensities selected for their opponents. The subjects given diazepam set, however, significantly higher shock intensities than placebo subjects. Thus, subjects given 10 mg of diazepam appeared more aggressive than placebo treated counterparts.

In initial studies employing the Cherek procedure, the drugs that were administered (nicotine, caffeine and alcohol) are eliminated very rapidly and successive doses could be administered over a relatively small number of sessions (10–16). Studies involving diazepam etc. involve materials, that are metabolized and excreted very slowly, requiring subjects to participate for 30 to 40 sessions. In the original procedure, the subject's aggressive responses had no effect upon subsequent provocations. Provoking point subtractions attributed to the other person were presented independently of the subject's aggressive responding. Under these conditions, aggressive responding decreased over sessions (extinction) and subjects very frequently would cease to respond aggressively. It was arranged that in the new procedure aggressive responding initiated specific periods during which scheduled point subtractions was not presented. Under these conditions, the subject's aggressive responding resulted in a temporary reduction in provocation. The contingency specified between aggressive responses and subsequent provocation was 'escape'. Under the escape contingency, at least one provoking point subtraction had to be presented to the subject before his aggressive responding produced an interval of time (either 125 or 500 sec) during which no provocations were presented. When this time interval elapsed, point subtractions were again scheduled and the subject would receive another point subtraction before his aggressive responses initiated another interval. Subjects are thus periodically provoked throughout the daily 50 min sessions at a frequency which varies inversely with the duration of the interval. Such a contingency maintains the subject's aggressive responding and mimics contingencies operating outside the laboratory which maintain aggressive behaviour.

Using this modified procedure, Cherek et al. (1986c) evaluated the effects of placebo and three doses of diazepam (2.5, 5 and 10 mg per 70 kg) on aggressive responding which involved repeated drug dosing and response measurements. Each subject received placebo and all three diazepam doses on four separate occasions. Seven of the eight subjects decreased their aggressive responding following the adminstration of the highest diazepam dose (10 mg per 70 kg). Three subjects also decreased aggressive responding after receiving 5 mg per 70 kg of diazepam.

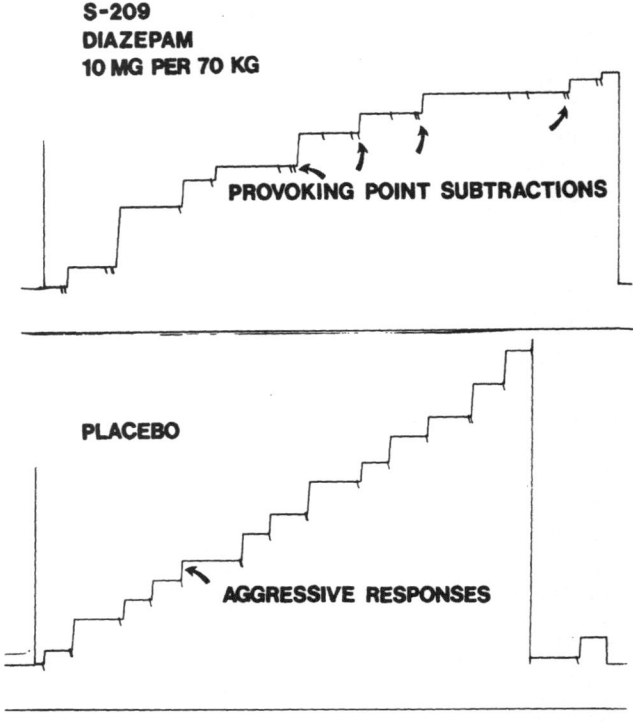

Fig. 1 Representative cumulative records of subject S–209 following administration of placebo (lower) or 10 mg per 70 kg of diazepam (upper portion). Aggressive responses are displayed on the stepper pen which moves vertically with each response. Provoking point subtractions presented to the subject are shown as diagonal slashes on the stepper pen.

This effect is shown in fig. 1 which displays sample cumulative records for subject S–209 following placebo (lower portions) and 10 mg per 70 kg of diazepam (upper portion). Aggressive responses are shown on the stepper pen which moves vertically each time the subject emits an aggressive response. Provoking point subtractions presented to the subject are shown as diagonal slashes on this stepper pen. Under placebo conditions, the subject immediately responds aggressively following provocation and emits a number of aggressive responses in succession. Following the administration of 10 mg per 70 kg of diazepam, the subject's aggressive responses are greatly diminished. This reduction in aggressive responses is the result of provocations no longer reliably resulting in aggressive responses (refer to arrows) and a fewer number of aggressive responses occurring when the subject does decide to retaliate. In this diazepam study, one subject (only) increased aggressive responding following the administration of diazepam but this occurred in the absence of provocation a feature not evident under placebo conditions.

In general, these two studies on diazepam provide conflicting results. Wilkinson (1985) found that 10 mg of diazepam produced more intense aggressive responding, whilst Cherek et al. (1986c) found dose–dependent decreases in aggressive responding in all but one subject. There are obvious methodological differences between these two studies which may account for these differences. The level of provocation may also be important. In the Wilkinson (1985) study the level of provocation presented to the subject increased throughout the experiment whereas in the Cherek et al. (1986c) study the frequency of provocation was relatively stable for individual subjects. Enhancement of aggressive responding by diazepam may only be observed in situations involving increasing levels of provocation.

NON–PHARMACOLOGICAL VARIABLES THAT ALTER THE EFFECTS OF DRUGS ON AGGRESSIVE RESPONDING

It seems important to briefly assess non–pharmacological variables that can alter the apparent effects of drugs on aggressive responsing. Two environmental variables, level or frequency of provocation and the consequences of aggressive responding, seem good candidates for such effects.

There is evidence that the increased aggressive responding observed following alcohol administration is enhanced in situations involving more intense or more frequent provocation. In a study utilizing Taylor's paradigm (Taylor et al. 1976), subjects were assigned to provoking or non–provoking conditions. In the provoking condition, the opponent set more intense shocks for the subject throughout the session. Subjects assigned to the non–provoking condition were informed that their opponent intended to set only low intensity shocks. Subjects in both conditions were given either placebo or 1.2 g/kg of alcohol. This alcohol dose increased the intensity of aggressive responses in subjects only under provoking conditions. Using the Cherek procedure (Cherek et al. 1985), subjects were assigned to a low (mean of 5 provocations per session) or a high (mean of 20 provocations per session) provocation condition. Both conditions showed increased aggressive responding following alcohol administration, but subjects assigned to the high provocation condition showed a greater and statistically significant increase in aggressive responding which was a function of alcohol dose. Three additional subjects were exposed to occasionally very high provoking conditions of 20, 25 or 30 provocations per session, while during intervening sessions they were given 2 to 4 provocations per session. Under these conditions, aggressive responding increased dramatically during the highly provoking sessions. Compared to placebo, a 0.5 g/kg alcohol dose resulted in a significant increase in the frequency of aggressive responding under conditions of very frequent provocation. Both the Taylor and the Cherek paradigms indicate that the tendency of alcohol administration to increase aggressive responding is greatly enhanced in situations involving intense and frequent provocation.

Within the experimental situation, the consequences of aggressive responding can also be manipulated by altering the frequency or intensity of aggressive responses directed to the subject which were ostensively initiated by the other person. Zeichner and Pihl (1979), as discussed earlier, modified the Buss procedure to include an ostensive counter–aggressive response which was available to the victim. The other person was able to present to the subject an auditory tone of varying intensity via headphones following each shock the subject presented to them. Subjects were assigned to two consequence conditions: (1) correlated in which the intensity of the tone presented to the subjects was directly related to the shock intensity the subject chose to administer to the other person, and (2) random in which varying intensities were presented to the subject independently of the shock intensity presented. Subjects under each consequence condition received either no drink, placebo drink or 1.2 g/kg of alcohol. The different consequence conditions had clear effects upon aggressive responding among subjects that received either no drink or placebo. These subjects administered less intense and shorter duration shocks when the consequences were correlated rather than random. However, the aggressive responding of the alcohol treated subjects was not affected by the different consequences conditions, and these subjects administered high intensity shocks to the other person in both consequence conditions.

A study (Cherek et al. 1986a) employing the modified Cherek procedure examined the effects of d-amphetamine on aggressive responding maintained by two different contingency conditions relating to counter-aggression. As discussed earlier, this modified procedure initiated aggressive responding by the subtractions of points exchangeable for money which were attributed to another person. Aggressive responding was maintained by the initiation of time periods during which scheduled provoking point subtractions were not presented. Subjects were assigned to an a) escape contingency where at least one provoking point subtraction had to be presented to the subject before his aggressive responding initiated a time interval (125 or 500 sec) during which no provocations were presented. Over the 50 min session, under this contingency, subjects were periodically provoked throughout the session b) avoidance contingency, which differed in that the subject's aggressive responses prior to any provocation or during an interval when provocations were not presented, initiated another new interval. In the avoidance contingency condition, it was possible for the subjects to avoid all scheduled provocations. D-amphetamine produced a dose-dependent decrease in aggressive responding among subjects assigned to the escape contingency condition but the effect of this drug in the avoidance contingency group was quite different. Under the avoidance contingency, aggressive responding was typically increased following the administration of the 10 mg/70 kg dose, but decreased to slightly below placebo levels after administration of the highest d-amphetamine dose (20 mg/70 kg). These differences indicate that the effects of d-amphetamine on aggressive responding can be affected by the contingency relationship between those aggressive responses and subsequent provocation.

Future research in this area with other drugs and determining the effect of other factors will undoubtedly provide numerous examples of non-pharmacological variables, which can alter the effect of even the same drug in the same individual under different environmental conditions.

FUTURE RESEARCH

Three areas of research seem likely to receive increased attention by researchers: (1) response validity, (2) specificity of drug effects and (3) individual differences. The question of response validity has received relatively little attention in studies of human aggressive responding in the laboratory. Studies should begin to determine if individual subjects with documented histories of aggressive behaviour respond differently in the laboratory from subjects without such histories. Questionnaire assessments of hostility and aggression such as the Buss-Durkee Hostility Inventory (Buss and Durkee 1957) could also be correlated with the frequency or intensity of aggressive responding.

The specificity of drug action is also a very important consideration in evaluating the effects of drugs on aggressive responding. A drug can result in decreased or increased aggressive responding as a result of non-specific, generalized stimulant or depressant effects which tell us little about drug effects upon aggressive behaviour. Researchers have repeatedly emphasized the importance of assessing the specificity of drug effects on aggressive behaviour (e.g. Miczek and Krsiak 1979; Thompson and Boren 1977). One method used to evaluate specificity of drug effects is to determine the effects on more than one response option available to the subject (Sidman 1959). The present Cherek

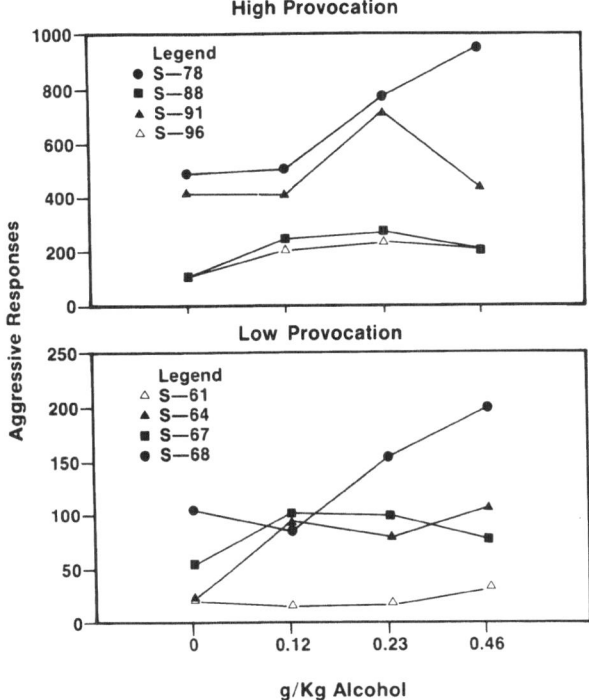

Fig. 2 The effects of placebo (0) and three doses (0, 0.12, 0.23 and 0.46 g/kg) of alcohol on the number of aggressive responses per session. Data points at each dose represent the mean of four different sessions. Subjects assigned to low provocation frequency (mean of 5 per session) are shown in the bottom panel, and subjects assigned to high provocation frequency (mean of 20 per session) are shown in the upper panel.

methodology allows assessment of specificity by comparing drug effects on two button pressing response options available to the subject. Experiments with stimulants (nicotine, caffeine and d–amphetamine) indicated that the dose–dependent decreased aggressive responding could not be attributed to a generalized, non–specific depressant action since these same doses resulted in dose–dependent increases in non–aggressive monetary–reinforced responding. (Cherek 1981; Cherek et al. 1983, 1986d). Similarily, alcohol produced dose–dependent increases in aggressive responding, while non–aggressive monetary–reinforced responding was unchanged or slightly decreased (Cherek et al. 1985). Such assessments of specificity of drug action are important in the interpretation of drug effects on aggressive behaviour.

The investigation of factors which may contribute to individual differences in their behavioural responses to particular drugs in the laboratory seems important. Frequently, research studies have given the impression that drugs consistently increase aggressive responding in all subjects. Cherek et al. (1985) reported that alcohol produced dose–dependent increases in aggressive responding among group of subjects assigned to low or high frequencies of provocation. However, if one plots individual dose–response curves (see fig. 2), large differences between subjects are evident. As alcohol dose is increased in both provocation conditions, some subjects show substantial increases in aggressive responding whereas others show only slight increases or no change in aggressive responding. These differences in response to alcohol are certainly recognized outside the laboratory and research will have to be carried out to explore current environmental factors and historical events which may account for them.

As discussed in the section on sedative drugs, diazepam administration produces increased aggressive responding in some subjects but decreased aggressive responding in most subjects. Following participation in the diazepam experiment, subjects completed the Buss–Durkee Hostility Inventory. Total hostility scores were highly correlated with changes in aggressive responding following administration of the highest diazepam dose ($r = 0.78$, $p < .01$). Subjects with relatively high total hostility scores increased aggressive responding following diazepam administration, and subjects with lower scores either had no change or decreased aggressive responding following diazepam administration. Subjects were also given placebo drinks and a single dose of 0.5 g/kg of alcohol after they completed the diazepam dose–response study. Again, there was a strong correlation between changes in aggressive responding following the 0.5 g/kg alcohol dose and changes following the highest diazepam dose ($r = 0.91$, $p < .001$). Thus, questionnaire assessment of aggression on hostility and changes in aggressive responding following alcohol administration predicts individual differences in changes in aggressive responding following diazepam administration.

Acknowledgements: The authors wish to thank Thoman H. Kelly, Ph.D. and Steve Matthews for their assistance. Our research was supported by grant DA 03166 from the National Institute on Drug Abuse.

REFERENCES

Abel EL (1977) The relationship between cannabis and violence: A review. Psychol Bull 84: 193–211

Bennett RM, Buss AH, Carpenter JA (1969) Alcohol and human physical aggression. Quart J Stud Alc 30: 870–876

Buss AH (1961) The psychology of aggression. J Wiley and Son, New York

Buss AH, Durkee A (1957) An inventory for assessing different kinds of hostility. J Consult Psychol 21: 343–349

Cherek DR (1981) Effects of smoking different doses of nicotine on human aggressive behavior. Psychopharmacology 75: 339–345

Cherek DR, Kelly TH, Steinberg JL (1986a) Behavioral contingencies and d–amphetamine effects on human aggressive and non–aggressive responding. In: Harris LS (ed) Problems of drug dependence, 1985, NIDA Research Monograph, Vol 67, U.S. Government Printing Office, Washington, D.C., pp 184–189

Cherek DR, Kelly TH, Steinberg JL, Friedman TT (1986b) Effects of acute alcohol or diazepam administration on human aggressive and non–aggressive responding. In: Shagass C et al. (eds) Biological Psychiatry 1985, Elsevier, New York, pp 497–499

Cherek DR, Steinberg JL (1986) Effects of drugs on human aggressive behavior. In: Burrows GD, Werry JS (eds) Advances in human psychopharmacology, Vol. IV, JAI Press, Greenwich CN (in press)

Cherek DR, Steinberg JL, Brauchi JT (1983) Effects of caffeine on human aggressive behavior. Psychiatry Res 8: 137–145

Cherek DR, Steinberg JL, Brauchi JT (1984) Regular or decaffeinated coffee and subsequent human aggressive behavior. Psychiat Res 11: 251–258

Cherek DR, Steinberg JL, Kelly TH (1986c) Effects of diazepam on human laboratory aggression: Correlations with alcohol effects and hostility measures. In: Harris LS (ed) Problems of drug dependence, 1986, NIDA Research Monograph, U.S. Government Printing Office, Washington, D.C. (in press)

Cherek DR, Steinberg JL, Kelly TH, Robinson DE (1986d) Effects of d–amphetamine on human aggressive behavior. Psychopharmacology 88: 381–386

Cherek DR, Steinberg JL, Manno BR (1985) Effects of alcohol on human aggressive behavior. J Stud Alc 46: 321–328

Meyerscough R, Taylor S (1985) The effects of marijuana on human physical aggression. J Pers Soc Psychol 49: 1541–1546

Miczek KA, Krsiak M (1979) Drug effects on agonistic behavior. In: Thompson T, Dews PB (eds) Advances in behavioral pharmacology, Vol 2, Academic Press, New York, pp 87–162

Schechter MD, Rand MJ (1974) Effect of acute deprivation of smoking on aggression and hostility. Psychopharmacologia 35: 19–28

Shuntich RJ, Taylor SP (1972) The effects of alcohol on human physical aggression. J Exp Res Pers 6: 34–38

Sidman M (1959) Behavioral pharmacology. Psychopharmacologia 1: 1–19

Smith EO, Byrd LD (1984) Contrasting effects of d–amphetamine on affiliation and aggression in monkeys. Pharmacol Biochem Behav 20: 225–260

Taylor SP (1967) Aggressive behavior and physiological arousal as a function of provocation and the tendency to inhibit aggression. J Pers 35: 297–310

Taylor SP, Gammon CB (1975) Effects of type and dose of alcohol on human physical aggression. J Pers Soc Psychol 32: 169–175

Taylor SP, Leonard KE (1983) Alcohol and human physical aggression. In: Green RG, Donnerstein EL (eds) Aggression: Theoretical and Empirical Reviews, Vol 2, Academic Press, New York, pp 77–101

Taylor SP, Schmutte GT, Leonard KE, Cranston JW (1979) The effects of alcohol and extreme provocation on the use of a highly noxious electric shock. Motivat Emot 3: 73–81

Taylor SP, Vardaris RM, Rawtich AB, Gammon CB, Cranston JW, Lubetkin AI (1976) The effects of alcohol and delta-9-tetrahydrocannabinol on human physical aggression. Aggr Behav 2: 153–161

Thompson T, Boren JJ (1977) Operant behavioral pharmacology. In: Honig WK, Staddon JER (eds) Handbook of operant behavior, Prentice-Hall Inc, Englewood Cliffs, NJ, pp 540–569

Tinklenberg JR, Stillman RC (1970) Drug use and violence. In: Daniels DN, Gilula MF, Ochberg FM (eds) Violence and the struggle for existence, Little Brown and Co, Boston, pp 327–365

Wilkinson CJ (1985) Effects of diazepam (Valium) and trait anxiety on human physical aggression and emotional state. J Behav Med 8: 101–114

Zeichner A, Pihl RO (1979) Effects of alcohol and behavior contingencies on human aggression. J Abn Psychol 88: 153–160

Zeichner A, Pihl RO (1980) Effects of alcohol and instigator intent on human aggression. J Stud Alc 41: 265–276

PSYCHOPHARMACOLOGY OF AGGRESSION IN HUMANS

Michael H. Sheard. Department of Psychiatry, Yale University Medical School, New Haven, CT 06508, U.S.A.

INTRODUCTION

This chapter reviews studies on the pharmacology of aggression published over the past two to three years. It is noteworthy that problems with aggressive behaviour span diagnostic categories and that it is essential to emphasize that aggression covers a wide range of behaviours, which in very basic ways, serve both individual and species survival. Consequently only maladaptive or pathological aggressive behaviour is an appropriate target for pharmacotherapy. Neurochemical studies in animals suggest that cholinergic and catecholaminergic mechanisms are involved in the induction and enhancement of predatory or instrumental aggression while serotonergic and gamma amino butyric acid–ergic mechanisms are inhibitory. In affective aggression, dopaminergic mechanisms are facilitatory while noradrenergic and serotonergic mechanisms are inhibitory. These findings have still not led to a completely rational anti–aggressive pharmacopeia.

In acute management, the method of rapid tranquilization has come under some revision. Previous work had indicated the potential usefulness of lithium, carbamazepine, and propranolol as anti–aggressive agents and additional reports on these will be mentioned. Unfortunately, only one or two control studies have been added to this literature. Some research studies on the relation between caffeine and alcohol to aggressive behaviour are newly included, as well as a section on hormones. The use of methyl phenidate and pargyline in aggressive behaviour and in the anti–social personality syndrome is reviewed and finally some initial results on a new anti–aggressive agent called sultopride.

ACUTE MANAGEMENT

The management of the violent or potentially violent patient remains a major issue for emergency psychiatric practice. The method of rapid tranquilization by repeated intramuscular injections of neuroleptics has become a widely publicized and utilized technique (Anderson et al. 1975). However, there remain several unanswered questions in regards to type of patient and dose. Thus one of the initial studies reported a rapid reduction in acute symptoms after three hours. Eleven of twenty–four patients improved in three hours and another five within seventy–two hours. However, Donlon et al. (1979) found no difference between a moderate (15 mg) and a very high (45 mg) dose of haloperidol followed by tapering (gradually reducing the dose) in acute schizophrenic patients. A recent review also challenges the necessity for high repeated doses (Linden et al. 1982). When anxiety and/or tension is the predominant affect, as in alcohol or drug withdrawal states, in the absence of psychosis, then sedation with a short–acting barbiturate or the administration of benzodiazepines may be indicated. If parenteral administration is necessary, then lorazepam is preferable given intramuscularly, since diazepam is poorly absorbed when given by intramuscular injection and intravenous injections are frequently not feasible in emergency situations. Sodium amytal remains a useful sedative given by intramuscular or intravenous administration, at doses up to

500 mg in a 2.5% solution, given at a rate of 1 ml a minute, until the patient becomes sleepy. This method requires the presence of resuscitation equipment and should not be undertaken in elderly or delirious patients or in cases where drug use is unclear.

Recent studies have shown a relationship between the severity of schizophrenic symptoms and violent behaviour (Yesavage 1982). This relationship was measured by the Brief Psychiatric Rating Scale and inpatient "danger related measures". In a further analysis, the relationship between serum levels of thiothixene and inpatient danger related events was studied in fifty–eight male schizophrenic patients over a twelve month period. The mean age of the patients was thirty with a range from twenty–two to fifty–three. Serum levels of thiothixene ranged between 3 and 45 ng/ml. Significant correlations were found with physical assault (–0.48), verbal assault (–0.37) and the total score of danger related events (–0.65). Of the nine patients who committed assault, eight had serum levels less than the mean for the group. This study suggests that schizophrenic patients who commit violent acts may not be receiving doses of medication which maintain adequate serum levels of drug It also follows that estimation of a single serum level of neuroleptic following an acute dose may have some usefulness as a predictor of violence.

Thus far there is no evidence that one antipsychotic agent is better than any other for the acute management of psychotic agitation. A differentiation can be made, however, between non–sedating (e.g. trifluophenazine, haloperidol, thiothixene) and sedating (e.g. chlorpromazine, thioridazine) types of antipsychotic agents. In general, for emergency situations, the non–sedating types seem preferable.

LITHIUM

A recent study was carried out in outpatients of the K.E.M. Hospital in Bombay (Bagdia et al. 1983). There were twenty patients, seven male and thirteen female with an age range of thirteen to thirty–eight and a mean of twenty–four. The diagnoses were thirteen mental retardation, four epilepsy, two schizophrenia, one personality disorder. All had received adequate trials with antipsychotics, ECT, anticonvulsants or other tranquilizers for relevant periods of time without control of aggressive behaviour. A relative kept weekly recordings of all overt aggressive behaviour for the six weeks trial. Dosage of lithium carbonate was regulated according to clinical judgement, side effects and serum level. An aggression scale was completed weekly on the basis of the relative's report on frequency and severity of outbursts. The results showed that fourteen out of twenty patients had moderate to marked improvement by the end of the six week period. The initial aggression score and that seen in the sixth week were significantly different on the basis of a t–test. There were four drop outs within the first two weeks for unknown reasons. This is the first large scale report on outpatients and while confirming the anti–aggressive effect of lithium in non bipolar patients it suffers from lack of controls. Also, the study fails to mention whether or not other treatments were stopped or continued during the lithium trial.

The use of lithium in children with severe aggressive behaviour problems is reported in a study from Hungary (Vetro et al. 1981). Seventeen children aged three to twelve were treated as inpatients. The hospitalization was necessitated by hyperaggressivity. Twelve were boys and five were girls. Ten of these children had previously experienced pharmacotherapy together with individual and family psychotherapy without improvement. Cyclothymic personalities or patients with a family history of bipolar or unipolar illness were excluded. Eleven patients had normal intelligence and six were retarded. There were minor neurologic abnormalities in four cases and one individual had a right side spastic hemiparesis from previous brain surgery. The duration of lithium administration was from four months to two years with an average of one year. The dosage of lithium maintained a serum level of 0.7–1.3 mmol/L. Measurement of the children with normal intelligence aged between seven and twelve was by means of the Rosenzweig Picture Frustation Test (P.F.T). This was administered before lithium and after four and six months. In children under seven or retarded, reliance was placed on reports received from nursery, school or parents. It was found that there was a significant reduction in explicit aggressive reactions and an increase in intrapunitive (self–punishing) responses. Thus the authors claim that the children's sociability and ability to adapt to the environment were improved. Both hetero and auto–aggression decreased. Reports revealed significant improvement in one case where the P.F.T. failed to show it. In the seven younger or retarded cases, no improvement occurred in two cases because of poor compliance with the administration of lithium, while a third showed no response despite an adequate serum lithium level. No serious side effects were reported. In summary, this study shows that lithium can be safely administered to young children with hyperaggressivity. Unfortunately, no controls were used.

A comparison of the effects of lithium and haloperidol in aggressive hospitalized school–aged children (Platt et al. 1981) has concentrated on measurements of cognition. This is of great importance, because while neuroleptics can reduce aggressiveness in children, the doses required often produce sedation and can interfere with cognition. As the authors point out, it is particularly important that children who are developing cognitive skills as well as generally having academic difficulties in association with psychiatric problems, receive drugs for behavioural problems which interfere minimally with cognition and learning. In this study the selection criteria were 1) disruptive, highly explosive aggressive behaviour which had resulted in hospitalization; 2) no evidence of psychosis and/or mental retardation; 3) diagnosis of conduct disorder, undersocialized–aggressive (DSM III); 4) age 6–12 years. Two independent raters concurred on these criteria. Thirty–seven children commenced the study but seven were dropped because their aggressive behaviour ceased on the ward. This left in the study twenty–eight boys and two girls. Children were drug free for four weeks and underwent a two week placebo baseline. Following this baseline, children were randomized to lithium, haloperidol or placebo treatments. Cognitive and behavioural assessments were performed at the end of placebo baseline period and again after four weeks. Performance and verbal I.Q. scores were similar across all three groups. The study was performed under double blind conditions. Dosages of lithium and haloperidol were individually regulated up to 16 mg/day haloperidol and 2000 mg/day lithium. All medications were given in identical capsules.

A large battery of measures were used including questionnaires and rating scales, e.g. the Timed Objective Rating Scale for Aggression, and an Aggression Checklist (Hardesty and Shopsin unpublished). The cognitive battery included 1) a reaction time test, 2) the Porteus Maze Test, 3) the Matching Familiar Figures Tesk, 4) a short term memory test, 5) a concepts attainment test and 6) the Stroop Test. The results showed that both lithium and haloperidol at doses which reduce severe behavioural symptoms do not have major negative effects on cognitive performance.

Lithium has been reported to worsen the clinical picture in temporal lobe epilepsy. In another recent clinical case report interictal aggressive behaviour was exacerbated by lithium (Schiff et al. 1982). A fifty–two year old male was admitted to the hospital for treatment of episodic violent behaviour. At age fifteen he suffered a severe head injury and was comatose for two days. Following this, he became hyperreligious, began to write extensive poetry, and was sexually hypoactive. He suffered from frequent episodes of violent behaviour lasting five to ten minutes after which he complained of exhaustion and fell asleep. There was no evidence of aura, automatism or convulsive activity. Computerized tomography scans showed dilatation of the temporal horns of the lateral ventricles. Daily depth and surface EEG recordings were made while on phenytoin and carbamazepine. In the right anterior amygdala there was a discrete focus of 2Hz sharp activity as well as an occasional run of positive spikes. Similar activity occurred in the left anterior amygdala. The ratio of right to left spike activity was 8:1. A decision was made to treat with lithium and a seven day trial caused a severe worsening of the spike and polyspike as well as a progressive loss of violent impulse control. The lithium was discontinued after seven days. The report does not make clear whether or not lithium was simply added to the anticonvulsant medication nor what effect carbamazepine alone had on the violent behaviour. The serum level of carbamazepine was, however, below the therapeutic range. In any case, it appears from this report and previous reports, that lithium is contraindicated in violence associated with temporal lobe epilepsy.

More double blind studies of lithium in carefully defined and diagnosed populations which show problems with aggressive behaviour are urgently needed.

CARBAMAZEPINE

There have been a few recent reports on the efficacy of carbamazepine in violent psychiatric patients. One of these (Lunchins 1983) studied seven patients, two female and five male. Six patients suffered from schizophrenia and one from a mixed personality disorder. The mean age was thirty-five with a range of twenty-three to forty-eight and the duration of hospitalization was twelve years with a range of five to nineteen. Three six week periods were compared: six weeks before, six weeks on and six weeks off carbamazepine. Episodes of physical aggression during these periods were counted by ward staff. There was a significant difference in the frequency of violent episodes during these three periods with fewer being seen when patients were on carbamazepine than in the other two conditions. All patients continued to receive their customary neuroleptic medication which was similar over the three periods. The dose of carbamazepine ranged between 400 to 1600 mg daily and maintained a mean serum level of 8.5 µg/ml. All these patients were described as having a normal EEG off carbamazepine but without the use of nasopharyngeal leads. These findings suggest that carbamazepine may be effective for the control of aggressive behaviour in patients who do not have an abnormal EEG.

Another report from Finland (Hakola and Laulumaa 1982) concerns the effect of carbamazepine on eight women with schizophrenia (of unspecified type) with violent episodic outbursts. The age range was twenty-one to sixty-four with a mean of thirty-eight. The mean duration of hospitalization was about ten years. EEG examinations revealed non-specific abnormalities such as generalized slowing as well as irritative activity with focal tendencies in three cases and without focal tendencies in four cases. All patients continued on their regular medication and received carbamazepine 400–800 mg, leading to daily serum levels of 17.3–31.2 µmol/l. The duration of treatment was two months to eleven years (mean 2.7 years). Under carbamazepine the violent behaviour almost completely disappeared in all patients. The authors also report that the final doses of neuroleptics were able to be considerably reduced. Although no controls were used, this study provides supportive evidence that carbamazepine can be effective for treating violent behaviour in schizophrenia.

A single case report (Yassa and Dupont 1983) is the use of carbamazepine in the treatment of a thirty-six year old male patient who had been diagnosed as suffering from paranoid schizophrenia at the age of sixteen. He had been treated with chlorpromazine and ECT. Interestingly, the ECT initially produced a good response. In the succeeding years he developed violent behaviour, showing generally unprovoked outbursts with clear consciousness but in response to paranoid ideation. A course of ECT improved the behaviour but it soon relapsed. He was subsequently kept on maintenance ECT, receiving 40 ECT sessions between 1979 and 1980. In August 1980, he was started on carbamazepine, 200 mg, which was gradually raised to 200 mg three times a day in addition to his regular medication which consisted of 1800 mg of chlorpromazine daily. The number of violent episodes gradually decreased, while the carbamazepine was increased and the chlorpromazine decreased. The final dose was 600 mg/day carbamazepine and 800 mg/day chlorpromazine and he was discharged from the hospital in July 1982 after eleven years of inpatient treatment. EEG's in this case were always reported as normal and brain scans and skull x-rays were also normal. Although this single case study is impressive, it lacks control and therefore it is interesting to report that there is an ongoing double blind placebo controlled study of borderline patients which has found carbamazepine superior to placebo in controlling behavioural dyscontrol (Gardner and Cowdry 1986). Clearly more such studies are needed.

The importance of considering the role of concomitant therapies for the control of hyperaggressive behaviour is pointed out by another case study (Rapport et al. 1983). This is a case of a 13.7 year old female with mental retardation and a seizure disorder secondary to measles encephalitis. She developed assaultive behaviour which remained uncontrolled by anticonvulsant medication including carbamazepine until she received behaviour therapy in addition.

PROPRANOLOL

Additional evidence for a useful anti-aggressive effect of propranolol has accumulated in the past two to three years. A report from Cuba (Leon et al. 1983) documents the use of propranolol in forty-five female patients suffering from mental retardation with and without epilepsy. Propranolol in a dose of 2.1 mg/kg produced a reduction in the number and intensity of aggressive episodes in the majority (35 cases). The dose ranged between 30 and 180 mg daily. In only one case was it necessary to stop the propranolol treatment and no significant side effects were reported.

The usefulness of propranolol for aggressive behaviour was also documented in three cases from the Massachusetts Mental Health Center (Ratey et al. 1983). First, a forty-two year old woman had a chronic history of schizophrenia with severe psychotic symptoms in addition to mild mental retardation. Medical and neurological examinations proved negative and she had received extensive trials with neuroleptics and antidepressants without benefit. Over the previous five years, she had become increasingly assaultive with sudden outbursts of rage. Propranolol was administered in addition to the neuroleptic medication and increased to 200 mg/day in divided doses. Over the course of the next three months, the degree of hostility and number of verbal and physical assaults decreased substantially. The second case was a forty-nine year old woman, also mentally retarded, who had been in and out of institutions since the age of thirteen. While usually pleasant, she exhibited periods of violent acting out for no apparent reason. She also had violent temper tantrums at changes of shift and when harassed by other patients. After propranolol at a dose of 30 mg three times a day, her temper outbursts gradually decreased until they became almost non-existent. While still easily upset, she did not give in to rage attacks. The third case was a twenty-four year old man who suffered severe neurological deficits after head and spinal cord injury received in a car crash. Despite no previous psychiatric history, he developed severe depression and made two suicide attempts, followed by two periods of paranoia, delusional ideation and assaultive episodes (usually when frustrated, but also sometimes unprovoked). Treatment with antidepressants and neuroleptics was unsuccessful. Because he had a severe tremor, propranolol was prescribed and the dose gradually increased to 300 mg/day. His behaviour improved very dramatically. He was discharged but had to be readmitted following a reduction in dose of the propranolol because of bradycardia. He improved once again when the dose of propranolol reached 300 mg/day.

The authors make the following recommendations for the use of propranolol:

1. Presence of discrete provoked or unprovoked rage episodes regardless of underlying psychopathology or neurologic impairment.
2. Presence of persistent aggressiveness or assaultiveness in patients who also have other forms of major mental illness.
3. Absence of treatable medical conditions, e.g. seizures.
4. No medical contraindications.
5. Ability to closely monitor the patient.

Another very interesting case report (Yudofsky et al. 1984) concerns the use of propranolol for the treatment of rage and violent behaviour associated with Korsakoff's Psychosis. In this case, a forty year old male patient with a twenty year history of alcoholism (with several episodes of delirium tremens and a seizure disorder) was being treated for delirium tremens. This was a difficult treatment problem as the patient showed rage outbursts which occurred more than ten times a day and were triggered by minor frustrations. He injured several aides and broke his own leg in the course of these outbursts. Over a two month period the patient required restraints despite all customary pharmacologic (he received up to 14 mg/day of haloperidol) and behavioural treatments. Propranolol was gradually added to this treatment until he reached a dose of 600 mg/day. Two weeks after this dose, the rage attacks were markedly reduced and restraints were no longer necessary. Attempts to reduce propranolol dose resulted in a resumption of the rage attacks. The authors make the point that substantially larger doses of propranolol are required to treat these violent episodes than are required for more usual treatment of hypertension. However, this and the earlier report suggest that the required dosage of propranolol is variable and can actually be quite low in some cases.

A preliminary report of a study designed to compare propranolol and carbamazepine (Mattes et al. 1984) in inpatients over the age of sixteen with uncontrolled rage outbursts is available for the first twenty-eight patients. The subjects met the first two DSM III criteria for intermittent explosive disorder. Namely, 1) several discrete episodes of loss of control of aggressive impulses resulting in serious assaults or destruction of property, and 2) behaviour that is grossly out of proportion to any precipitating psychosocial stressor. Patients with diagnoses requiring other treatments were excluded. Subjects were randomly assigned to carbamazepine (to a therapeutic level) or propranolol (up to 640 mg/day). If they had contraindication to either drug or could not tolerate one, they were given the other. Seven patients received an adequate trial of propranolol and twenty of carbamazepine. Most patients improved on either medication. It is interesting to note that no patient characteristics, (including abnormal EEG), were useful in predicting response. This included diagnoses which broke down as follows: 9 Attention Deficit Disorder, 12 Intermittent Explosive Disorder, 5 Antisocial Personality Disorder, 3 Unsocialized Conduct Disorder, Aggressive, 11 Drug Abuse, 12 Alcohol Abuse, 3 Borderline Personality Disorder and 7 Organic Disorder. It is clear from this that any subject may have more than one diagnosis. These results suggest that both propranolol and carbamazepine may be helpful for any patient with uncontrolled temper outbursts.

CAFFEINE

The use of new methodology which elicits aggressive responding from research subjects by subtracting money from them and attributing this to a fictitious person has demonstrated that caffeine, as compared to placebo, decreases aggressive responding but increases non-aggressive responding (Cherek et al. 1983). In these experiments, aggression is operationally defined as pressing a button which delivers to them a blast of sixty dB white noise. A third button produced a monetary reward for points which were given for every one hundred presses. This enabled a measure to be obtained simultaneously for non-aggressive responding. Since this increased under the influence of caffeine, the decrease in aggressive responding could not be attributed to a generalized non-specific depressant effect. In another experiment, the authors showed that it was indeed the caffeine in the coffee which produced the effect since it was not observed with the decaffeinated coffee (Cherek et al. 1984a).

ALCOHOL

The same methodology as described above for the assessment of the effects of caffeine upon aggressive behaviour has been used to assess the effects of alcohol (Cherek et al. 1984b). The results indicated substantial increases in the number of aggressive responses with alcohol doses of 0.25 g/kg and higher. This amount of alcohol is equivalent to that consumed in one drink. At the same time these doses did not alter the monetary reinforced responding, supporting the notion that the increase in aggressive responding was not due to any generalized stimulant action. Previous studies had only shown effects at much higher doses of alcohol.

METHYLPHENIDATE, PEMOLINE AND PARGYLINE

Stimulants such as methylphenidate and pemoline have been used with success in the Attention Deficit Disorder of children (previously known as the hyperactive child syndrome or minimal brain dysfunction). It has also been reported that this disorder can persist into adulthood with persistent symptoms of attentional deficit and impulsivity. In particular, these subjects are described as having hot tempers and poor control over angry responses. In adults, this syndrome is now called Attention Deficit Disorder, Residual Type.

Drug studies in this population have shown that both pemoline and methylphenidate are significantly better than placebo. However, there are disadvantages to stimulant medication, including a) it is effective in only about two thirds of the population, b) Methylphenidate has the potential for abuse and c) both drugs have a short lived action. For these reasons, a trial of pargyline (a long-acting monoamine oxidase inhibitor whose substrate is primarily phenylethylamine and dopamine) was undertaken (Wender et al. 1984). An open trial was conducted with twenty-two individuals who received 10 mg pargyline daily with slow increases every five to seven days if tolerated. Six patients dropped out because of side effects resulting in thirteen male and three female subjects. Mean ages were 31.6 ± 6.1 with a range of 21–43. The trial lasted six weeks and median daily doses were 30 mg with a range of 10–50 mg. There was a clinically and statistically significant improvement in symptoms, including anger, which commenced about seven to ten days after starting the drug. In all, 68 percent of the sixteen patients showed a moderate to marked improvement.

A trial of methylphenidate in two cases who met a diagnosis of antisocial personality disorder in addition to having evidence of attention deficit disorder in childhood also has been reported (Stringer and Yosef 1983). Both were males in their early twenties who were treated as inpatients in an open trial. After a few weeks of baseline observation, the hyperaggressive behaviour of both patients while on the methylphenidate (20 mg twice a day) decreased. However, the disadvantages of methylphenidate (its short duration of action, and particularly its potential for abuse), should be carefully weighed in such cases.

HORMONES

Treatment with antiandrogenic agents such as cyproterone acetate (Schering) has largely replaced surgical castration or estrogen therapy for the treatment of sexual offenders in Europe. In one recent report (Berner et al. 1983), twenty one inmates of a prison facility in Vienna volunteered to take cyproterone acetate in addition to supportive psychotherapy. The drug was given as a daily oral dose of 100 mg for one to ten years. Taking the drug had no effect on release from the prison. Follow-up was only available for a relatively small percentage of cases. Moreover, of those followed up between one and eight years, the authors report over two thirds did not repeat offences. The re-arrest rate for sexual offences was twenty-eight percent. Unfortunately, no comparison was performed with subjects treated with psychotherapy alone.

Medroxyprogesterone remains the most common pharmacological approach in the USA. The use of depo-medroxyprogesterone acetate has been previously described for the treatment of aggression in temporal lobe epilepsy. In a recent paper, its use as an adjuvant in three aggressive patients is described (O'Connor and Baker 1983). The patients were three males, aged forty, twenty-two and twenty-eight, all diagnosed as having chronic schizophrenic symptoms, although very few clinical details are given. The first two patients had chronic assaultive behaviour while the third kicked the furniture, banged his head on the wall, often injuring himself. Sterile water or depo-medroxyprogesterone acetate (in increasing doses of 25 to 75 mg per week) was given. The doses were changed at one to two month intervals and given in a random order, and crossed over so that neither patient nor staff evaluating behaviour knew what the injection was. Evaluations were once a week with the Brief Psychiatric Rating Scale and all episodes of violent behaviour were counted. Serum FSH, LH, plasma testosterone and cortisol were measured just prior to the drug injections. The inter-relationships of rating scales and hormone levels with drug treatment was tested by analysis of variance. There was significant dose drug-related improvement in the first two cases but not in the third.

These findings suggest that synthetic steroids may be useful adjuvants in the treatment of male patients with schizophrenia and violence. It should be noted that synthetic steroids appear to have some central psychotropic action in addition to their endocrine effects.

SULTOPRIDE

An initial report on five human subjects (Blondel 1982) reports excellent results with sultopride in reduction of aggression and agitation across a spectrum of psychiatric diagnoses, bipolar manic, mental retardation, post encephalopathy, paranoid schizophrenia and a severe hysterical character disorder. Doses used ranged about 1200 mg/day by mouth and 600–1200 mg by intramuscular injection. The drug appears to be well tolerated and is compatible with other neuroleptic medications.

SUMMARY

There are now several agents available of potential promise for the treatment of violent aggressive behaviour. These include lithium, carbamazepine, propranolol, methylphenidate and hormones such as cyproterone acetate and depo-medroxy progesterone acetate. All require considerable further study with careful controls under double blind conditions, in different psychiatric populations, either alone or as adjuvants to standard pharmacologic therapies. Finally it is interesting to point out the development of a reliable rating scale for the objective rating of verbal and physical aggression (Yudofsky et al. 1986) for use in these studies.

REFERENCES

Anderson WH, Kuehnle JC, Cantanzano DM (1975) Rapid treatment of acute psychosis. Am J Psychiatry 133: 1076–1078

Bagadia VN, Lakdawala PD, Pradhan PV, Mundra VK, Desai NK, Shah LP (1983) Lithium in aggression. Indian J Psychiatry 25: 107–109

Berner W, Brownstone G, Sluga W (1983) The cyproteroneacetate treatment of sexual offenders. Neurosci Biobehav Rev 7: 441–443

Blondel F (1982) Sultopride et Agitation. Sem Hop Paris 58, 44: 2615–2616

Cherek DR, Steinberg JL, Brauchi JT (1983) Effects of caffeine on human aggressive behavior. Psychiatry Res 8: 137–145

Cherek DR, Steinberg JL, Brauchi JT (1984a) Regular or decaffeinated coffee and subsequent human aggressive behavior. Psychiatry Res 11: 251–258

Cherek DR, Steinberg JL, Vines RV (1984b) Low doses of alcohol affect human aggressive responses. Biol Psychiatry 19: 263–267

Donlon PT, Hopkin J, Tupin JP (1979) Overview: Efficacy and safety of the rapid neuroleptization method with injectable haloperidol. Am J Psychiatry 136: 273–278

Gardner DL, Cowdry RW (1986) Positive effects of carbamazepine on behavioral dyscontrol in borderline personality disorder. Am J Psychiatry 143: 519–522

Hakola HPA, Laulumaa VA (1982) Carbamazepine in treatment of violent schizophrenics. Lancet 1: 1358

Leon Z, Alonso RJL, Basterrechea L (1983) Accion del propranolol en la agresividad de los oligofrenicos encefalopaticos. Jornada Interna de Defectologia en el Hogar de Impedidos Fisicos y Mentales 4: 431–436

Linden R, David JM, Rubenstein J (1982) High versus low dose treatment with antipsychotic agents. Psychiatry Ann 12: 769–781

Luchins DJ (1983) Carbamazepine for the violent psychiatric patient. Lancet 2: 766

Mattes JA, Rosenberg J, Mays D (1984) Carbamazepine versus propranolol in patients with uncontrolled rage outbursts. Psychopharmacol Bull 20: 98–106

O'Connor M, Baker HWG (1983) Depo–medroxy progesterone acetate as an adjunctive treatment in three aggressive schizophrenic patients. Acta Psychiatr Scand 67: 399–402

Platt JE, Campbell M, Green WH, Perry R, Cohen IL (1981) Effects of lithium carbonate and haloperidol on cognition in aggression hospitalized school–age children. J Clin Psychopharmacol 1: 8–13

Rapport MD, Sonis WA, Fialkov MJ, Matson JL, Kazdin AE (1983) Carbamazepine and behavior therapy for aggressive behavior. Behav Modif 7: 255–265

Ratey JJ, Morrill R, Oxenkrug G (1983) Use of propranolol for provoked and unprovoked episodes of rage.Am J Psychiatry 140: 1356–1357

Schiff HB, Sabin TD, Geller A, Alexander L, Mark V (1982) Lithium in aggressive behavior. Am J Psychiatry 139: 1346–1348

Stringer AY, Josef NC (1983) Methylphenidate in the treatment of aggression in two patients with antisocial personality disorder. Am J Psychiatry 140: 1365–1366

Vetro A, Pallag P, Szentistvanyi LI, Vargha M, Szilard J (1981) Treatment of childhood aggressivity with lithium. Aggressologie 22: 27–30

Wender PH, Wood DR, Reimherr FW, Ward M (1983) An open trial of pargyline in the treatment of attention deficit disorder, residual type. Psychiatry Res 9: 329–336

Yassa R, Dupont D (1983) Carbamazepine in the treatment of aggressive behavior in schizophrenic patients: a case report. Can J Psychiatry 28: 566–568

Yesavage JA (1982) Inpatient violence and the schizophrenic patient: An inverse correlation between danger–related events and neuroleptic levels. Biol Psychiatry 17: 1331–1337

Yudofsky SC, Silver JM, Jackson W, Endicott J, Williams D (1986) The overt aggression scale for the objective rating of verbal and physical aggression. Am J Psychiatry 143: 35–39

Yudofsky SC, Stevens L, Silver J, Barsa J, Williams D (1984) Propranolol in the treatment of rage and violent behavior associated with Korsakoff's psychosis. Am J Psychiatry 141: 114–115

INDEX